Immunology and Immune System Disorders

nova
Medicine & Health
New York

Immunology and Immune System Disorders

The Innate Immune System in Health and Disease: From the Lab Bench Work to Its Clinical Implications. Volume 1
Jorge Morales-Montor (Editor)
2022. ISBN: 978-1-68507-507-1 (Hardcover)

Vaccines: Operation Warp Speed, Regulation and Safety
Oliver Huerta (Editor)
2021. ISBN: 978-1-53619-059-5 (Softcover)
2020. ISBN: 978-1-53619-096-0 (eBook)

IgG4 Related Disease: Illustrated Pathology of Novel Systemic Disease
Syuichi Koarada, MD (Editor)
2020. ISBN: 978-1-53618-050-3 (Hardcover)
2020. ISBN: 978-1-53618-051-0 (eBook)

Basics and Fundamentals of Immunology
Manzoor Ahmad Mir, PhD (Editor)
2020. ISBN: 978-1-53616-639-2 (Hardcover)
2019. ISBN: 978-1-53616-640-8 (eBook)

Th17 Cells in Health and Disease
Tsvetelina Velikova, MD, PhD (Editor)
2020. ISBN: 978-1-53617-152-5 (Hardcover)
2020. ISBN: 978-1-53617-153-2 (eBook)

Cytokines and Their Therapeutic Potential
Manzoor Ahmad Mir, PhD (Editor)
2020. ISBN: 978-1-53617-017-7 (Softcover)
2020. ISBN: 978-1-53617-018-4 (eBook)

More information about this series can be found at
https://novapublishers.com/product-category/series/immunology-and-immune-system-disorders/

Jorge Morales-Montor
Editor

The Innate Immune System in Health and Disease

From the Lab Bench Work to Its Clinical Implications. Volume 1

nova
Medicine & Health
New York

DOI: https://doi.org/10.52305/KCVE8303

Library of Congress Cataloging-in-Publication Data

ISBN: 978-1-68507-507-1

Published by Nova Science Publishers, Inc. † New York

Contents

Preface

The aim of this book, *The Innate Immune System in Health and Disease: From the Lab Bench Work to Its Clinical Implications – Volume 1*, is to provide updated information to scientists and clinicians that is valuable in their quest to gather information, carry out new investigations, and develop novel drugs that are more effective and do not cause adverse effects targeting the innate immune system. This book is of high priority to people interested in an update on innate immunity. Volume 1 examines the participation of innate immunity in diseases such as infections by viruses, cancer, or parasites. Specifically, this volume discusses innate immunity as it relates to chronic HIV infection, type 2 diabetes, and SARS-CoV-2, as well as its relationship with vitamin D. The dual role of neutrophils in cancer development and the plasticity of natural killer cells are investigated. Also, the immune response to infections from helminth and *Toxoplasma gondii* are described.

The Innate Immune System in Health and Disease: From the Lab Bench Work to its Clinical Implications – Volume 1 promises to be a must-have book on the shelf of all people who want to know about the role of the basic functioning of the innate immune system in several diseases of actual relevance to human health.

Key Features:

- Provides basic and clinical evidence based on molecular interactions to address the role of innate immunity in viral diseases, such as HIV and SARS Cov-2
- Discusses the role of vitamin D as modulator of the immune response during viral infections
- Addresses the immunometabolic interactions in type 2 diabetes and SARS-CoV-2 infection (COVID-19)
- Analyzes the dual role for neutrophils and myeloid-derived suppressor cells in the development of colorectal cancer

- Provides evidence of the plasticity of natural killer cells in cancer and its opportunities for therapy
- Delves into the dual role of neutrophils in cancer development
- Presents updated information on innate immunity against *Toxoplasma gondii* infection
- Discusses in depth the role of the innate immune response during helminth infections

We hope that readers enjoy this book and that it may be of use to all people involved, from scientists in the lab to clinicians in the field.

Chapter 1 - The innate immune system plays an essential role in controlling viral infections and represents the first line of defense against pathogens. Human immunodeficiency virus type 1 (HIV) still remains as a global disease burden. Despite an effective combined antiretroviral therapy (cART) to control viral replication and to maintain a chronic infection, it does not eradicate HIV. During HIV replication, RNA genome undergoes reverse transcription and integrates the resulting double-stranded HIV DNA into the cellular genome, constituting HIV proviral DNA, that persists indefinitely in a small reservoir of latently infected cells and cannot be detected by the immune system nor targeted by cART. In acute HIV infection, viral load rises exponentially, and sentinel innate immune cells recognize a wide range of pathogen-associated molecular patterns through host pattern recognition receptors (PRRs) as well as cytosolic sensors. The sensing of HIV occurs in CD4+ cells, including innate immune cells and CD4+ T cells, and activates intracellular signaling pathways to induce host antiviral response. This response includes transcription of type I interferons (IFNs) and IFN-stimulated genes (ISGs) that restrict virus infection, production of proinflammatory cytokines and chemokines, and increased levels of costimulatory molecules which initiate innate antiviral immunity to control virus spread. Proinflammatory molecules are critical for confining HIV replication by activating innate immune cells and recruiting to the site of entry and to local lymphatic tissue. In chronic phase of HIV infection, an exacerbated and sustained activation of the innate immune system, in particular the type I interferon (IFN-1) system, is detrimental to the host since it may be involved in immunosuppression and progression to acquired immunodeficiency syndrome (AIDS). In this chapter the authors summarize recent findings regarding the main challenges for translational research on HIV chronic infection and innate immunity: the persistence of HIV reservoir and chronic inflammation. The authors highlight the role of HIV reservoir to maintain a chronic and latent infection and how a sustained immune activation promotes a robust inflammatory response that contributes to innate control of viral infection and simultaneously recruits HIV target cells and promotes disease progression. The authors discuss the molecular basis by which innate immunity is regulated in controlled HIV infection, and how systemic effects of PRRs, cytosolic receptors, intracellular signaling pathways, specific viral-restriction factors and innate immune cells define the antiviral immune response. The close relationship between HIV reservoir and the inflammatory response is reviewed. Recent strategies targeted to eradicate reservoir and anti-inflammatory interventions are described.

Chapter 2 - The active form of Vitamin D is 1,25-dihydroxyvitamin D3 (1,25$(OH)_2$D3), a metabolite that is involved in the absorption of minerals such as calcium, magnesium, and phosphate. Vitamin D can be obtained either from the diet or by synthesis induced by UVB radiation in the lower layers of the epidermis. Binding of 1,25 $(OH)_2$D3 to its nuclear receptor enhances the expression of proteins required for calcium absorption in the intestine and for maintaining its homeostatic levels in other tissues. Also, the active form of vitamin D3 has been shown to have immunomodulatory properties *in vitro* and *in vivo*. Several cells of the immune system can synthesize the active form of Vitamin D from its precursor, and express the Vitamin D receptor. Thus, vitamin D or its precursor can influence the immune response by up-regulating the production of anti-microbial and other proteins in cells of the innate immune system, while it can also modulate the activity of cells of the adaptive immune response. Adequate levels of Vitamin D are important in viral infections such as that produced by SARS-CoV-2. Low serum levels of Vitamin D and its metabolites have been found to be associated with increased risk of infection by SARS-CoV-2, as well as severe presentation and mortality in COVID-19. This has led to an interest in the potential prophylactic and/or therapeutic use of Vitamin D in this disease. Here, the authors summarize the known actions of Vitamin D on cells of the immune system, with emphasis on its effects in viral diseases, and review several studies on the effect of supplementation with Vitamin D in patients with COVID-19.

Chapter 3 - Coronavirus disease 2019 (COVID-19) is caused by the severe acute respiratory syndrome coronavirus-2 (SARS-CoV-2) that has affected the lives of millions of people worldwide. Type 2 diabetes (T2D) acts in synergy with SARS-CoV-2 infection to increase the progression, severity, and mortality of COVID-19. For this reason, there is an urgent need to understand the immunometabolic bases of the main etiological factors of T2D that affect the severity of COVID-19. Herein, the authors revise how hyperglycemia contributes to the cytokine storm commonly associated with COVID-19 by stimulating monocytes and macrophages to produce interleukin (IL-) 1 beta, IL-6, and tumor necrosis factor alpha (TNF-alpha) in the airway epithelium. The main mechanisms through which hyperglycemia promotes reactive oxygen species (ROS) release, inhibition of T cell activation, and neutrophil extracellular traps (NETs) in the lungs of patients with severe SARS-CoV-2 infection are also discussed. The authors further study the molecular mechanisms by which proinflammatory cytokines induce insulin resistance, and their deleterious effects on pancreatic β-cell exhaustion in T2D patients critically ill with COVID-19. The authors revise the effect of excess glucose on advanced glycation end product (AGE) formation and the role of AGEs in perpetuating pneumonia and acute respiratory distress syndrome (ARDS). The authors examine the contribution of preexisting endothelial dysfunction secondary to diabetes in the development of neutrophil trafficking, vascular leaking, and thrombotic events in patients with severe SARS-CoV-2 infection. Finally, the authors discuss the role of other mechanisms potentially involved in COVID-19 evolution such as hyperglucagonemia, adiposity, dyslipidemia, and glomerular damage. As the authors have outlined here, T2D acts in synergy with SARS-CoV-2 infection to increase the progression, severity, and mortality of COVID-19. The authors think a better understanding of the immunometabolic factors involved in T2D that contribute to the severity of COVID-19 will improve our ability to recognize patients at much higher mortality risk.

Chapter 4 - Worldwide, colorectal cancer (CRC) is the third most common cancer in men and the second in women, while it is the third leading cause of cancer death in men and the fourth in women. Only 15% of CRCs arise from inherited diseases such as Lynch syndrome and polyposis associated with APC and MUTYH genes. The remaining 85% arises sporadically and is associated with different factors such as obesity, sedentary lifestyle, alcohol and tobacco consumption, low fiber consumption, high consumption of red/processed meats, alterations in the microbiota, and inflammatory bowel diseases. Inflammation is now recognized as a hallmark for tumor progression, especially in CRC. Inflammation implies many cell populations of the immune system and soluble factors, molecules, and cells that are not part of the immune system. Thus, an outstanding balance of the inflammatory response is critical to restrain or not the development of CRC. Even more, a disbalance of the inflammatory response may participate in restricting or favoring CRC development. Cells from the innate immune system and adaptive immune response play a significant role in orchestrating immunomodulation during

CRC. In this chapter, the authors analyzed concrete evidence supporting that neutrophils and myeloid-derived suppressor cells (MDSCs) have an essential role in promoting tumor development during CRC and regulating its growth and metastasis. Most of the studies on MDSCs focus on how these cells mediate an increasing suppressive capacity, which controls the effector immune response generating protection for tumors. Controversial and more recent reports indicate that MDSCs may play a role in controlling excessive immune responses and consequently inflammation. Similarly, neutrophils have been considered a bad prognostic marker in patients suffering from CRC, but as recently described, the N1 type neutrophils may play an opposite role. These cell populations are now recognized and may go through a process called "phenotypic plasticity," where they co-express transcription factors and membrane molecules that promote an inflammatory profile. The authors focus on evidence that describes the interaction between the different effector populations of the immune response and its modulation by neutrophils and MDSCs adapted to the tumor microenvironment. Including the mechanisms used by neutrophils and MDSCs to modulate putative protective or deleterious immune responses, as well as the different subpopulations of both cell types participating in tumor progression, generating susceptibility or resistance during CRC development. Finally, the authors discuss whether innate immune cells and neutrophils or MDSCs might or might not be a therapeutic target for effectively reducing the morbidity and mortality caused by CRC.

Chapter 5 - Natural killer (NK) cells are vital components of the innate immune system. NK cells are implicated in pleiotropic processes such as transplant rejection, cell killing after viral infections, inflammatory diseases, and cancer. This kind of cellular immune effector is an important switch for the adaptive immune response mediated by the activation of T lymphocytes. It has been reported that the percentage of NK cells in the tumor microenvironment is low. However, recent works have pointed out that different lineages of NK cells, which are characteristic components of the hyper-secretory profile of cytokines and granzymes, led to the proliferation of malignant cells and the formation of distant metastases. Moreover, the disturbed metabolism of NK cells encouraged by tumor cells can promote the cancer process. Until now, little information about the relationship between different subsets of NK cells, their function, and regulation of cancer promotion is known. Thus, the present chapter focuses on the deep knowledge of the regulation of these cellular mediators in the cancer context to expand the scientific evidence of their possible use as a therapeutic strategy that improves clinical prognosis.

Chapter 6 - Neutrophils, the most abundant leukocytes in blood, are the first cells to be recruited to sites of inflammation and infection. At these sites, neutrophils display an array of anti-microbial functions, including phagocytosis, degranulation, production of reactive oxygen species, and formation of neutrophil extracellular traps (NET). In addition, neutrophils are also found infiltrating many types of tumors. Neutrophils seem to have an important role in progression of malignant disease. In fact, increasing clinical evidence suggests that elevated neutrophil counts in blood correlate with poor prognosis. The tumor microenvironment controls neutrophil recruitment and in turn tumor-associated neutrophils (TAN) help tumor progression. However, it is well established that neutrophils also possess potent anti-tumor effector functions. Hence, TAN can be detrimental or beneficial to the host. In this chapter, the authors describe these two aspects of neutrophil biology during cancer progression. The authors discuss recent studies that provide insight into the mechanisms for neutrophil migration to the tumor, for neutrophil promotion of tumor progression, and for neutrophil activation to enhance their anti-tumor functions.

Chapter 7 - *Toxoplasma gondii* is a protozoan parasite that belongs to Apicomplexa phylum, it can infect almost all warm-blooded animals worldwide and one-third of human population is infected with this parasite. In immunocompromised individuals, such as pregnant women, AIDS patients, chemotherapy recipients and immunosuppressed tissue transplant patients, infection and reactivation can lead to miscarriage, malformations in the newborn, blindness and lethal encephalitis. Since 50's decade a mixture of sulfadiazine-pyrimethamine has been used as conventional therapy against toxoplasmosis. Although sulfadiazine – pyrimethamine are synergic it is known that present severe side effects and has a limited effectiveness, mainly on chronic stage disease, and there is not an available vaccine for human use. As an obligate intracelular opportunistic pathogen, *T. gondii* celular invassion is essential for its survival, which occurs almost immediately after the degradation of ingested cysts. For this reason the study of innate immune response has been difficult, especially in humans. In this chapter the authors analyzed the innate immune response produced by *T. gondii*, focusing in the main pathway infection, ingestion of contaminated food with sporulated cysts.

Chapter 8 - Helminthosis not only affect a large number of people in the world (more than 1.5 billion people are estimated to be infected by these parasites), they also affect a large number of animal species, and many of them are zoonotic. This implies great economic losses, not only in the cost of the different treatments, but also due to the decrease in the different productive and reproductive parameters in other animal species. Although many of them are asymptomatic infections, in some cases serious health problems can occur, as well as a predisposition to different comorbidities or co-infections. In controlling a helminth infection, the immune response plays an important role. In particular, the innate response is one of the first and main defense mechanisms of the organism that are activated when coming into contact with these parasites. It has long been known that different cells of the innate immune response participate through different mechanisms, in the control and elimination of helminths. However, both new cells and new strategies implemented by the innate immune system to combat these infections have recently been described. In this chapter, the authors will discuss the participation of cells of the innate immune response, as well as the different effector mechanisms that are activated during helminth infections, and how they can act, either directly on parasites or by modulating the future adaptive immune response.

Section 1. Innate Immunity and Viruses

Chapter 1

Innate Immunity in Chronic HIV Infection: Relationship Between Viral Reservoir and Inflammatory Response and Its Clinical Implications

Perla Mariana Del Río-Estrada[1], Enrique Espinosa[2] and Mónica Viveros-Rogel[3,*]

[1]Instituto Nacional de Enfermedades Respiratorias (INER) Ismael Cosío Villegas, Centro de Investigación en Enfermedades Infecciosas (CIENI), Mexico City, Mexico
[2]Instituto Nacional de Enfermedades Respiratorias (INER) Ismael Cosío Villegas, Laboratory of Integrative Immunology, Mexico City, Mexico
[3]Instituto Nacional de Ciencias Médicas y Nutrición Salvador Zubirán (INCMNSZ), Department of Infectious Diseases, Mexico City, Mexico

Abstract

The innate immune system plays an essential role in controlling viral infections and represents the first line of defense against pathogens. Human immunodeficiency virus type 1 (HIV) still remains as a global disease burden. Despite an effective combined antiretroviral therapy (cART) to control viral replication and to maintain a chronic infection, it does not eradicate HIV. During HIV replication, RNA genome undergoes reverse transcription and integrates the resulting double-stranded HIV DNA into the cellular genome, constituting HIV proviral DNA, that persists indefinitely in a small reservoir of latently infected

* Corresponding Author's E-mail: monica.viverosr@incmnsz.mx.

In: The Innate Immune System in Health and Disease
Editor: Jorge Morales-Montor
ISBN: 978-1-68507-507-1

cells and cannot be detected by the immune system nor targeted by cART. In acute HIV infection, viral load rises exponentially, and sentinel innate immune cells recognize a wide range of pathogen-associated molecular patterns through host pattern recognition receptors (PRRs) as well as cytosolic sensors. The sensing of HIV occurs in CD4+ cells, including innate immune cells and CD4+ T cells, and activates intracellular signaling pathways to induce host antiviral response. This response includes transcription of type I interferons (IFNs) and IFN-stimulated genes (ISGs) that restrict virus infection, production of proinflammatory cytokines and chemokines, and increased levels of costimulatory molecules which initiate innate antiviral immunity to control virus spread. Proinflammatory molecules are critical for confining HIV replication by activating innate immune cells and recruiting to the site of entry and to local lymphatic tissue. In chronic phase of HIV infection, an exacerbated and sustained activation of the innate immune system, in particular the type I interferon (IFN-1) system, is detrimental to the host since it may be involved in immunosuppression and progression to acquired immunodeficiency syndrome (AIDS). In this chapter we summarize recent findings regarding the main challenges for translational research on HIV chronic infection and innate immunity: the persistence of HIV reservoir and chronic inflammation. We highlight the role of HIV reservoir to maintain a chronic and latent infection and how a sustained immune activation promotes a robust inflammatory response that contributes to innate control of viral infection and simultaneously recruits HIV target cells and promotes disease progression. We discuss the molecular basis by which innate immunity is regulated in controlled HIV infection, and how systemic effects of PRRs, cytosolic receptors, intracellular signaling pathways, specific viral-restriction factors and innate immune cells define the antiviral immune response. The close relationship between HIV reservoir and the inflammatory response is reviewed. Recent strategies targeted to eradicate reservoir and anti-inflammatory interventions are described.

Keywords: innate immunity, viruses, HIV, AIDS, PRRs

Introduction

Over four decades of research, major scientific advances have been made to develop strategies to prevent and eradicate HIV infection. It is undeniable the impact of the combined antiretroviral therapy (cART) on the life expectancy and quality of life of infected individuals, making HIV infection a chronic treatable disease (Gulick et al., 2019). Furthermore, the expectations of a sterilizing cure are increasingly close. However, there is still no effective vaccine for HIV/AIDS and additional issues involving the emergence of drug resistant viral variants, long-term side effects of the life-long cART and the lack of universal access to antiretrovirals have prevented ending the HIV epidemic.

After virus entry into CD4+ target cells, HIV RNA is converted to DNA and integrates into the host genome to form HIV proviral DNA or provirus constituting the viral reservoir (Katlama et al., 2013). HIV remains latent in a pool of resting CD4+ cells and uses the host cell machinery to produce its proteins and ensure the viral persistence, replication and transmission. Despite the advances of cART, the main obstacle to eradicate HIV is the establishment and persistence of long half-life viral reservoirs, containing transcriptionally inactive, but replication-competent HIV proviruses (Reviewed in Bruner et al., 2019; Cohn et al., 2020). In addition to the existence of viral reservoirs, HIV+ patients present a chronic immune activation that begins in early infection and is characterized by increased levels of proinflammatory mediators promoting a persistent inflammation state (Brites-Alves et al., 2018) and thymus dysfunction (Manjati et al., 2016). Chronic immune activation is a hallmark of chronic HIV infection and although with cART the activation level decreases, does not reach the same level of uninfected individuals. Chronic immune activation is considered as the driving force of CD4+ T cell depletion and progression to AIDS, and even at residual immune activation may lead to a systemic inflammation largely responsible for the increased non-AIDS morbidity and mortality observed in treated HIV+ individuals (Reviewed in Paiardini et al., 2013). Many studies have shown that patients on effective long-term cART may develop non-AIDS-related complications, such as cardiovascular disease, kidney disease, neurocognitive disorders, liver disease and several types of cancer that can be attributed in part to the chronic immune dysfunction, including chronic immune activation and immunosuppression (Deeks et al., 2013). Furthermore, even patients undergoing long-term cART do not show a complete reconstitution of their anti-HIV immunity (Goicoechea et al., 2006).

Due to the impediment of cART and host immune response to eradicate the viral reservoir, a controlled chronic infection is established until treatment is interrupted and a viral rebound is observed because of reactivation of HIV replication, reaching viremia comparable to pre-therapy levels (Cohen et al., 2018).

As in other viral infections, innate immune responses against HIV acts immediately as the first line of defense system and natural protection against the virus, through a process of pathogen sensing and activating signaling pathways during acute stage of infection. Once HIV pathogen-associated molecular patterns (PAMPs) have been recognized by the host pattern recognition receptors (PRRs) of sentinel innate immune cells (mononuclear phagocytes as monocytes, macrophages, DCs and NKs), cell-innate immune response initiates leading to the increase of antiviral defenses. The activation of innate immune response has been associated with a stronger adaptive immune response in different animal and human models, however the consequences in the course and outcome of HIV-disease are less well understood. Indeed, HIV-innate immune response has a major role in defining the function of the subsequent adaptive immune response and in the control of viral infection.

In the context of HIV infection, a natural clearance of infection apparently does not exist, indicating that innate immune response is insufficient to eradicate the virus. However, unlike the initial belief that HIV could not elicit a robust innate immune response, there is clear evidence of partial control of viral replication in primary infection (Fidler et al., 2016). In this stage, a high-level plasma viremia is observed which is subsequently limited by host immune responses and confers a partial immunological recovery for the most of untreated HIV individuals (Robb et al., 2016). Furthermore, several worldwide cohort studies have shown that there is a small group of HIV individuals (5-10%), who remained asymptomatic for more than 10 years in the absence of cART. These individuals have been extensively studied and are classified as long-term non-progressors (LTNPs), elite controllers, and viremic controllers (Gurdasani et al., 2014; Lopez-Galindez et al., 2019; Jiang et al., 2020). There is another group of individuals known as high-risk exposed HIV seronegatives (ESNs) characterized by multiple and prolonged exposures to the virus but remains asymptomatic and seronegative for infection (Horton et al., 2010). These groups of infected individuals represent natural models to control HIV infection in the absence of cART. Numerous studies of these individuals have led to the identifaction of natural resistance mechanisms to HIV, some related to the virus as the size of the inoculum and the molecular characteristics of the HIV strain, others to the host immunogenetic background or environmental factors as gut microbiota (Tamalet et al., 2021). However, the causes of natural resistance to HIV in these groups of patients have not yet been fully elucidated.

In this chapter we present recent advances related with two main challenges for innate immunity in controlled chronic HIV infection: the persistence of HIV reservoir and the systemic chronic inflammatory response.

A general scenario of innate immunity in the control of chronic HIV infection is described including its crucial components and the molecular basis by which innate immune response is regulated to elicit an antiviral response and inflammation. We highlight the role of HIV reservoir to maintain a chronic and incurable infection that simultaneously evades and exploits the host immune response to achieve latency and drive chronic inflammation. Current strategies targeted to eradicate reservoir and control inflammatory response in HIV infection are described.

Innate Immune Response in the Control of HIV Infection

Innate Immunity

The innate immune system is considered the first line of defense against infection or tissue injury, its components include physical barriers, soluble recognition and effector molecules, and cellular components. Innate immune cells include mononuclear phagocytes such as monocytes, macrophages, dendritic cells (DCs), polymorphonuclear cells (neutrophils, basophils, and eosinophils), natural killer cells (NK), and a special subset of T cells with limited antigen receptor diversity (γδ T cells). All these cells represent key components of the innate immune system due to its ability to sense the presence of pathogens and their ability to start an immediate and specific immune response. Activation of the immune system is initiated by pattern recognition molecules that can be expressed on innate cells or circulate in the blood. Accordingly, cells of the innate immune system must sense the presence of pathogens in the extracellular compartments to establish contact and either kill or engulf them. Soluble components that help in this process include members of the pentraxin family like C-reactive protein, lipopolysaccharide binding protein (LBP), the soluble form of CD14 (sCD14), and a set of proteolytic enzymes known as the complement system which is considered one of the most important enzymatic systems involved in the innate immune response. The complement system can be activated by separate pathways, one activated by antibodies (classical pathway), a second pathway dependent on plasma glycoprotein properdin, and a third one activated by mannose-binding protein (MASP) (Beutler B, 2004).

Activation of the innate immune system occurs very early after HIV infection. Days after HIV has entered the body, the levels of cytokines and interleukins increase in plasma in conjunction with the increase in the activity of innate immune cells, increase activity of antiviral restriction factors and of the components of the inflammasome, and consequent activation of the adaptive immune system. The events that occur during the first days of HIV infection are known to determine the progression to AIDS. Importantly, antiretroviral treatment can modify many of these events.

Sensing of HIV through PRRs

Innate immune cells recognize molecular structures shared by microorganisms called pathogen-associated molecular patterns (PAMPs) through pattern recognition receptors (PRRs). Innate cells also recognize damage-associated molecules released after tissue damage or cell death. Among PRRs are Toll-like receptors (TLRs), cytoplasmic DNA sensors, lectin receptors, cytoplasmic receptors like Nod-like receptors (NRLs) or RIG-I-like receptors (RLRs), and others, all located in different cellular compartments (Cai et al., 2019; Xia et al., 2021). PRR signaling activate downstream transcription factors such as IRF3, IRF7, NF-κB that drive the induction of antiviral genes. Upon activation, PRRs can also induce phagocytosis, maturation of antigen presenting cells (APCs) like DC or macrophages, or the production of inflammation molecules.

Cytosolic Sensors

Cytosolic sensors detecting HIV include: cGAS (DNA sensors cyclic GMP-AMP synthase), IFI16, PQBP1, the RNA sensor DDX3, the cDNA sensor NONO, among others. cGAS and IFI16 are the sensors that have been more described in the context of the HIV infection and have been shown to induce strong immune responses against HIV (Jakobsen et al., 2015). cGAS can detect the accumulation of double-stranded DNA (dsDNA), the reverse transcription product of HIV, by a length-dependent mechanism and not sequence-dependent (Luecke et al., 2017). IFI16, member of the AIM2-like receptor family, share some characteristics with cGAS. Both sensors are interferon-inducible, bind dsDNA, and have similar expression among different cell types and tissues. cGAS is mainly cytosolic while IFI16 needs to be deacetylated to translocate from the nucleus to the cytosol (Krapp et al., 2018). Importantly, the function of IFI16 is associated with the severity of disease produced by the pathogen thus, can influence in HIV disease progression (Booiman et al., 2014). cGAS activation generates the second messenger cGAMP (cyclic guanosine monophosphate-adenosine monophosphate) (Wu et al., 2013) and induces the activation of the adaptor protein STING. STING is transmembrane protein attached to the ER that, upon cGAMP binding, recruits the serine kinase TBK1 (Tanaka et al., 2012) to recruit IRF3 protein. Finally, IRF3 activation drives the expression of IFN-stimulated genes (ISG) (Liu et al., 2015). IFI16 detects HIV single-stranded DNA (ssDNA) and induces IFN-β production using the same STING–TBK1–IRF3/7 pathway, signaling cascade in macrophages (Jakobsen et al., 2013). In addition, IFI16 has been shown to induce pyroptotic death of infected resting tonsillar CD4 T cells by a mechanism involving the activation of capase-1 and not STING activation, as well as the death of bystander cells due to the production of IL-1β (Monroe et al., 2014). The differential expression of these proteins on CD4 T cells or macrophages make these proteins a good target for designing new molecules that can be used as alternative treatments for the elimination of HIV reservoir.

Toll-Like Receptors (TLRs)

TLRs are expressed mostly on myeloid cells but also in hematopoietic stem and progenitor cells and in non-immune cells such as epithelial or endothelial cells (Schmid et al., 2011; Carty et al., 2021). TLRs are glycoproteins structurally composed of three domains, a ligand-recognition leucin-rich repeat (LRR), a transmembrane domain, and an intracellular domain with either ligand-binding or signaling motifs (TIR, Toll/IL-1 receptor) (Gay et al., 2014). Signaling through TLRs requires cytoplasmic adaptors, mainly MyD88 or TRIF proteins (Beutler et al., 2004; Aluri et al., 2021). TLRs can be located either on the cell surface or within endosomes (Martinsen et al., 2020).

Toll-like receptor activation during HIV infection occurs not only in infected cells but also in non-permissive cells like pDCs. Several studies have described TLR activation by HIV, envelope proteins gp41 and gp120 can bind and activate TLR-2 (Reuven et al., 2014); Gag proteins p17 and p24 bind and activate TLR-2/TLR-1 and TLR2/TLR-6 heterodimers (Henrick et al., 2015). Importantly, TLR2 activation has been associated with high risk to acquire HIV infection and with faster disease progression (Laplana et al., 2020; Vidyant et al., 2017). In addition to the microbial translocation products produced due to damage of the gut epithelial cells during HIV infection, TLR-4 can be activated by the HIV gp120 protein (del Corno et al., 2016; Nazli et al., 2013). HIV trigger TLR-7 and -8 activation with the consequent production of IFN-α, TNF-α and IL-1β (Guo et al., 2014), contributing to the activation of pDCs and macrophages (Bernard et al., 2014). All these data suggest that constant TLR-4, -7, and -8 activation together with the constant production of pro-inflammatory cytokines due to TLR activation might contribute to the persistent immune activation and inflammatory environment observed in people living with HIV, even in people on cART (Marchetti et al., 2013). Despite more information about the role of TLRs in people with controlled viremia due to cART is needed, TLRs have been considered important targets to control or eliminate HIV.

Antiviral Innate Effector Cells: NK Cells, Dendritic Cells, Monocytes and Macrophages

Cells of the innate immune system initiate immune responses induced directly by viral particles or by soluble factors produced by infected cells. Host restriction factors expressed in innate effector cells can directly inhibit viral replication and usually results in a transient expression of inflammatory cytokines by infected cells. Finally, cells of the innate immune system activate and coordinate cells of the adaptive immune system to mount a complete immune response leading to the elimination of the pathogen.

Natural Killer Cells

Natural killer cells (NK) play a key role in the host defense due to their cytotoxic activity against virus-infected cells or tumor cells. Natural killer cells can be divided in two major subsets, $CD56^{bright}CD56^{neg}$ cells that are 10% of all NK cells and are the most common NK cells in lymphoid tissues, and $CD56^{dim}CD16^{+}$ cells that have higher amounts of cytolytic granules, constitute 90% of the total of NK cells, and can migrate to inflamed tissues. A third group of NK cells that lack the expression of CD56 but are $CD16^{high}$ has been described during HIV infection. These cells are dysfunctional NK cells with low cytotoxic capacity and low production of cytokines upon activation (Alter et al., 2005). Importantly, a subset of NK cells that produce IL-22, a member of the IL-10 family that regulates inflammation at the gastrointestinal tract (GI) and mucosal-associated lymphoid tissues, has been associated with natural resistance to HIV in exposed seronegative individuals (Missé et al., 2006).

In the context of HIV infection, NK cells have an important role during the acute phase of HIV infection due to rapid IFN-γ production and strong cytotoxic activity that has been associated with a higher CD4 T cell set point. However phenotypic and functional abnormalities of NK have been described since the acute phase of HIV infection. HIV causes an expansion and activation of the whole NK cell pool but with abnormal distribution of its subsets, a reduction of pro-inflammatory subpopulation NKCD56^{bright} and increases in CD56^{dim}CD16^{+} and of a dysfunctional NK cell subpopulation CD56^{neg}CD16^{+} have been reported in people living with HIV (Alter et al., 2005 and 2009). HIV infection reduces the expression of MHC molecules in infected cells, the ligands of NKG2A/C and KIRs (Zulu et al., 2017). Also, it has been described different expression of KIR receptors, high activation markers, and uncontrolled cytokine production in NK cells of HIV+ people, compared to NK cells of HIV negative people (Zulu et al., 2017). Alteration in the expression of receptors include a reduced expression of NCRs, NKp30, NKp44, ad NKp56 (Junior et al., 2016), essential for the containment of HIV; and an upregulation of the NKG2D receptor that leads to lysis and death of these cells (Bayigga et al., 2014). CD56^{neg}CD16^{+} NK cells are characterized by low expression of NKG2A, NCR and high NKG2C and have low cytokine expression and low cytolytic activity (Milush et al., 2013). A reduction of intracellular perforin and granzyme A granules in total NK cells was associated with the reduced ability to ADCC killing in people living with HIV (Altfeld et al., 2011). Recently, it has been shown that the high inflammatory cytokine environment during viremic HIV infection drives to the expansion of CD94^{+}CD56^{high} subset of NK cells, characterized by its increased cytotoxicity and degranulation against infected HIV CD4 T cells, which supports its important role to control HIV viremia (Wang et al., 2020). Further, HIV vaccine trial RV144 emphasized the importance of ADCC activity in the acquisition of HIV as authors found correlates of ADCC with lower risk of infection (Tomaras et al., 2013). There have been proposed three methods by which NK cells control HIV infection: DC editing, which implies the lysis of immature DCs via NKp30 and IFN-γ with the preservation of mature DCs (Flores-Alvarez et al., 2018); help to increase the recruitment of cytotoxic cells after TLR-7 or -8 activation (Vibholm et al., 2017); or by the production of IFN-γ, TNF-α and β-chemokines that are ligands of CCR5 impeding HIV attachment to this co-receptor (Vibholm et al., 2017). Antiretroviral treatment induces a general recovery of NK cells at different times after cART initiation but with defects in the subset distribution and IFN-γ persistent production (Alter et al., 2009).

Recent evidence suggests that NK cells also have an important role in the containment of the latent HIV reservoir. NK cells and not CD8 T cells, were correlates of the decrease in HIV-DNA in HIV+ people on ART (Olesen et al., 2015). Several *in vitro* models of HIV latency with cells of people on cART showed that NK cells can kill infected CD4 T cells after reactivation (Desimio et al., 2018; Garrido et al., 2019). It has been hypothesized that killing activity of NK cells in HIV reactivation models is dependent on the expression of ligands of the NKG2D receptor. The increase of NKG2D in latently infected CD4 T cells treated with combinations of latency reversal agents (LRAs) has been observed in *in vitro* models (Covino et al., 2021). Therapies targeting NK cells with checkpoints blockade, CAR-NK cells, NK intracellular signaling modulation with small molecules, can be promising treatment for HIV infection, as has been for some cancers (Andre et al., 2018; Liu et al., 2018). CAR-NK cell therapy, consisting of CARs, made with single chain fragments of the variable region of an antibody that recognize a specific target and gamma or zeta subunits of an Ig or TCR, reduced tumors in cancer patients without side effects (Liu et al., 2020) suggesting that CAR-NK cells as good candidates to improve immune responses against HIV infection (Zenere et al., 2019).

Dendritic Cells

Dendritic cells (DCs) link innate and adaptive immunity. The main functions of these cells are capture, process, and present antigens to induce activation of other cells of the innate and adaptive immune system (Collin et al., 2018). Dendritic cells are usually found in immature state distributed in the body with special localization in mucosal and lymphoid tissues. Immature DC can sense and recognize pathogens through PRRs and produce pro-inflammatory cytokines to initiate inflammation and activation of the innate or adaptive immune responses. An increase in the expression of co-stimulatory proteins, in the antigen-presenting capacity, and migration to secondary lymphoid tissues are part of the maturation process of DCs. Once mature, DCs can activate cells of the adaptive immune system. All DCs lack the expression of lineage markers (CD3, CD19, CD56) and express high levels of HLA-DR and can be divided in two main groups: plasmacytoid DCs (pDCs) and conventional DCs (cDCs). New types of blood DCs have been described with single cell technology (Villani et al., 2017) that will require further study to know their exact role in HIV control.

Conventional Dendritic Cells

Conventional DCs (cDCs) are a specialized group of APCs that can sense and process antigens. These are short-lived cells that give rise to resident or migratory DCs in tissues like secondary lymphoid tissues. Conventional DCs of the lymphoid tissues localize in the T- and B-cell zones and are prepared to activate these cells after its maturation. Immature cDCs have high phagocytic abilities and, when mature, also have increased capacity to produce cytokines (IL-6, IL-12, IL-15, IL-23, TNF-α, IL-1β). Conventional DCs can be divided in two subsets based on the expression of BDCA-3 (blood DC antigen 3) and BDCA-1 (blood DC antigen 1) (Schlitzer et al., 2015). BDCA-3+ cDCs can activate NK cells and have a high capacity for cross-presentation to CD8 Tc to induce Th1 responses. Also, these cells can produce type I and III IFNs in response to TLR-3 and TLR-9 activation (Hemont et al., 2013). BDCA-3+ cDCs reside in peripheral tissues, lungs, skin, and peripheral blood and can migrate to LN where are localized in paracortical regions. These cells express the transcription factor IRF8 (Kristoff et al., 2019). BDCA-1+ cDCs are the most abundant cDCs in peripheral blood and express different markers among the different tissues where these cells reside. BDCA-1+ cDCs differ from BDCA-3+ cells in their expression of multiple TLRs 1 – 8, the expression of IRF4 and not IRF8, and by their capacity to induce Th1, Th2, Th17, immune responses, and Tregs after sensing intracellular pathogens, extracellular bacteria, parasites or fungi (Schlitzer et al., 2015; Sittig et a., 2016).

Plasmacytoid Dendritic Cells

Plasmacytoid DCs (pDCs) are long-lived cells that produce high amounts of type I IFNs upon viral infection, pDCs differ from conventional dendritic cells (cDC) in morphology, phenotype, and function (Reizis B., 2019). Plasmacytoid DCs can be recognized by the expression of BDCA-2 and BDCA-4 (blood dendritic cell antigen 2 and 4), CD123 (IL3Ra), ILT7 (immunoglobulin-like transcript 7), and high expression of TLR-7 and TLR-9. Immature pDCs can be recruited to inflamed tissues through signaling of chemokines, once in the tissue pDCs can uptake and recognize pathogens and signal through TLRs.

Plasmacytoid DCs are important players in the primary response against viruses due to their substantial production of IFNs type I and III through sustained TLR-7 and -9 signaling (Yin et al., 2012). Once activated, pDCs undergo changes in their morphology upregulating HLA-DR and activation markers that allow the antigenic presentation to T cells (Colonna and Cella, 2007). pDCs are responsible for over 95% of the type I IFN produced in response to viruses, constitutively express IRF4, IRF7, and IRF8 which are transcriptional factors essential for IFN-1 production (Siegal et al., 1999). Types of IFNs that can be produced by pDCs depend on the TLR that sense the pathogen (IFN-α, β, κ, ω, τ and IFN type III (Yin et al., 2012; Hemman et al., 2017). Interferons bind to IFN-α/–β receptors expressed in almost all nucleated cells and activate signaling cascades that stimulate interferon-stimulated genes (ISGs) that encode many antiviral proteins. Type I IFNs block viral infection, resulting in pathogen clearance (Galicia et al., 2014) and stimulate cDCs and other cells like NK and CD8 T cells, contributing to their effector functions against viruses (Gerosa et al., 2005). Type I IFNs also promote the differentiation of B cells into Ab-secreting cells (Jego et al., 2003) and activation of caspase-1 via inflammasome and the production of IL-1β and IL-8, cytokines that have been associated apoptotic cell death (Anand et al., 2011). Furthermore, pDCs can also produce other inflammatory cytokines such as IL-6, TNF-α, IL-8, and IP-10, possibly contributing to pathologic inflammatory process.

HIV infection induces irreversible changes in DC numbers and function. The anatomical distribution of DCs changes during HIV infection, a notable reduction in the number of DCs in blood occurs since early stages of HIV infection (Sabado et al., 2010) probably due to the redistribution to lymphoid tissues (Dillon et al., 2008; Carranza et al., 2016; Lehman et al., 2010). Dendritic cells express high levels of the restriction factors making them refractory to HIV infection despite their expression of CD4 and CXCR4 or CCR5. The main restriction factor limiting HIV replication in cDC and monocyte-derived DCs (MDDCs) is SAMHD1 (Ryoo et al., 2014), other restriction factors expressed in these cells are cyclophilin A (Manel et al., 2010) and TRIM5a (Pertel et al., 2011). Therefore, the possibility that cDC cells become infected and produce low amounts of virus, along with the fact that these cells have long half-life, suggest that they can be a cellular component of the latent HIV reservoir.

The role of pDCs during HIV infection can be beneficial or detrimental. On one hand, pDCs control HIV through controlled IFN-α production, functioning as APCs (Fonteneau et al, 2003), or by killing infected cells through TRAIL pathway (Barblu et al., 2021). Effective immune responses linked to enhanced IFN responses have been described in DCs from HIV controllers, people with a natural control of HIV replication (Barblu et al.,

2012, Martin-Moreno and Muñoz-Fernandez, 2019). On the other hand, IFN-α production by pDCs can boost CD4 T cell infection making them more susceptible to infection and, at the same time, facilitating the recruitment of more CD4 T cells to sites of high HIV replication (mechanisms reviewed in Martin-Moreno and Muñoz-Fernandez, 2019). Furthermore, exacerbated production of IFN-α by pDCs, probably due to hyper-response to TLR-7 (Sabado et al., 2010), stimulates strong antiviral responses that may contribute to the chronicity of HIV infection (O´Brien et al., 2013), to the chronic activation of immune cells observed during HIV disease (Benlahrech et al., 2011), and impedes cDC differentiation from monocytes (Kodama et al., 2010) altering its antigen presenting capacity. Alternative treatments that modulate the production of cytokines, especially of IFN-α, might contribute to the reduction of detrimental inflammation and activation. Chloroquine has been assessed in HIV+ individuals on cART but did not improve CD4 T-cell recovery or reduce immune activation or inflammatory markers despite being well tolerated (Routy et al., 2015).

Monocytes and Macrophages

Monocytes are mononuclear cells that protect against viral, bacterial, fungal and parasitic infections by killing pathogens via phagocytosis or the production of reactive oxygen species (ROS), myeloperoxidases, or inflammatory cytokines (Saha et al. 2011). Monocytes can differentiate into macrophages or DCs, depending on the inflammatory microenvironment, but are not the only precursors of these cells. The main markers to recognize monocytes are CD14, a protein that function as an accessory protein of TLR-4, and CD16 (Fcγ receptor III involved in ADCC). Monocyte subsets can be identified based on the expression of CD16 and CD14: classical monocytes (CD14++CD16-), constitute 90% of all monocytes; intermediate monocytes (CD14++ CD16+) that represent around 5% of circulating monocytes, and non-classical monocytes (CD14+ CD16++) (Ziegler et al., 2010; Kapellos et al., 2019). Recruitment of monocytes to the site of inflammation depends on the expression of chemokine receptors like CCR2 or CX3CR1 (fractalkine receptor), once in the tissue, monocytes differentiate into macrophages. Functionally, monocyte subsets are slightly different, all monocytes can produce TNF-α, IL-1, and IL-6 in response to LPS stimulation being the non-classical monocytes the ones that produce higher amounts of TNF-α and the classical monocytes that produce higher amounts of IL-10. Classical monocytes have higher phagocytic and myeloperoxidase activity and non-classical monocytes have been involved in the pathogenesis of various diseases because can produce high amounts of TNF-α.

Blood monocytes are an important source of tissue macrophages during an inflammation process, a second type of macrophages that originate from yolk sac progenitors during embryogenesis also constitute the pool of tissue macrophages (Perdiguero et al., 2016; Hoeffel et al., 2018). Activation of monocytes and classic activation of macrophages requires exposure with certain cytokines like IFN-γ that increases its responsiveness against pathogens (Bosisio et al., 2002). During acute HIV infection, macrophages polarize into M1 (pro-inflammatory) and eventually shift to an M2 (anti-inflammatory) in chronic infection (Burdo et al., 2015; Murray et al., 2017). The proportion of the pool of tissue-resident and monocyte-derived macrophages vary depending on the homeostatic or inflammation state of the tissue (Wang et al., 2016). Inflammatory signals, like the ones seen during HIV infection, may induce a shift toward macrophage tolerance in the response to PRR ligands that might result in increased cytokine production (Biswas et al., 2009).

Phenotypic characteristics of monocytes change during HIV infection, a high expression of activation markers, adhesion molecules, and chemokine receptors have been described even in people on cART. Classical monocytes lose the expression of CCR2, their phagocytic activity, and their capacity to induce CD8 T cell activation and proliferation. An expansion of intermediate monocytes has been observed in viremic HIV+ individuals that is normalized during cART (Hearps et al., 2012; Westhorpe et al., 2014; Booiman et al., 2017) and decreased with Maraviroc intensification scheme (Ndlovu et al., 2014). Importantly, intermediate monocytes have been demonstrated to have HIV DNA in HIV+ treated individuals (Ellery et al., 2007).

Macrophages and monocyte play different roles in HIV infection. During the first days of infection, macrophages help in the initiation of the adaptive cellular immune response against HIV but can also help to establish infection at sites of viral entry. Furthermore, monocytes can help to the dissemination of HIV in the body due to its high migratory capacity to inflamed tissues (Valcour et al., 2010). Infected tissue-resident macrophages have been detected in lymph nodes during AIDS (Orenstein et al., 1997), in the liver (Mosoian et al., 2017), lungs (Jambo et al., 2014; Cribbs et al., 2015), and CNS (Ko et al., 2019) of HIV+ people, as the corresponding specialized macrophage of each tissue. Macrophages are considered long-lived component of the HIV reservoir as these cells are resistant to cytotoxic CD8 T cells (Clayton et al., 2018) and can reside in anatomical sanctuaries that will support viral persistence during cART (Estes et al., 2017).

Monocytes and macrophages are probably the innate immune cells more associated with the chronic inflammation observed during controlled HIV infection (Mosoian et al., 2017). Increased activation and differentiation of monocytes have been observed during treated HIV infection (Hearps et al., 2012; Booiman et al., 2017). Importantly, a large set of evidence suggest that some HIV-related inflammatory comorbidities in people on cART can be linked to monocyte/macrophage activation (reviewed in Jaworowski et al., 2019; Westhorpe et al., 2014; Sandler et al., 2011). It has been shown that inhibiting tissue factor activity on monocytes decrease significantly immune activation in a non-human primate (NHP) model of HIV infection (Schechter et al., 2017). Activated monocytes due to microbial translocation can cross the blood brain barrier and differentiate into pro-inflammatory macrophages in the brain causing neuronal death (Williams et al., 2014) suggesting that systemic inflammation also play a role in HIV-associated dementia (HAD).

Activation in Innate Immune Cells: Cytokines and Chemokines

Cytokines involved in development, differentiation, and function of innate and adaptive cells, include interleukins (ILs), interferons, hematopoietic growth factors (O'Shea et al., 2002). Cytokine receptors signal through two types of pathways: the JAK-STAT pathway and the activation of the MAP kinase cascade (Berenson et al., 2006). Among the cytokines that activate the JAK-STAT pathway are IFN-γ that activates JAK1/STAT1; IL-6, IL-10 that activates JAK1/STAT3, IL-10 that also activates STAT3, and IL-2, IL-4, IL-6, IL-23 that also activates STAT3, STAT4, and STAT6 (Murray et al., 2007). Importantly, the differentiation of T cells is influenced by the signaling of some interleukins, Th1 cells by IL-2; Th2 cells by IL-4, and Th17 cells by IL-6 and IL-23. Treg differentiation is influenced by IL-2/STAT5 signaling and Tfh differentiation by the signaling of IL-21 and STAT (Volgezang et al., 2008).

Chemokines are small peptides that facilitate the migration of immune and non-immune cells from the circulation into the tissues. There are two classes of chemokines in mammals, the C-C and CXC types and two minor subfamilies (CX3C and XC) that use G-coupled proteins for transducing their signal and normally promote pro-inflammatory responses or homeostatic signals involved in tissue responses, extracellular matrix remodeling, angiogenesis, and others. Chemokine receptors can be divided in four types: those that activate nuclear factor (NF)-κB and MAP kinases like TNF-α receptors, IL-1 family receptors, and the IL-17 family; receptors that activate the JAK-STAT pathway (most cytokines belong to this family); receptors of the TGF-β family that activates SMAD transcription factors; and growth factor receptors that signal via ERK kinases. Of note, some chemokine receptors are also coreceptors for HIV, among these CXCR4 and CCR5 are the most important. It has been described that some HIV variants can use other chemokine receptors (CCR1, CCR3, CCR8, CX3CR1, CXCR4, CXCR6). CCR5 HIV coreceptor is important for the establishment of HIV infection and CXCR4 has been linked to a switch of variants in late stages of HIV infection in around 40% of HIV+ people. The expression of CXCR4 and CCR5 in different cells may be determinant in the establishment of the HIV reservoir.

In the context of HIV infection, cytokines can modulate HIV replication in CD4 T cells and macrophages; IFN-α, IFN-β, IL-10, IL-13 and IL-16 inhibit viral replication, and TNF-α, IL-1, IL-6 stimulate HIV replication (Kedzierska et al., 2001). Proinflammatory cytokines, TNF-α, IL-1, IL-6 and IL-8, are mostly synthetized by macrophages, some cytokines are considered markers of inflammation during HIV infection: TNF-α, sTNFR-1 and -2, IFN-γ, IL-1, -6, -8, -10, -17, and -23, MCP-1 (monocyte chemotactic protein-1), MIP-1α and 1β (macrophage inflammatory proteins -1a and -1b); IP-10. Some of these cytokines have been associated with markers of clinical progression or comorbidities of HIV, and cART.

IP-10

IP-10 or CXCL10 is produced mainly by monocytes but other cells like NK cells, T cells, endothelial or stromal cells can also produce it (Wu et al., 2017). Its receptor CXCR3 is expressed on T lymphocytes and NK cells principally. IP-10 secretion is driven by IFN-γ and increase STAT, NF-κB, MAPK and PI3K kinase signaling (Dhillon et al., 2007), process mediated by HIV Tat protein during HIV infection (Williams et al., 2009). In the setting of HIV infection, IP-10 levels are increased and can be detected since day 6 after plasma viral load is detectable. This cytokine is strongly associated with the time to reduction of the CD4 T cell counts during Fiebig stages III-V of the acute phase of HIV infection (Stacey et al., 2009; Jiao et al., 2012) and with plasma viral load during the chronic phase of infection (Mhandire et al., 2017). IP-10 levels are reduced with cART implementation but not in people that do not respond to the treatment, event that is influenced by the nadir of CD4+ T cells (Valverde et al., 2018). Increases in IP-10 levels disrupt T cell functions like the production of IFN-γ (Ramírez et al., 2014) and have been associated with T cell activation during cART (Cortes et al., 2018). Importantly, it has been postulated that IP-10 may promote latent HIV infection due to changes in the phosphorylation status of filamentous actin and accumulation of HIV DNA in macrophages. Importantly, IP-10 levels can be used to monitor HIV disease progression (Pastor et al. 2017) or to detect co-infections as its levels in plasma are high in HIV individuals with co-infections with HCV or TB (Garcia-Basteiro et al., 2017). It is overproduced in the CSF of HIV+ people and correlates with HIV RNA of CD4+ T cells found in the CSF (Andersone t al., 2021).

IL-6

IL-6 is considered one of the most prominent pro-inflammatory cytokines, regulates cellular activities in various organs like brain, helps in the regulation of the metabolism and in the development of some diseases like cancers, HIV infection and severe COVID19 (Narazaki et al., 2018; Gubernatorova et al., 2020). It is produced by myeloid cells after TLR stimulation together with IL-1β and TNF-α. IL-6 can activate a membrane-bound receptor (IL-6R) and a second receptor protein gp130. The fact that all cells of the human body express gp130 suggests that this pathway may be involved in some pathologies in which IL-6 is present. IL-6 concentration in plasma increases 1,000 folds during severe inflammatory pathologies (Tanaka et al., 2014). During HIV infection, IL-6 is considered a marker of inflammation along with other markers like high-sensitivity C-reactive protein, D-dimer, and others (Nixon et al., 2010; Neuhaus et al., 2010). Elevated levels of IL-6 in plasma of HIV+ people are considered a predictor of mortality and of the appearance of opportunistic infections (Rodger et al., 2009). Despite IL-6 levels are reduced by cART, IL-6 levels have been associated with cardiovascular disease after cART initiation (Neuhaus et al., 2010). Along with TNF-α, higher IL-6 levels have been associated with major depression disorders (Musinguzi et al., 2018). Importantly, IL-6 levels increase after controlled treatment interruption (El-Sadr et al., 2006).

TNF-a

TNF-α is a pleiotropic cytokine produced by activated immune cells including monocytes and macrophages, T and NK cells in response of TLR activation.

TNF-α initiates signaling through the activation of two receptors: TNFR1 or TNFR2. TNFR1 is expressed in all human tissues while TNFR2 is mainly expressed in immune cells. Activated TNFR1 signaling involves the activation of NF-κB and MAPK kinases to induce tissue inflammation, cell survival and proliferation (Brenner et al., 2015) and is essential to induce cytotoxic and proinflammatory responses against pathogens. Of note, the excessive production of TNF-α can be harmful and lead to diseases like autoimmune diseases. TNF-α has been associated with HIV disease progression, with increased viral load set point and shorter times to reach less than 500 CD4 T cells/μL (Vaidya et al., 2014). The effect of cART on TNF-α is not clear as studies reported contrasting results. TNF-α has a protective role against HIV infection in *in vitro* models, it inhibits HIV entry into monocyte-derived macrophages and in alveolar macrophages by reducing CCR5 expression on these cells (Herbein et al., 1996). In contrast, TNF-α increases HIV transcription in monocytes and lymphoid cells that have been already infected and tested for reactivation with LRAs, probably due to the activation of NF-κB (Banerjee et al. 2012). Although it has been proposed the use of TNF-α signaling modulators as HIV treatment, the detrimental effect of TNF-α must be considered as some studies using TNF treatment have shown reactivation of tuberculosis (Keane et al., 2001).

IL-10 Family

The IL-10 family of cytokines are a group of molecules that can be divided in three groups, based on their functions: IL-10 group, exerts immunosuppressive function to reduce inflammatory responses during the resolutions phase of infections or inflammation processes (Engelhardt et al., 2014); the IL-20 family (IL-19, IL-20, IL-22, IL-24, and IL-26) that acts mainly on epithelial and stromal cells to protect barrier tissues and involved in tissue remodeling and repair (Sonnenberg et al., 2011); and the IL-28 family that includes IL-28A, IL-28B, and IL-29 (collectively named as type III Interferons) (Laezar et al., 2019). Of these, IL-10 is the most well-characterized, it is produced in response of PRRs activation in activated T cells, macrophages, DCs, NK and B cells. Optimal production of IL-10 in Tregs depends on TGF-β signaling (Josefowicz et al., 2012). IL-10 signal through the JAK-STAT signaling pathway and acts on myeloid cells to inhibit its proinflammatory cytokine production and antigen presenting capabilities (Fiorentino et al., 1991a, 1991b). IL-10 can directly suppress the activation of T cells and the production of INF-γ and IL-2 (Donnely et al., 1999) and promote the survival of Tregs (Chaudhry et al., 2011; Coomes et al., 2017). IL-22 has an important role in innate pathogen defense, it has been linked with severe inflammation and is highly expressed on activated CD4 T cells and ILC3s (group 3 innate lymphoid cells) (Rutz et al., 2014; Sonnenberg et al., 2011).

Despite IL-10 is linked to modulation of the immune response, it is also well known that, if IL-10 signaling is compromised or dysregulated, it may contribute to the development of disease or chronic infection (Oldstone, 2015). IL-10 have an important role in HIV pathogenesis, high levels of this interleukin have been observed in HIV+ people (Stylianou et al., 1999) and may be associated with susceptibility to infections caused by mycobacteria (Muller et al., 1998). Notably, IL-10 has been linked to the development of the Immune Reconstitution Inflammatory Syndrome (IRIS), a clinical condition observed in a subset of HIV+ individuals that starts treatment, in which the immune system begins to recover, but then responds to a previously acquired opportunistic infection with an exacerbated inflammatory response that makes the symptoms worse. (Takodera et al., 2013). The levels of IL-10 and IL-22 are increased in people that developed IRIS due to tuberculosis reactivation, being macrophages the main cells that produce these cytokines during IRIS (Takodera et al., 2013).

Interferons

Interferons (IFNs) are cytokines produced in response to pathogenic host invasion or inflammation by a variety of cells. pDCs are the main producers of IFN-α after viral infections (Siegal et al., 1999). In humans, IFNs can be divided in three families based on their structure, genetics, functional characteristics, and biological activity according to receptor they use (Zhou et al., 2014). Type I family, signal through IFNα/β receptor (IFNAR), include 12 types of IFN-α, IFN-β, IFN-κ, IFN-ε, and IFN-ω (Schreiber et al., 2015) and are involved in innate and adaptive responses during viral infections. Type II IFN, include only IFNγ produced by T and NK cells and have immunoregulatory and inflammatory properties (Gallin et al., 1995). Type III IFNs composed of IFNλ1, IFNλ2, IFNλ3, and IFNλ4, important in epithelial and mucosal tissues (Hemman et al., 2017). Type I IFN subtypes differ in localization and their affinity for IFNARs that determine its biological activity. To date, several members of the type I interferon family are known:

- *IFN-α:* produced by leukocytes, has different subtypes with different biological activities and receptor binding affinities. Mainly produced by pDCs and is critical for the activation of NK cells enhancing its cytotoxic activities and IFN-γ production. Induce the transcription of interferon-stimulated genes (ISG) with anti-viral activities.
- IFN-β: produced mainly by fibroblasts, IFN β inhibits production of the inflammatory Th1 cytokines, IFN γ and IL-12, by human dendritic cells (Nagai et al., 2007), promotes the production of the anti-inflammatory IL-10
- *IFN-ε*: expressed in the lungs, brain, skin, intestine, uterus, vagina, and ovarian tissue, has weaker antiviral and NK cell induced cytotoxicity than IFN-α and -β. Induces more genes of the TNF-α pathway and phagocyte activation to block HIV replication than IFN-α (Li et al., 2017). IFN-ε has anti-HIV activity mediated by the expression of host restriction factors like MX2, BST2, IFITM3, and APOBEC3G (Garcia-Minambres et al., 2017). IFN-ε cytokine production may contribute to anti-HIV responses in cervical ephitelium (Abdulhaqq et al., 2016).
- *IFN-ω*: contribute to the expression of cytokines released by innate immune cells, it has been proposed to be used in patients who are resistant to IFN-α, because it produces fewer adverse effects than IFN-α.
- *IFN-κ*: expressed in keratinocytes and lymphoid cell populations like monocytes and monocyte-derived DCs (DeCarlo et al., 2010).
- *IFN-τ*: seem to have the highest antiviral activity with no evidence of toxicity, effectively suppress HIV replication in monocyte-derived macrophages and enhance the expression of β-chemokines but not of IL-1β and TNF-α (Rogez-Kreuz et al., 2005).

Interferons are produced very early after HIV infection as a rapid innate immune response but also during the chronic phase of the infection, levels of IFN-α correlate positively with HIV viral load and negatively with CD4 T cell counts (Stacey et al., 2009, Hardy et al., 2013). It has been proposed that IFNs have a dual effect during HIV infection, non-pathogenic IFN signaling has been associated with HIV replication control (Barblu et al., 2012) but progression to AIDS have been associated with a persistent production and increased IFN responses. The beneficial effect of interferons during HIV infection has been related to the activation of ISGs that inhibit HIV replication (reviewed in Bosinger et al., 2015). IFN-I signaling interferes with HIV replication as it favors the expression of restriction factors such as SAMHD1, APOBEC3, TRIM5a, and Tetherin (Nchioua et al., 2020). The production of different IFNs, IFN-α, -β- and $-\gamma$, helps to inhibit HIV infection of macrophages and CD4 T cells (Goujon et al., 2010). Type I IFNs also stimulate the activation of NK cells, support their survival and enhance its cytotoxic activity and the expression of Il-15 receptor (Stackaruk et al., 2013) to help with the elimination of infected cells. Conversely, the constant IFN-related signaling shift the immune response toward a detrimental signal (Utay et al., 2016), sustained IFN-α production has been associated with increased HIV viral load, progressive CD4 T cell depletion, immune dysfunction, and immune activation (Stacey et al., 2009; Martinson et al., 2010) leading to immune suppression and AIDS (Sandler et al., 2014).

Novel Strategies to Control HIV Infection

Dendritic Cells in Immunotherapies

Due to their role as APC, DCs have been proposed to be used as vaccine adjuvants or immunotherapies against HIV. Dendritic cell-based immunotherapies have been successful in mice models where MDDCs induce cytotoxic responses against tumors (Kalinski et al., 2011). This therapy involves loading of autologous DCs with antigens and re-introduction of these cells into the patient. Immunotherapies have the advantage that the researcher can choose which antigens to load in the DCs. It has been proposed that the loading DCs with pools of peptides of the HIV envelope protein might generate an immune response that produce broadly neutralizing antibodies (Mohamed et al., 2020). However, the efficacy of DC-based HIV vaccines or immunotherapies have been low (38%) (reviewed in Coehlo et al., 2016) pointing to new methods to enhance the immunogenicity and antiviral efficacy of DC-based vaccines against HIV infection, probably a combination DC-based vaccine in combination with TLRs agonists. The use of DCs to support the "shock and kill" strategy to eliminate the HIV reservoir have been proposed since DCs can stimulate NK or CD8 T cell responses to recognize and kill infected CD4 T cells. Clinical trials on people with cART using DCs charged with different HIV antigens have shown increased CTL responses, including increases in HIV-specific responses (Jacobson et al., 2016; Gandhi et al., 2016; Van Gulk et al., 2012). In recent years, DCs have been proposed for its use to reactivate HIV provirus working as LRAs themselves (Ren et al., 2017). Tissue resident and blood cDCs have been shown to reactivate virus from latently infected CD4+ T cells (Van der Sluis et al., 2015). Loading of immature DCs with SEB super antigen resulted in reactivation of HIV in CD4+ T cells infected *in vitro* (Marini et al., 2008). An interesting study showed that co-culture of MDDC1 with latently infected cells trigger activation of HIV through PI3K-AKT-mTOR signaling (Kristoff et al., 2019) supporting the idea that certain types of DCs can function as natural LRAs. Further studies will be needed to explore the potential of DCs as LRAs and their use to eliminate latently infected cells.

Monocytes and HIV Reservoir Elimination

Despite the mechanisms for HIV latency have not been completely understood, it is influenced by multiple factors like sequestration of cellular restriction factors, epigenetic regulation, among others. HIV latency in macrophages have been studied in animal models of HIV infection. Replication-competent SIV virus have been recovered in cultures of macrophages from macaques infected with SIV and treated with tenofovir (Abreu et al., 2019). Sustained viral replication in tissue macrophages was observed after CD4 T cell depletion in a NHP infected with SHIV (Igarashi et al., 2001) supporting their role in the maintenance of the HIV reservoir. HIV latency mechanisms might differ in monocytes/macrophages, compared to latency mechanisms of CD4 T cells. Viral latency in monocytes can be related to the low levels of CycT1 which is a regulatory unit of the complex p-TEFb, a critical cofactor for HIV Tat protein. Low levels or CycT1 expression in monocytes have been related to low HIV transcription and thus in viral latency in these cells (Liou et al., 2002). Macrophages can contain unintegrated HIV DNA and persist longer periods than CD4+ T cells (Kelly et al., 2008) contributing to viral persistence. Accordingly, LRAs compounds have been shown to have different effects on CD4 T cells and macrophages, HDAC inhibitors can reactivate HIV from patients on ART but have not decreased proviral DNA in CD4 T cells (Archin et al., 2017). Importantly, *in vitro* that HDAC inhibitors can reactivate and decrease HIV release through autophagy pathway in monocyte-derived macrophages (Campbel et al., 2015). Thus, it is important to identify the cellular factors and pathways that regulate HIV latency in macrophages and CD4 T cells to improve the efficacy of LRAs or its combination as "shock and kill" strategy to deplete the replication-competent HIV reservoir.

More recently, it has been proposed a second strategy to eliminate the replication-competent HIV reservoir through a "prime, shock and kill strategy" (Badley et al., 2013). Taking into account that some cells that harbor HIV DNA, like macrophages, are resistant to apoptotic stimulus, this approach postulates that first, cells must become sensitive to apoptotic mechanisms before HIV reactivation. Several anti-apoptotic proteins are expressed in latently infected cells, either CD4 T cells or macrophages, including BCL-2, X-linked inhibitor of apoptosis protein (XIAP), BCL-XL, among others (Lopez-Huertas, 2013). Targeting anti-apoptotic pathways in macrophages before or during proviral reactivation with LRAs may promote cell death. Importantly, it has been reported that killing of infected macrophages by CD8 T cells takes more time than killing of CD4 T cells. Long synapses between CD8 T cells and infected macrophages stimulate the secretion of IFN-γ with the consequent production of pro-inflammatory chemokines by macrophages (Clayton et al., 2018), linking the macrophage component of the HIV reservoir with inflammation observed in treated HIV-patients. Further therapies aiming to eliminate the macrophage component of the HIV reservoir through killing with cytotoxic cells must overcome the probable increase in the production of pro-inflammatory cytokines by macrophages as part of its interaction with cytotoxic CD8 T cells.

Modulation of TLR Signaling

The modulation of TLR signaling have been proposed as an important target as therapeutic agents of HIV infection due to the important role of TLRs in inducing the expression of antiviral genes or promoting T and B cell responses against HIV. As TLRs can activate the transcription of factors that promote HIV gene expression, TLR agonists have been studied in *in vitro* and *ex vivo* models as probable LRAs. Some of these compounds have been used in clinical trials, although results on the reduction in HIV reservoir or latently infected cells have not been successful.

Nowadays, plenty of TLR agonists have been tested as LRAs in different *in vitro* HIV latency models, some of them already tested in clinical trials with contrasting results (reviewed in Macedo et al., 2019 and in Martinsen et al., 2020). As examples, the use of TLR-3 agonists in people on cART showed no changes in usRNA and total DNA in CD4 T cells (Saxena et al., 2019); TLR-7 agonists, used as LRAs in peripheral blood mononuclear cells (PBMCs) from HIV+ people on cART, showed an increase in HIV RNA of PBMCs and improved the killing of infected CD4 T cells (Tsai et al., 2017). Non-human primate models of SIV infection in treatment interruption (TI) studies showed that TLR-7 agonist induced small increases in SIV viremia ("blips") after four doses but had no effect on the time of viral rebound after TI, suggesting no reduction in latently infected cells (Lim et al., 2018). MGN1703, a TLR-9 agonist, increased unspliced RNA (usRNA) in PBMCs of HIV+ people on cART (Offerssen et al., 2016) but, contrasting this report, other study showed no reduction on vDNA or replication competent virus after TLR-9 agonists administered subcutaneously in HIV+ people (Vibholm et al., 2019). Pam3CSK4, a TLR-1/TLR-2 agonist, increased the transcription of HIV and p24 protein in a central memory CD4+ T cell model (Novis et al., 2013). It has also been described that TLR-2 agonists increased the permissiveness of CCR6+ CD4 T cells via NF-κB (Bolduc et al., 2017). Further studies must be done to clarify the interaction of these cells in supporting residual viral replication after TLR-2 activation. Importantly, LPS, a TLR-4 agonist, has been shown to reactivate HIV in macrophages of cART-treated HIV people (Ganor et al., 2019, Macedo et al., 2019) suggesting a close relation between bacterial translocation and the activation of TLRs supporting HIV replication. Alternatively, TLR agonists have been tested to boost CD8 or NK cell cytotoxic activities after replication of competent HIV reservoir. Immunomodulatory role of TLRs on pDCs, NK cells, CD4 and CD8 T cells, has been demonstrated after TLR-9 stimulation (Vibholm et al., 2017 and 2019, Krarup et al., 2018). Activation of immune cells could boost the clearance on latently infected cells after treatment interruption. The TLR-9 agonist MGN1703 was shown to stimulate NK and T cell ability to recognize and kill infected cells (Vibholm et al., 2019).

As results from clinical trials with TLR agonists have not been encouraging (reviewed in Macedo et al., 2019), it has been proposed the use of new combinations of TLR agonists with broad neutralizing antibodies (bNabs) as therapeutic use to control HIV replication after controlled treatment interruption. A combination of the TLR-7 agonist with the bNab PG121 in SHIV-infected monkeys (a non-human primate model of HIV infection) showed a delay in viral rebound after treatment interruption and, importantly, lowered the levels of DNA in LNMCs (Borducchi et al., 2018). Controlled clinical studies with combinations of TLR agonists with other immunomodulatory molecules must be performed. It will be important to evaluate how modulation of these treatments affects the expression of activation and inflammation markers in people on cART and its relation with HIV reservoir.

Modulation of Interferon Signaling

Type I IFNs have been used as therapeutic tool for HIV (Bossinger et al., 2015). Early studies of the role of IFN-α in HIV infection were performed before the cART era. Some studies that used IFN-α as monotherapy showed less severe CD4 depletion, lower HIV burden, fewer opportunistic infections, and slower progression to AIDS but other reports showed no significant improvements (reviewed in Bosinger et al., 2015). Despite it was shown that IFN-α was well tolerated by HIV+ people the side-effects of its administration overcame its beneficial effects. Other schemes of use of IFNs, like the use of IFN-β in combination with the use of antiretrovirals, were also tested but did not show better results than the use of only antiretrovirals. Importantly, adding IFN-α to cART scheme did not improve CD4 T cell reconstitution (Bossinger et al., 2015). Reports of the supplementation with IFN-α in people on cART differed in the timing, doses, and scheme of IFN-α implementation, its results emphasized the importance of the timing of IFN administration. A study in NHP showed that administration of IFN–α during the rectal challenge with SIV prevented infection (Sandler et al., 2014) proposing that enhancing IFN-α signaling during the first days of infection may be critical and clinically beneficial. The development of pegylated-IFN (Peg-IFN) showed better pharmacodynamics, studies using Peg-IFN as monotherapy underlined the fact that people with higher ISG expression reached a state of tolerance against IFN-α (Iyer et al. 2017), supporting the fact that timing of IFN-α administration is important.

The use of IFN as a cART adjuvant have also been considered in studies of analytical treatment interruption (ATI) studies. The use of Peg-IFN showed a reduction in the time of viral rebound after ATI but had no changes on HIV reservoir (HIV-DNA) (Dianzani et al., 2008), even if cART was given during primary HIV infection (Goujard et al., 2012). Another study showed that treatment of HIV-infected subjects under cART with Peg-IFN-α2a was associated with the suppression of HIV viral load and decreased HIV-DNA, if given as an add-on therapy before ATI (Azzoni et al., 2013). IFN-α treatment might have beneficial effects on HIV+ people but if given at the incorrect time, might lead to contribution to persistent activation and inflammation, tolerance, or negative regulation of the IFN-IFNAR.

IFN-α in combination with cART and LRAs may help to eliminate HIV reservoirs (Hua et al., 2018; Noel et al., 2018; Sandstrom et al., 2017). However, this new combination of treatments must achieve a control between the immune response and the induction of pathogenic systemic inflammation.

An important study that used recombinant IL-21 and peg-IFN-α2a limited residual inflammation and delayed viral rebound after withdrawal of antiretroviral therapy in NHP models of SIV infection (Harper et al., 2021). The use of subtypes of IFN-α may be a good option as these have different binding affinity to its IFNAR and different biological activities that modulate differently signal transduction (George et al., 2018). This has been shown *in vitro* in human intestinal cells tested for the anti-HIV activity of different IFN-α subtypes, showing that IFN-α8, -α6, -α14, and -α17 reduced the infection of the cells (Harper et al., 2015), and in mice models where IFNα14 effectively suppressed HIV replication through a mechanism involving increased expression of inhibitor factors and NK cell activity without inducing uncontrolled inflammation (Lavender et al., 2016). In humans, the use of hidroxychloroquine, an inhibitor of TLR-7 and TLR-9 that reduce IFN production *in vitro* (Martinson et al., 2010), has been used for the modulation of excessive IFN-signature in the context of HIV infection. Despite some reports showed that hydroxychloroquine reduces activation of T cells and plasma levels of LPS (Piconi et al., 2011), its effect in people receiving ART has not been clear (Routy et al., 2015). Interfering with IFN signaling can be another method to counteract the excessive immune activation associated with chronic IFN stimulation. Two studies with humanized mice models of HIV infection with ART have tested IFNAR antagonists, showing ISG expression in conjunction with a reduction in CD8 T cells exhaustion (Zhen et al., 2017) and, more importantly, a decrease in cell-associated HIV DNA and in inducible HIV reservoir (tested with viral outgrowth assay) (Cheng et al., 2017). However, such encouraging results were not observed in a study with NHP treated with IFNAR antagonists where it was shown reduced antiviral gene expression but increase SIV reservoir size and pronounced CD4 T cell loss with concurrent progression to AIDS (Sandler et al., 2014; Nganou-Makamdop et al., 2018). Inhibitors of the JAK/STAT pathway induced by IFN have also been tested, JAK1/2 inhibitors limited HIV replication and reactivation in an *in vitro* model using primary lymphocytes and macrophages (Gavegnano et al., 2014) but, importantly, it was shown limited production of the pro-inflammatory cytokines IL-6, TNF-α, IL-1*b* without affecting the reactivation capacity of different LRAs (Spivak et al., 2016). Some of these JAK1/2 inhibitors are now being tested in phase 2 trials in people on ART for safety evaluation and its effects on viral suppression and CD4 T cell counts. Further studies with controlled schemes of IFN-α administration are needed to clarify the therapeutic use of IFNs for HIV control or elimination.

Moreover, the use of different IFN-α subtypes might help to redirect IFN-α signaling to certain tissues or cells considered as components of the HIV reservoir.

HIV Reservoir

Definition of HIV Reservoir

HIV infection remains incurable despite the efficient combined antiretroviral therapy (cART), due to the early establishment of long-lived viral reservoirs that are mostly settled in resting CD4+ memory T cells constituting the latent provirus (Chomont et al., 2009; Rausch et al., 2020). Interruption of cART leads to a relapse of viremia and disease progression, suggesting the existence of a latent and replication-competent viral reservoir, hence strict adherence to a life-long treatment regimen is mandatory for controlling HIV infection (Rothenberger et al., 2015). The rapid advances in the field of cART have shown that the early administration of treatment reduces the size and diversity of the latent reservoir compared to those individuals treated during the late stages of infection, however, does not prevent its establishment (Colby et al., 2018; Josefsson et al., 2013; Pedro et al., 2019).

The HIV reservoir is defined as a pool of CD4+ latently infected cells that exist in a long-term resting state containing full-length transcriptionally inactive, but replication-competent HIV proviruses in individuals undergoing long-term cART with sustained viremia suppression, and that is established within few days after infection. It is worth highlighting the difference between the latent reservoir including intact HIV provirus that is transcriptionally silent, but capable of being induced to produce replication-competent viruses, and active reservoir defined as persistently infected cells that continue to transcribe HIV RNA and presenting HIV antigen (Sung et al., 2018).

Once HIV latent reservoirs have formed in different cell types or anatomical sites, they are refractory to cART and due to their resting phenotype, unrecognized by the host immune system, leading to HIV persistence for life (Dufour et al., 2020). Several studies have described the decay dynamics of the latent reservoir and the best characterized in individuals on cART with sustained viremia suppression for a long time is a pool of resting CD4+ memory T cells, which has been estimated to have a half-life of approximately 44 months (Crooks et al., 2015). Due to this long-term stability of the reservoir, eradication of HIV would take over 70 years on lifelong cART alone. Therefore, it is imperative to develop new therapeutic agents and molecular strategies to target the latently infected cells to reduce and eradicate the viral reservoir.

The Latent Reservoir

Cellular Reservoirs: Resting CD4+ T Cells Subtypes

There is vast evidence that resting CD4+ memory T cells are the major contributors to the latent HIV reservoir both during cART and untreated infection. CD4+ memory T cells can be classified into five classic memory groups: T stem cell memory (T_{SCM}), T central memory (T_{CM}), T transitional-memory (T_{TM}), T effector-memory (T_{EM}), and a variety of terminally differentiated helper T cells known as tissue-resident memory cells (T_{RM}) (Reviewed in: Cohn et al., 2020). HIV infected memory CD4+ T cells subsets display different transcriptional profiles, proviral inducibility and contribution to the pool of cells harboring intact and replication-competent HIV genomes. Diverse studies have demonstrated the persistence of HIV DNA in all subsets of CD4+ memory T cells, even in naïve cells, suggesting that direct infection of resting CD4+ T cells can also contribute to the establishment of latency (Roche et al., 2020; Venanzi et al., 2019). However naïve cells constitute a small fraction of the latent reservoir and other studies have found HIV DNA from naïve CD4+ T cells but failed to harbor significant levels of replication-competent latent proviruses (Soriano-Sarabia et al., 2014). On the other hand, it has been shown that the contribution of infected CD4+ memory T cells subsets to the pool of cells harboring intact HIV genomes is different and depends basically on transcriptional activity and induction of proviral HIV DNA (Grau-Expósito et al., 2019; Kwon et al; 2020; Pardons et al., 2019). It has been described that about 1 of 10^3 CD4+ memory T cells harbors HIV DNA in successfully treated individuals, and about 1 of 10^6 cells produces a replication-competent proviruses (Avettand-Fenoel et al., 2016).

Each group of CD4+ memory T cells has a particular phenotype determined by the membrane expression of specific chemokine and homing receptors (Mueller et al., 2013; reviewed in Rausch et al., 2020; Ventura et al., 2020) and these cellular characteristics are related among other factors, with the half-life of HIV proviruses in memory T cell subsets: it is 277, 144, 133, and 88 months for T_{SCM}, T_{CM}, T_{TM} and T_{EM}, respectively. The half-life of HIV provirus is correlated with the rate of decay of the reservoir as well as the tissue compartment that influences the trafficking behavior of the tissue-resident memory cells (Wang et al., 2017; Wong et al., 2016). Tissue-resident memory cells are another distinct T cell subset that resides in the barrier tissues and also within other non-lymphoid organs as brain, kidney and joints (Carbone et al., 2015; Damouche et al., 2015).

Among T cell subsets, T_{CM} cells generally constitute the largest fraction of the CD4+ memory T cells that contribute to the viral reservoir in individuals on cART, because the persistence of HIV DNA is probably facilitated by the long half-life of these cells (Chomont et al., 2009; Lee et al., 2016;). Although HIV DNA can be found also in the T_{TM} cells, the majority of the replication-competent and inducible reservoir is largely restricted to the T_{CM} cells (Soriano-Sarabia et al., 2014). To highlight the relevance of T_{CM} cells in maintaining the latent reservoir, in one study among HIV controllers with a long-term virological remission after the interruption of early cART, were demonstrated to have very low levels of HIV infected T_{CM} cells. These results may have relevant implications for a functional HIV cure (Sáez-Cirión et al., 2013).

T_{SCM} cells makes a significant contribution to the reservoir of treated patients because of their long-lived, poorly differentiated stem cell-like characteristics and a strong propensity for homeostatic proliferation. The contribution of T_{SCM} cells to the total viral reservoir in CD4+ T cells has been measured by DNA as approximately 8% and is increased over time on suppressive cART. These cells have been found to harbor integrated HIV-DNA within their genome in cART-suppressed HIV-infected individuals and are less vulnerable to the viral cytopathic effects, promoting the establishment of the latent reservoir (Buzon et al., 2014). In addition to the fact that T_{CM} and T_{SCM} cells (defined by expression of CCR7 and CD27) have been identified as the most relevant reservoirs that harbor HIV DNA, they can differentiate into T effector-memory (T_{EM}) cells, which can expand significantly and have been shown to harbor more intact HIV genomes compared to T_{CM} cells (Hiener eta al., 2017).

Several CD4+ memory T cells may also be categorized by specific immune functions defined by their antimicrobial properties, cytokine secretion patterns, expression of specialized transcription factors and maturational aspect that influence cellular longevity and long-term survival (Muranski et al., 2013; Lee et al., 2016). According to these criteria, six helper CD4+ T cell lineages with the functional polarization have been identified: Th1, Th2 and Th17 that target specific classes of pathogens, Th9 (regulation of allergic inflammation, anti-tumor and anti-parasites immunity), T regulatory cells (T_{reg}; regulation of antimicrobial immunity and maintaining self-tolerance), and T follicular helper cells (T_{FH}; support of humoral immunity) (Rausch et al., 2020; Geginat et al., 2014; Saravia et al., 2019). These various CD4+ T cell subsets have different degrees of plasticity or stability in tissues and consequently, they can acquire new characteristics and functions in chronic HIV controlled infection. In addition to these categorizing CD4+ T cell lineages based on their functional properties, recent studies highlighted the presence of a new CD4+ T cell population, γδ T cells that have an important role for recognition of lipid antigens independently of antigen presentation by MHC complexes and may be relevant for the composition of the HIV reservoir in patients on cART (Kalyan et al., 2013). In a rigorous study, inducible, latent but replication-competent HIV reservoir was recovered from peripheral γδ T cells in some cART-suppressed, HIV individuals. The results revealed that the frequency of replication-competent proviruses in these CD4+ T cell subset was relatively high (1 in just 5000 γδ T cells), indicating that despite of γδ T cells is only a small percentage of the total T cell population, they may indeed contribute to persistence of fully functional virus despite cART (Soriano-Sarabia et al., 2015).

Current knowledge indicates that IL-17 producing helper T cells (Th17 cells) represent highly heterogeneous, unstable and long-lasting lymphocytes showing strong propensity of homeostatic proliferation. These characteristics are common among cells that contribute to the latent reservoir (Geginat et al., 2014; Muranski et al., 2013). Some cells from this T cell subset produce several effector cytokines besides IL-17, as IL-22, a cytokine that promotes epithelial proliferation and barrier function (Liang et al., 2006). Th17 cells were found to harbor high levels of HIV DNA in individuals on long-term cART, particularly in the gut tissue, a cellular microenvironment favorable for HIV replication (Sun et al., 2015; Pardons et al., 2019; Wacleche et al., 2016). Moreover, in *ex-vivo* assays, Th17 cells from HIV individuals undergoing long cART seem to be more susceptible to HIV infection as they harbor high levels of HIV DNA (Gosselin et al., 2010).

T follicular helper cells (T_{FH}) play a relevant role in the persistence of latent HIV reservoir and represent a major site for HIV replication and production on and off cART, although there is still a debate about whether these cells serve as a long-term reservoir for HIV in the setting of optimal cART (Fukazawa et al., 2015; Perreau et al., 2013). T_{FH} cells are essential for the formation and maintenance of germinal centers defined as anatomic regions where B cells are selected and differentiated into memory B cells or long-lived plasma cells. T_{FH} cells characterized by surface expression of CXCR5 and PD-1, can enter germinal centers evading CD8+ T cell mediated cytotoxicity. These cells reside in lymph node follicles providing critical helper signals and support the germinal center reaction for generation of effective humoral immunity. Latently infected T_{FH} cells can persist for a long time avoiding the adaptive immune response against HIV. (Mintz et al., 2020). In tissues, the B cell follicle is a major viral reservoir where T_{FH} cells are highly susceptible to infection with HIV, and high levels of replication-competent HIV provirus have been detected, particularly during untreated HIV infection and in the first years on cART. Currently there are not data to determine the long-term stability of infected T_{FH} cells during controlled HIV infection by cART.

T regulatory cells (T_{reg} cells) maintain self-tolerance to avoid autoimmunity, however their potential role in chronic HIV infection has also been described. These cells express FoxP3 as a master transcription factor and show a dual effect in HIV infection: a detrimental effect by inhibiting beneficial anti-HIV immunity, and a beneficial effect by limiting pathologic immune activation. Interestingly, it has been shown that T_{reg} cells from cART-treated patients harbor high levels of HIV DNA compared to other CD4+ T cell subsets, with a half-life of about 20 months. However, there is still unknown whether T_{reg} cells contain replication-competent virus that contributed to the HIV reservoir because the lack of standardized assays to measure levels and function of T_{reg} cells (Tran et al., 2008).

Finally, tissue-resident memory cells (T_{RM}) are a subset of CD4+ and CD8+ memory T cells that can be found in virtually every human tissue, including mucosal tissues, barrier surfaces and in other non-lymphoid tissues such as the lung, liver, gut, and skin; in addition, some T_{RM} cells are present in lymphoid sites, such as the lymph nodes and spleen. T_{RM} cells can be defined by a constitutive expression of CD69 in addition to tissue-dependent coexpression of CD103, CD49a, and other receptors. These CD4+ memory T cells patrol limited areas within non-lymphoid tissues such that, this tissue-based immunosurveillance dictates the course of HIV infection (Shacklett et al., 2020; O'Neil et al., 2021). Recent studies have highlighted the role of T_{RM} cells as key targets for HIV infection and persistence. Despite T_{RM} cells dominate the T cell response in tissues and can harbor HIV provirus, it may be difficult to characterize because these tissues are difficult-to-access; thus, there have been limited studies focused on T_{RM} cells in HIV infected individuals (Cantero-Perez et al., 2019). Recent studies show that in adipose tissues T_{RM} cells may support HIV infection and are able to represent a site for viral persistence, however the contribution of these cells to HIV reservoir remains unclear (Couturier et al., 2015; Damouche et al.,2015).

Other Cell Populations as HIV Reservoirs

HIV persists also in regulatory T cells (Treg), myeloid cells as monocyte-macrophages, microglia, and dentritic cells (DCs) however, the contribution of non-CD4+ T cells as a HIV latent reservoir is still controversial (Wallet et al., 2019; Chatzidimitiou et al; 2020; Cohn et al., 2020; Rausch et al., 2020).

The relevance of the macrophage reservoir in HIV persistence has generated great debates since their role as phagocytes may account for the presence of HIV nucleic acids and proteins in these cells (Calantone et al., 2014). Moreover, the drastic reduction in reservoir size due to the administration of cART, limited access to tissue samples and technical limitations of current detection assays has failed to convincingly demonstrate the presence of a replication-competent monocyte/ macrophage HIV reservoir in patients with controlled viremia. There is accumulated evidence showing the presence of HIV DNA in macrophages and monocytes in virologically suppressed individuals and may be important in the establishment of chronic infection at the site of virus exposure. However, *in vivo* evidence for infected macrophages is limited, suggesting the reservoir is small. Furthermore, it remains unclear whether these cells persist during cART because their short half-lives (Calantone et al., 2014). Most macrophages reside in tissue compartments that are not easily accessible hence, their relevance as a latent reservoir during treatment remains undefined. Conversely, growing evidence support the role of macrophages in HIV persistence due to some biological reasons. Firstly, macrophages are non-dividing cells, extreme heterogenous in response to the local cytokine microenvironment, showing a reversible phenotype and functional response (Gordon et al., 2017). This diversity allows macrophages to play a variety of roles in innate immune response and phagocytosis. Macrophage polarization affects their susceptibility to HIV and can range in acute and chronic stages of the infection (Murray et al., 2016).

Macrophages express receptor CD4 and chemokine coreceptors CCR5 and CXCR4, on their cell surface, allowing for productive HIV infection both *in vivo* and *in vitro*. While HIV infected macrophages have been found in various tissue compartments *in vivo* as the central nervous system (CNS), lymph nodes, gut lymphoid associated tissue (GALT) and lung, how they contribute to the viable HIV reservoir is still largely unknown (Wallet et al., 2019; Balcom et al., 2019; Jambo et al., 2014; Avalos et al., 2016). Heterogenicity of macrophages allows them to present different functions depending on the tissue they reside. In the setting of HIV infection, macrophages resident in the GALT are resistant to infection, whereas macrophages in the rectum and alveolar are easily permissive (McElrath et al.,

2013; Jambo et al., 2014). Macrophages susceptibility to HIV infection has been shown to be drastically different among tissues and mechanisms that may control latency, suggesting some macrophage phenotypes provide better reservoirs (Hendricks et al., 2021). Contrary to the initial assumed idea that macrophages have a moderate life span and lack self-renewing potential, several studies have shown that these cells are capable of self-renewal and harbor HIV provirus despite cART, suggesting their potentially relevant role as long-lived HIV reservoirs. However, is still unknown whether such cells harbor inducible replication-competent HIV. Developing *in vitro* assays to identify and quantify inducible HIV latent reservoir in macrophages has been a great challenge. Animal models, particularly humanized myeloid-only mouse (MoM) model may provide more definitive evidence regarding HIV persistence in tissue macrophages during cART (Honeycutt et al., 2013). The "MoM" model only has myeloid but not T cells and is currently used to identify macrophages enriched of replication-competent HIV.

Monocytes are derived from pluripotent hematopoietic stem cells which differentiate into common CD34+myeloid progenitors, granulocyte/monocyte progenitors and committed monocytes progenitors in the bone marrow (Geissmann et al., 2010). The migration behavior of monocytes subsets is influenced by their chemokine receptor expression profiles and has implications for their ability to seed tissue reservoirs. Monocytes express CD4 and the coreceptor CCR5 in very low levels, resulting in inconsistent data regarding their susceptibility to HIV infection *in vitro* (Carter et al. 2019; Wong et al., 2019). In HIV infected individuals, intermediate monocytes population is expanded during all stages of infection, especially in individuals with detectable viral load (Booiman et al., 2017; Kruize et al., 2019). Additionally, some studies have demonstrated that in a peripheral blood, a small monocytes population (<0.1%) harbor replication-competent HIV provirus both untreated and treated HIV infection (Ellery et al., 2007).

Dentritic cells (DCs) are critical and very diverse initiators of innate immune response, which include cell lineages that can serve as antigen-presenting cells: plasmacytoid (pDCs), myeloid (mDCs) and Langerhans cells, and a follicular dentritic cells (FDCs) that lack this function. Both pDCs and mDCs also express CD4, CCR5 and CXCR4 receptors involved in HIV infection. Immature DCs capture HIV that enter peripheral mucosal tissues and migrate to secondary lymphoid organs where they transfer the HIV to resting autologous CD4+ T cells during the antigen presentation process via an infectious synapse. The unique capacity of immature DCs to act as vehicles for HIV transmission is pivotal for early pathogenesis. Although many studies have demonstrated that DCs and monocytes can be infected to some extent *in vivo*, these cells are unlikely to contribute directly to HIV reservoirs because of their short half-lives (Pace et al., 2011).

Astrocytes are the most abundant cell type in the brain and their role in HIV latency is highly controversial (Li et al., 2016). On one hand, it has been shown its participation in cell-to-cell viral transfer route from infected CD4+ T cells allowing HIV successfully establishes latency in these cells (Luo et al., 2015). On the other hand, it has been found that astrocytes were evidently resistant to infection of cell free HIV *in vitro* or cell-to-cell routes but are able to engulf infected macrophage material (Russell et al., 2017). *In vivo*, HIV does not productively infect astrocytes, neurons, oligodendrocytes, or brain endothelial cells. Further studies need to be conducted to define the role of astrocytes in HIV infection.

Other newly identified as probable HIV reservoirs in the brain are the pericytes of the blood-brain barrier (BBB). Pericytes are localized wrapped around endothelial cells of the capillaries of the brain and are necessary for proper formation and maintenance of the BBB. Importantly, pericytes can be infected by HIV and support productive virus replication. Further, BBB-pericytes are susceptible to establish a latent infection, which can be reactivated. Recent studies in mouse model and in human post-mortem samples of HIV infected brains, have confirmed the presence of HIV infected pericytes (Bertrand et al., 2019).

Finally, there is growing evidence that suggest the possible long-term HIV persistence on non-T cell reservoirs in individuals on cART, such as liver and renal epithelial cells, fibrocytes and NK cells (Kandathil et al., 2017).

Tissue Reservoirs

Tissue reservoirs are usually defined as a tissue or organ containing cells that continue to harbor HIV in treated patients, independently of the mechanism maintaining them. They are also referred as anatomic sanctuaries for HIV and include central nervous system (CNS), lymphoid tissue, adipose tissue, and gut-associated lymphoid tissue (GALT). Due to their localization and distribution within the body, tissue sanctuaries have low cART penetration which may promote limited HIV provirus propagation at minimal levels that allow the emergence of drug resistant variants. Experimental *in vitro* and *in vivo* models with non-human primates and human autopsy studies, have shown that lymph nodes and gut mucosa contain the largest amount of HIV and infected cells. Studies in SIV (Simian Immunodeficiency Virus)-infected macaques and human autopsy cohort demonstrated that the majority of the total body reservoir was found in the gut (Banga et al., 2016; Estes et al., 2017).

The CNS is an important viral reservoir that is considered as an immune privileged site due to the presence of the blood-brain barrier and the reduced antiretroviral drug efficacy and penetration in the brain. There are three types of CD4+ cells in the CNS: CD4+ T cell, macrophages and microglia which can be infected by HIV. *In vitro* data have shown the presence of macrophage-tropic HIV lineages in late disease in the CNS, these results were confirmed by quantitative viral outgrowth assay (QVOA) demonstrating the presence of latently infected macrophages and microglia which contain replication-competent provirus. CD4+ T cells are present in very low concentration in the brain and the contribution to the HIV reservoir is limited (Asahchop et al., 2017; Avalos et al., 2017). In the CNS, HIV persist mainly in resident macrophages, like microglial cells and perivascular macrophages (Ko et al., 2019).

Although HIV cannot infect preadipocytes and adipocytes *in vivo* due to their extremely low expression of CD4, CCR5 and CXCR4 receptors, adipose tissue contains a plethora of target innate immunity cells as CD4+ T cells, CD8+ T cells, T_{reg} cells, NK cells and B cells which express an activated memory phenotype (CD69+). These cells can be stimulated by adipocytes and promote HIV replication. Recent studies demonstrated the presence of replication-competent virus in the adipose tissue (Couturier et al., 2015; Damouche et al., 2015).

The B cell follicle within lymphoid tissue has unique properties that make it ideal in HIV latency and persistence, in part due to the suboptimal antiretroviral drug penetration. In states of inflammation induced by HIV, lymph nodes contain activated T_{FH} cells and FDCs which are well described reservoirs. Furthermore, HIV-specific CD8+ T cells are excluded from the follicle and lymphoid tissue have reduced cytolytic activity, thus providing HIV with an immune sanctuary (Fukazawa et al., 2015; Reuter et al. 2017). In lymphoid tissues like spleen and lymph node, resting CD4+ memory T cells are the predominant reservoir, although, latently infected resident macrophages have also been detected.

Gut-associated lymphoid tissue (GALT) is the largest component of the lymphoid system that contains the highest concentration of CD4+ T cells. HIV target cells in the gut include a vast variety of innate immunity cells as: macrophages, T_{reg} cells, interleukin 17 producing helper T cells (Th17 cells), interferon gamma and interleukin 17 producing helper T cells (Th1/Th17), interleukin 22 producing helper T cells (Th22), T_{FH} cells, T_{TM} cells, T_{RM} cells, gd T cells, T_{SM} cells and T_{CM} cells. Therefore, GALT is probably the main source of ongoing HIV replication given the presence of high amount of virus and infected cells, its extensive distribution, and its role in immune function (Thompson et al., 2017). Moreover, the early establishment of HIV in GALT compartment promotes viral persistence. Clinical evidence has shown that rectal macrophages express higher levels of CCR5, which suggests that the most distal parts of the gut are particularly susceptible to HIV infection. Additionally, it has recently reported that PD-1 can serve as a marker of HIV persistence in rectal tissue (Khoury et al., 2017; McElrath et al., 2013). Several groups have been identified high levels of HIV RNA and DNA in gut compared with PBMCs, and their distribution is different in the GI. On the other hand, genetic analysis indicates that viral gene expression in GALT is maintained even in plasma samples with undetectable viral loads. Together, these data support persistent HIV replication within GALT.

The genital mucosal tissues are the main HIV entry portal on sexual transmission. Genital tracts are rich in tissue macrophages and might be potential immune sanctuaries due suboptimal antiretroviral drug penetration. Recent studies found evidence that suggests that human urethral tissue macrophages from treated HIV infected individuals constitute a principal HIV-reservoir. Additionally, tissue-resident memory CD4+ T cells (T_{RM}) in the cervix of female genital tract were enriched with HIV DNA (Cantero-Perez et al., 2019; Ganor et al., 2019).

The Active Reservoir

Active reservoir represents a pool of CD4 + cells that continue to support productive virus transcription despite cART and might contribute to virus rebound after interruption of treatment. During effective cART, a minority of HIV infected cells (approximately 10% of circulating infected cells) are transcriptionally active, producing detectable levels of HIV RNA, HIV proteins and intact virions. This fact indicates that HIV reservoir can be found since a profound latent state with no or very little viral RNA produced, to an active state of virus production. There is a little to no evidence of sustained active viral replication and productive infection in the presence of cART. Remarkably, a recent study showed that the frequency of cells expressing HIV RNA is similar in untreated infected individuals and on cART, although significant lower levels of viral RNA were detectable during cART (Wiegand et al., 2017). Some studies have found that the active reservoir may contain cells phenotypically distinct from those harboring the latent reservoir (Abdel-Mohsen et al., 2018; Hogan et al., 2018).

T follicular helper cells (T_{FH}) are the best described active reservoir that is localized in the B cell follicle of the lymph node. T_{FH} cells are highly susceptible to HIV infection and replication due to colocalization with follicular dentritic cells which promotes cell-to-cell transmission of virus (Heesters et al., 2015).

More than 95% of the integrated viral genomes or HIV proviruses detected in PBMCs from infected patients on cART, are defective because of internal deletions and lethal mutations. Although these defective proviruses had long been thought to have a little significance to HIV pathogenesis, as they do not encode replication-competent viruses, it has been recently demonstrated that defective viral genomes are not silent and can also produce viral proteins (Gag and Nef proteins) that are immunogenic and can be found at all stages of HIV infection. The long-term production of viral proteins produced by these defective proviruses in absence of viral replication may have a deleterious effect on the host that may be associated with persist immune activation/inflammation during cART (Imamichi et al., 2020).

Markers of the HIV Reservoir

Cellular Markers of Latent HIV Reservoir

The identification of markers of the HIV reservoir remains controversial. The discrepant results are likely because different definitions of viral reservoir and technical limitations. Cellular markers of resting CD4+ memory T cells containing replication-competent HIV proviruses would allow selective reservoir targeting. Additionally, specific subsets of CD4+ T cells may differ in propensity of clonal expansion or latency reversal. Currently, specific markers of latent HIV infection have not been defined. Three subsets of CD4+ memory T cells are defined by the cell surface proteins CD45, CCR7, and CD27: central memory cells (T_{CM}: CD45RO+CCR7+CD27+), transitional memory cells (T_{TM}: CD45RO+CCR7-CD27+), and effector memory cells (T_{EM}: CD45RO+CCR7-CD27-). Naïve cells (T_N:CD45RO-CCR7+CD27+) home to secondary lymphoid organs to differentiate into memory cells (Farber et al., 2014; Kwon et al., 2020). Depending on microenvironment and localization in the body, resting CD4+ memory T cells can be found in quiescent, activated (CD69+ or HLA-DR+), proliferating, or exhausted (expressing immune checkpoint receptors, such as program cell death-1 (PD-

1), TIGIT, and LAG-3) (Bruel et al., 2018). Thus, an accurate identification of clinically relevant markers of HIV reservoir is needed. If a specific phenotype can be identified, it facilitates the detection of persistently infected cells *in vitro* and in animal models, then HIV cure strategies can be targeted to eliminate the cells containing the latent reservoir.

Markers of Activation of the HIV Reservoir

The low frequency of infected resting CD4+ T cells and low levels of HIV RNA in samples of patients undergoing cART greatly limit the ability to identify and quantify the latent reservoir. Several molecules have been proposed as latency markers however, the identification of unique cellular biomarkers that distinguish latently infected cells from healthy uninfected CD4+ memory T cells as well as its proportion in tissues remain undefined.

HIV proviruses are enriched in cells that express markers of activation: HLA-DR (MHC class II), CD25 (the alpha chain of the IL-2 receptor and a constitutive marker of T_{reg} cells) and CD69 (a marker of tissue T_{RM} cells). However, it is important to study total CD4+T cell population rather than resting CD4+ T cells to address the issues of long-term persistence of HIV during cART. An important fraction of latently CD4+ infected cells express immune checkpoints with phenotypes that may vary during cART. Immune checkpoint molecules are defined as ligand-receptor pairs that exert an inhibitory or stimulatory effects on immune response. Most of these molecules are expressed on cells of the innate and adaptive immune system. Findings from several studies have shown that HIV DNA has found to be modestly enriched in cells expressing immune checkpoint receptors, programmed cell death-1 (PD-1), cytotoxic T-lymphocyte-associated protein 4 (CTLA-4), T cell immunoglobulin and ITIM domain (TIGIT), lymphocyte activation gene 3 (LAG-3), T cell immunoglobulin and mucin 3 (TIM-3) and CD160 (Cohn et al., 2020; Zhang et al., 2020). A comprehensive study of latently CD4+ infected cells showed that a combination of PD-1, TIGIT and LAG-3 immune checkpoints, correlates with the presence of integrated HIV DNA in blood cell from HIV infected patients. It was demonstrated that CD4+ T cells expressing these three markers display ten times more HIV DNA than total CD4+ T cell population (Fromentin et al., 2016). Other markers of the CD4+ T cell HIV reservoir have been also described although their role in viral persistence in not fully clear: CD30, CD2 and IFITM1, CD32a (Bruel et al., 2018).

Total HIV DNA a Clinically Relevant Marker

Several markers have been proposed to quantify and monitor the HIV reservoir but there is no consensus method. Total HIV DNA is the most widely studied marker of the HIV reservoir, having been quantified in many patient cohorts and clinical trials. In addition, total HIV DNA has been proposed as biomarker of viral reservoir dynamics because of some relevant characteristics: it is easy to measure in blood and different tissues by simple, sensitive and reproducible methods, is predictive of progression to AIDS and death, is predictive of the response to cART and treatment efficacy, and it is also predictive of the presence and severity of HIV associated end-organ disorders. Furthermore, the total HIV DNA load is indicative of viral pathogenesis and persistence because includes all viral forms, integrated and unintegrated HIV DNA (that includes episomal 1 long terminal repeat [1-LTR], 2-LTR, and linear HIV DNA) that participate in cell activation and HIV pathogenesis (Avettand-Fenoel et al., 2016). All viral forms coexist in infected cells. The most prevalent viral form in infected cell is labile, linear HIV DNA that constitute a preintegrated form of latency and in quiescent CD4+ T cells (Blankson et al., 2000), 1-LTR and 2-LTR episomal forms represent less than 10% of unintegrated HIV DNA (Kilzer et al., 2003). Intracellular HIV RNA load has been proposed as another reservoir marker which quantifies HIV transcription and reflects the activity of the HIV reservoir. This reservoir marker can be useful for evaluating residual viral load in patients on cART (Pasternak et al., 2013).

The Latent HIV Reservoir: Formation and Compartmentalization

Several mechanisms are involved in the formation of HIV reservoirs. One mechanism is the direct infection of latently CD4+ T cells. This idea is supported by the fact that resting CD4+T cells are the predominant infected cells during early infection when reservoirs are forming. *In vitro* studies have demonstrated the direct infection of resting CD4+ T cells (Dai et al., 2009; Vatakis et al., 2009). Another idea is that latently infected CD4+ T cells may arise from activated CD4+ T cells that return to a resting state after becoming infected. Additionally, other factors as cytokine-rich environment, and the presence of DC cells and macrophages, may enhance reservoir formation by different mechanisms (Wu et. al., 2006). However, it has been a great challenge to determine if these mechanisms of reservoir formation occur *in vivo* and if latent CD4+ T cells were infected while they were in resting or in activated state.

Despite cART, in most HIV-infected individuals the establishment of viral reservoirs occurs in multiple tissues and anatomic compartments/sanctuary sites within a few weeks post-infection resulting in the generation of distinct viral populations. (Thompson et al., 2011; Sturdevant et al., 2015). HIV population during acute infection is typically homogenous, until virus compartmentalization occurs across different anatomic sites (Svicher et al., 2014). HIV compartmentalization refers to the presence of distinct viral populations across different tissues or across different cell types within an infected individual, as a consequence of tissue-specific genetic differentiation and a biological barrier that restricts the viral migration between anatomic sites. Several studies have shown that viral compartments facilitate HIV evolution and divergence between peripheral blood and tissue compartments giving rise to tissue-adapted variants, which may contribute to HIV pathogenesis (Pillai et al., 2006). HIV compartmentalization can be detected by phylogenetic methods, based on sequence determination of HIV envelope. Viral compartments have been described in HIV infected individuals, including lymph nodes, GALT tissue, CNS, kidney and lower genitourinary (GU) tract. There is still controversy regarding the compartmentalization in lymphoid tissues, gut, and GU tract. HIV compartmentalization between the central nervous system (CNS) and the peripheral blood, it has been shown to be associated with HIV-associated neurodegenerative disorder (HAND) in up to 50% of HIV infected patients. Such viral compartmentalization in CNS has been attributed to viral tropism in the two compartments (Bale et al., 2019; Bednar et al., 2015; Gianella et al., 2016; Wallet et al., 2019). Understanding the viral dynamics in these compartments is imperative to develop functional cure strategies.

HIV Integration into the Host Genome

The key protagonists in HIV entry are the envelope glycoprotein (Env) and main receptor CD4 and co-receptors CCR5 and CXCR4 on the plasma membrane of the target cell. HIV entry begins with the adhesion of the virion to the target cell and ends with the fusion of the cell and viral membranes followed by delivery of the viral core into the cytoplasm, where two RNA genome strands are reverse-transcribed into a complementary DNA (cDNA) copy and the formation of an integration-competent pre-integration complex

(PIC) which is implicated in virus nuclear import. The PIC consists of a large nucleoprotein complex of host proteins as well as processed cDNA associated with HIV integrase, which catalizes the endonuclease-mediated priming of the 5' and 3' ends of the cDNA for full integration into the host genome (Engelman et al., 2018). Integration is a crucial step for HIV replication and for the establishment of the reservoir that persists in HIV individuals. Integration on the host genome is non-random, HIV interacts with cellular factors to target integration to interior regions of transcriptionally active genes within gene dense regions of chromatin (Schröder et al., 2002). Once integrated, the HIV cDNA is termed a provirus and can be accessed by the host cell transcriptional machinery to produce viral proteins. Results from several comprehensive studies demonstrate that integration of HIV provirus into the host genome can lead to three scenarios: 1. apoptosis triggered by double-strand DNA breakage caused by the integration process, 2. active transcription of the provirus (active reservoir), 3.and the transcriptional repression of the proviral sequence (hallmark of latent reservoir). These latently infected cells constitute the persistent HIV reservoir which eventually resurges when cART is interrupted (Ventura et al., 2020).

Establishment of HIV Latency

There is still a debate regarding about when and how HIV latency is established after virus integration into the host genome. Since human studies of early reservoir establishment are complicated to conduct because ethical challenges, several groups have used non-human primate models of HIV infection (rhesus macaques infected with Simian Immunodeficiency Virus, SIV) to determine the duration of untreated infection needed to establish persistent infection. The results of these studies indicate that the persistent reservoir was established in less than 7 days after SIV inoculation and although early cART may not prevent the establishment of the reservoir, it may accelerate its decay. On the other hand, less than 2% of the total body T cells are circulating in peripheral blood, therefore, it is expected that HIV infected cells are found in various tissues after years of cART (Chaillon et al., 2020). Altogether, these results suggest that the anatomical location and the timing of the emergence of persistently infected cells contribute to the establishment of HIV latency. Other factors involved in the establishment of HIV-1 latency are: transition from activated cells to quiescence, condition that is favorable to HIV latency (Shan et al., 2017), the presence of immune checkpoint molecules that decrease T cell activation by silencing HIV transcription, and cell-to-cell transmission from activated, productively infected CD4+ T cells to resting CD4+ T cells leads to a non-productive infection that may result in viral latency (Kumar et al., 2015; Rezaei et al., 2018). Resting CD4+ T cells may also be directly infected, and latency may be established due to cell-cell interactions that increase their susceptibility (Zack et al., 1990). Some cytokines as IL-7, CCL19 and CCL20 are also known to increase the susceptibility of resting CD4+ T cells to latent infection in tissue reservoirs (Coiras et al., 2016).

Currently, the most widely accepted model of latency establishment is that provirus is initially active and later become to a resting memory state due to a cellular relaxation, a process called effector-memory-transition (EMT) (Shan et al., 2017). This model suggests a major contribution of host cell state on HIV entry into latency and induce a late latency (Shukla et al., 2020). Alternative, there has been described an early latency model that proposes the direct infection of resting CD4+ T cells in which the proviral genome remains transcriptionally inactive immediately after integration into unfavorable genomic areas (Chavez et al., 2015). In addition, because of some features involved in transcriptional regulation including immune cell state, integration landscape, and complex transcription circuit architecture, it has become difficult to define their contribution to proviral fate. Together, these studies indicate that both viral and host cell factors influence the establishment and maintenance of proviral latency.

HIV Persistence during Antiretroviral Therapy

After initiation of cART, HIV plasma viremia exhibited a multiphasic decay pattern in which the majority of plasma viremia (HIV RNA) and infected cells (HIV DNA) declined rapidly during first two phases, indicating the efficiency of antiretroviral drugs (Hilldorfer et al., 2012).

The majority of persistently CD4+ infected cells do not produce replication-competent proviruses during cART. However, the reactivation of the persistent-latently infected cells is considered as the main source of the new HIV RNA because these cells are replenished by homeostatic proliferation. Besides the natural longevity of different subsets of resting memory CD4+ T cells, it has been proposed that IL-7 and IL-15 mediated homeostatic proliferation of these cells can explain the reservoir stability over time, increasing level of infected cells *in vivo* (Chomont et al., 2009; Vandergeeten et al., 2013).

Clonal expansion of resting CD4+ memory T cells has been well documented and convincingly demonstrated by next-generation genome sequencing. Several studies in recent years have shown that the clonal expansion of memory T cells can contribute up to 50-60% of the entire reservoir during HIV infection (Murray et al., 2016).

Low levels of HIV replication in tissues during cART may indirectly maintain the latent reservoir and facilitate reservoir persistence. With the highly sensitive real-time PCR, residual viremia has made evidenced in many studies conducted in long-term successfully suppressed individuals on cART. Importantly, recent studies reported lower levels of antiretrovirals found in different anatomical sanctuaries, allowing ongoing viral replication (Fletcher et al., 2014). Additionally, sequencing analysis of HIV from lymph nodes and blood of treated individuals at several time-points, showed that viral evolution and trafficking between tissue compartments continue in patients with controlled plasma viremia. Consequently, HIV continues to replicate and reseed the viral reservoir despite cART (Lorenzo-Robledo et al., 2016). Based on these results, HIV persistence is due to viral latency and not drug resistance mutations or discernible evolution.

Another cause of HIV persistence during cART is the induction of an immune hyperactivation state. Residual viremia that increases immune activation thus further stimulates HIV replication, generating a vicious cycle. Additionally, residual viremia may induce a low grade of inflammation that could replenish the reservoir to a small extent. Therefore, HIV persistence during suppressive cART is supported by homeostatic and antigen-induced proliferation of latently infected cells, ongoing replication in the sanctuary sites as well as by increased immune activation and inflammatory response (Kim et al., 2016).

Tools to Quantify the Latent Reservoir

The main reasons why latency represents an obstacle for HIV cure are the absence of specific cellular biomarkers that would distinguish latently infected CD4+ cells from healthy uninfected CD4+ cells, and the need for effective techniques to accurately measure the latent reservoir *in vivo*. The size of the reservoir is considered as predictor of the time of viral rebound after interruption of cART. Therefore, quantification of latent reservoir is a priority in order to estimate the size of the reservoir, the time of viral rebound and to evaluate the efficacy of antiretrovirals.

Current methods to detect replication-competent proviruses include polymerase chain reaction (PCR)-based assays, and quantitative virus outgrowth assays (Q-VOA). The Q-VOA is currently the gold standard method which aims to identify the cells harbor replication-competent virus by measuring the relative release of infectious virions upon cellular stimulation. It is used to quantify drug-resistant infected cells from resting CD4+ T cells isolated from treated individuals. However, there are multiple drawbacks as time consuming (14-21 days to be completed), requires large amounts of samples and a biosafety level 3 laboratory, is expensive and has been proven to underestimate the size of reservoir because only a portion of CD4+ cells with replication-competent provirus are activated with a single round of stimulation (Wang et al., 2018).

Quantitative PCR (qPCR)-based assays are used to measure HIV cDNA or HIV RNA present in PBMCs or plasma obtained from infected individuals. The main limitations of these assays are that they are unable to distinguish between actively or latently infected cells, nor if the proviruses are replication-competent or defective, thus they overestimate the size of the reservoir. For the DNA-based qPCR assays, the most widely used is the AluPCR, which exclusively detects integrated proviral DNA. For the RNA-based qPCR, relative levels of extracellular RNA over cell-associated RNA are measured. However, this method also fails to distinguish latently infected cells (Castro-Gonzalez et al., 2018; Rouzioux et al., 2018). Tat/Rev-induced limiting dilution assay (TILDA) measures by qPCR the relative amount of cell-associated multiply-spliced HIV RNA after T cell stimulation to detect the presence of replication-competent proviruses (Procopio et al., 2015). Droplet digital PCR (ddPCR) is also used although presents false-positive results. Next generation sequencing (NGS) is currently used to genetically characterize proviruses and predict replication competency (Hiener et al., 2017).

Cure Strategies Targeting the Latent Reservoir

The primary objective for HIV cure is the eradication of the reservoir or to induce a sustained and effective immune control against infected cells, allowing life-long remission of virus without cART. Targeting HIV reservoirs is of vital importance to achieve either a sterilizing or a functional cure. However, targeting the latent reservoirs located particularly in anatomic sanctuaries including CNS, lymphoid tissue and GALT has been a great challenge. Poor access of antiretroviral drugs into the tissue reservoirs contributes to the persistence of HIV. In the case of CNS considered as an immune privileged site, the access of cART is limited due to the presence of the blood-brain barrier. Another limitation in this particular sanctuary is the existence of residual neuroinflammation which is responsible for the occurrence of HAND.

The current focus of HIV cure strategies can be divided into four broad areas: 1. Reducing the reservoir (including "Shock and Kill" strategy), 2. Immunologically control of viral rebound (including broadly neutralizing antibodies and chimeric antigen receptor T cells [CART-T]), 3. Silencing the reservoir ("Block and Lock" strategy), and 4. Gene therapy (CRISPR/Cas9 system, zinc finger nucleases [ZFN] and the transcription activator-like effector nucleases [TALENs]) (Cohn et al., 2020; Shukla et al., 2020;). All these strategies have strengths and weaknesses.

Shock and Kill Strategy

The Shock and kill strategy based on the reactivation of the latent provirus aims to "shock" latently infected cells into a reactivated state, using latency-reversing agents (LRAs) to activate virus transcription and then "kill" these reservoir cells via cytopathic effects of the reactivated virus, host immune responses (cytotoxic T cells [CTLs]) or other mechanisms (Margolis et al., 2020). LRAs agents target cellular factors involved in HIV latency. Many of LRAs are now used *ex vivo* and in clinical trials. There are currently >100 compounds used as LRAs which belong to two main families (Abner et al., 2019). One family of LRAs target epigenetic mechanisms that occur in latency and comprises: HDAC inhibitors (HDACi), histone methyl transferases inhibitors (HMTi) and DNA methylation inhibitors. Second family of LRAs induce the expression of positive cellular factors and/or their release from inactive complex. However, these drugs used alone failed to fully reactivate viral expression *ex vivo* and their lack of specificity. Many other compounds including TLR agonist, TNFα, cytokines, and antibodies are also used as LRAs.

Immunotherapy

Immunotherapies include the use of broadly neutralizing antibodies and chimeric antigen receptor T (CAR-T) cells to target viral envelope on the cell surface for killing. Currently, HIV-specific CARs are being applied to peripheral T cells, NK cells, and stem cells to boost recognition and killing of

HIV infected cells. The use of antibodies and CAR-T cells shows potential limitations as both require that the epitope be expressed on the cell surface at high densities, and both target the highly variable envelope glycoprotein (Mu et al., 2020; Herzig et al., 2019). These therapies may only work in combination with LRAs or during a treatment interruption.

Block and Lock Strategy

An alternative approach to achieve long-term control over HIV is to induce a long-lasting inhibition of HIV gene expression. Block and lock approach aims to enhance the latent virus state in an irreversible way, by blocking HIV transcription and locking the HIV promoter in a deep latent state via epigenetic silencing. Pharmacologic inhibitors called latency-promoting agents (LPAs) inhibit viral or cellular proteins involved in HIV transcription. The best-described LPA is didehydro-cortistatin A (dCA), a Tat inhibitor that decreases viral RNA in cells and tissues and delay viral rebound when cART is interrupted in a humanized mouse model. Cellular miRNAs are also potential molecules to be used in the Block and Lock strategy (Ahlenstiel et al., 2020).

The Gene Therapy

Gene therapy strategies have been widely used to permanently inactivate or eradicate HIV genome from latently infected cells. These strategies are based on nuclease-mediated gene editing tools such as the ZFN, TALENs and the CRISPR/Cas9 system. Since ZFN and TALENS are costly and time consuming, CRISPR/Cas9 system has become the molecular tool of choice. This technology uses a guided RNA (gRNA) to induce directed mutations and a Cas9 nuclease that excise DNA sequences of target specific cellular factors to eliminate the HIV provirus (Xiao et al., 2019). Despite CRISPR/Cas9 approach was initially promising, it resulted in a long-term escape caused by HIV DNA mutations in the targeted sequences, preventing single-guide RNA recognition. To prevent escape by resistance mutations, it is necessary to use several gRNAs, although when using multiplex gRNAs targets, the system can induce off-target cuts (Yoder et al., 2016). However, the self-regulation of CRISPR/Cas9 system by HIV may be a safe and accurate strategy for eliminating HIV infection and whose effect will be restricted to infected cells (Vergara et al., 2020). Cell/Gene therapy using stem cell transplantation has been successfully proved only with two patients: the Berlin and London patients. In this approach naturally HIV-resistant donor CCR5delta32 stem cells were used for transplantation in patients being treated for associated malignancies. However, this approach is not currently feasible in the general population of HIV individuals due to the high risk that represents an allogenic stem cell transplantation (Gupta et al., 2019).

Inflammatory Response in Controlled HIV Infection

Chronic inflammation is a hallmark of HIV infection that persists even after long-term successful cART and plays a relevant role in disease pathogenesis. Chronic inflammatory response can be brought about by HIV that is able to evade host immune system and persists in tissues for an extended period. During controlled HIV infection, upregulation and release of many pro-inflammatory cytokines and dysregulation of the Type I interferon (IFN-1) response, contribute to systemic inflammation leading to immune exhaustion, immunosene-scence and development of non-AIDS co-morbidities.

Clinical Implications of Systemic Inflammation and Reservoir in HIV-Infected Individuals

After initiation of cART, many factors associated with inflammation are normalized however, other markers of inflammation did not return to baseline levels and the causes are still unknown. In the setting of HIV infection, it is assumed that the aetiology of persistence inflammation is multifactorial and includes residual viremia, co-infection with other pathogens (cytomegalovirus, hepatitis C virus, Epstein-Barr virus), microbial translocation and dysbiosis caused by the disruption of the gut mucosa, loss of immunoregulatory function, and cumulative cART toxicity. These factors avoid the full control of inflammation increasing the risk of developing HIV associated non-AIDS complications (HANA), such as metabolic syndrome and cardiovascular diseases (CVDs) (Deeks et al., 2013). Additionally, there is clear evidence of innate immune activation with increased levels of soluble inflammatory and coagulation markers including IL-6, IL-8, IL-10, IL-12, IL-15-IL-18, IL-23, IL-1β, TGF-β, C-reactive protein (CRP), D-dimer, MCP-1, sCD14, sCD163, TNF-α, ICAM-1, VCAM-1, and LPS (Figure 1). These markers have been described as predictors of mortality in HIV infected individuals and have been targets of anti-inflammatory interventions (Kamat et al., 2012).

Most cells and tissues of the immune system are significantly impaired by chronic inflammation, accordingly, the function of some organs including gut, lymph nodes, liver, kidney, lungs, brain, and adipose tissue is also damaged despite effective cART. One of the first tissues to become infected by HIV after exposure is GALT the largest component of the lymphoid system that contains the highest concentration of CD4+ T cells, making this tissue an ideal site for HIV infection and for the establishment of viral reservoir. Many studies have evidenced the relevance of the loss of gut mucosal integrity in chronic immune activation and persistence inflammation. After a few weeks post-infection, HIV replicates markedly in this tissue showing a significant depletion of CD4+ memory T cells followed by disruption of the intestinal epithelium. The disruption of the gut may not be fully restored and lead to both an imbalance of the intestinal microbiota composition (dysbiosis) and to the release of bacterial products in the circulation (microbial translocation) that induce a chronic immune activation and inflammation. Microbial translocation is considered one of the main causes of morbidity and mortality in HIV infection. The mechanisms associated with the breach of gut mucosal include the loss of Th17 cells and alteration in Type I IFN responses. Residual replication in this compartment may propagate the infection to other organs and contribute to sustained immune dysregulation. (Reviewed in: Thompson et al., 2017 and Zicari et al., 2019).

Chronic immune inflammation also affects the lymph nodes which are a major tissue reservoir and sanctuary for HIV-infected CD4+ T cells. In treated-HIV patients, chronic activation of lymph nodes leads to upregulation of transforming growth factor β (TGF-β), which stimulates collagen production. The collagen modifies the structure and function of lymphoid tissue with the loss of naïve T cells. This devastating fibrotic effect is reduced by cART but lymphoid tissue does not recover entirely may be due to persistent inflammation and low grade or virus replication (Zeng et al., 2011).

HIV can lead to a liver and kidney diseases which have an inflammatory component and responses to TLRs from the gut breach. The persistence inflammation also affects the functionality of the thymus leading to long-term thymic dysfunction and clonal exhaustion of T cells (Reviewed in Lewis et al., 2019). HIV-induced chronic systemic inflammation is associated with increased risk of lung infections and chronic lung diseases. In HIV infected patients on cART, higher levels of systemic inflammatory markers, as CRP, have been associated with increased bacterial pneumonia risk. Furthermore, HIV can lead to deleterious pulmonary inflammatory responses (Cribbs et al., 2020).

Obesity in an increasingly factor in HIV infected individuals, that is influenced by several factors, including some viral products and antiretroviral regimens. Obesity leads to excessive adipose tissue that is a known HIV reservoir in addition to increasing systemic inflammation, development of metabolic syndrome and the risk of HANA. Adipose tissue is an important immune site characterized by accumulation of activated pro-inflammatory cells including macrophages, T cells, NK and NKT cells and granulocytes. The long-term elevated levels of inflammatory cytokines as IL-6 and TNFα by adipocytes exacerbate adipose inflammation. This inflammatory environment favors the establishment of HIV reservoir. Adipose tissue is characterized by a shift to pro-inflammatory M1-type macrophages, T helper 1 cells (Th1) and Th17 cells, with high levels of IL-1β and IL-18 downstream of inflammasome activation. This cellular shift contributes to inflammation and immune dysfunction (Damouche et al., 2015).

After invasion of the CNS, HIV infects microglia and tissue macrophages to establish reservoirs in the brain. Parenchymal microglia and perivascular macrophages are productively infected with HIV in brains of patients with or without neurological diseases. Once these immune cells are activated, release neurotoxic host factors such as pro-inflammatory cytokines (IL-6, IL-8, TNFα, IFNα, IL-1β) and chemokines (CCL2 and CCL5). HIV production in the brain is associated with the emergence of virus resistance and with HIV-associated neurocognitive disorders (HAND) in humans and in animal models such as the macaques. These disorders have a different clinical severity that ranges from asymptomatic neurocognitive impairment, mild neurocognitive disorder and the most severe HIV-associated dementia. About 50% of treated HIV patients develop HAND and chronic inflammation is postulated to be the driver of neuronal damage. The pathogenesis of HAND appears to be multifactorial: HIV replication in CNS, drug neurotoxicity of cART, CNS inflammation and co-infection with other pathogens (Wallet et al., 2019, reviewed in Zicari et al., 2019). Since inflammasome activation has been found in brain injury and neurodegenerative diseases, it was no surprising that elevated levels of IL-18 were found in cerebrospinal fluid in HIV patients with neurologic symptoms (Zicari et al., 2019; Nasi et al., 2017; Heron et al., 2017).

Metabolic syndrome (MetS) is a common comorbidity in HIV+ patients and has been linked to cardiovascular diseases (CVD), neurocognitive impairment and has been also associated indirectly with the effects of the combination of cART and inflammatory response. Persistent inflammation may be the major determinant in the pathogenesis of metabolic disorders and atherosclerosis. HIV+ patients present metabolic abnormalities such as hypertriglyceridaemia, low- and high- density lipoprotein cholesterol and insulin resistance.

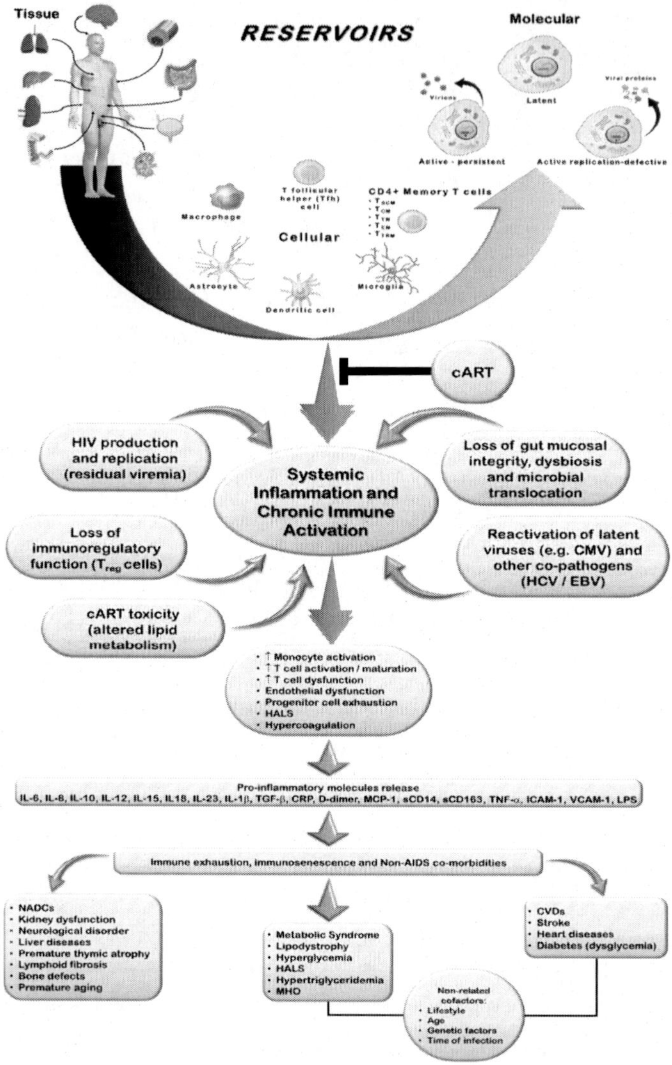

Figure 1. The relationship between viral reservoir and systemic inflammation in chronic HIV-infection and clinical implications. Despite effective long-term combined antiretroviral therapy (cART), HIV reservoirs (latent, active-persistent and active replication-defective reservoirs) persist mainly in CD4+ memory T cells subsets and in some innate immune cells as macrophages, monocytes, dentritic cells, microglia and astrocytes. HIV latency and chronic immune activation cause a persistent inflammation and immune dysfunction in HIV+ patients. The aetiology of systemic inflammation includes residual viremia, microbial translocation due to gut breach, the presence of co-pathogens, damage of immunoregulatory function, and antiretrovirals toxicity. HIV-associated inflammatory environment causes immune exhaustion, immunosenescence and increased risk of developing non–AIDS co-morbidities identified as metabolic syndrome (MetS), cardiovascular diseases (CVDs), non-AIDS-defining cancers (NADCs), HIV-associated lipodystrophy syndrome (HALS), and metabolically healthy obesity (MHO).

Morphological disorders include peripheral lipoatrophy and isolated or co-existing lipoaccumulation. Visceral obesity is a source of many inflammatory proteins and is considered a risk factor for CVD and neurocognitive disorders (Reviewed in Li Vecchi et al., 2014).

The effects of chronic immune activation and inflammation remain the major contributor to CVD risk development. Virus-related mechanisms in CVD include pro-inflammatory effects of HIV proteins released from the virus, CD4+ depletion, increased intestinal permeability, microbial translocation, and altered cholesterol metabolism. Recent studies suggest that monocytes and macrophages are the major source of inflammatory mediators that promote CVD, even in patients on cART. High levels of some pro-inflammatory molecules as sCD14, sCD163, CCL2 (MCP-1), IL-6 and D-dimer are associated with atherosclerotic cardiovascular disease (e.g., myocardial infarction), coronary stenosis and a greater risk for fatal CVD. Additionally, CD4+ T cell depletion among HIV-infected people is associated with higher rates and risk for incident acute infarction, heart failure, peripheral artery disease, and ischaemic stroke (Reviewed in: So-Armah et al., 2020).

Current Anti-Inflammatory Interventions in HIV Infection

Since HIV-associated chronic inflammation is not fully controlled with long-term cART, novel therapies targeting systemic inflammation are currently being tested to reduce its negative impact on the immune system and the risk of developing non-AIDS co-morbidities.

A strategy based on antiretroviral regimen simplifications (dual-drug or even single-drug) is currently being used in several clinical trials introduced as a first line therapy in adults to minimize the toxicity and the elevated inflammation and immune activation associated with the use of antiretrovirals. The follow-up studies are still ongoing with some controversial results. (Reviewed in: Zicari et al., 2019).

A successful anti-inflammatory therapy in HIV infection is based on the use of immunomodulators including statins, purinergic P2X receptor inhibitors, and chloroquine and its analogue hydroxychloroquine among others (Reviewed in: Bandera et al., 2017). The statins have immunomodulatory effects through different mechanisms including cytokine production, expression of HLA class II antigens on macrophages, and inhibition of the expression of co-stimulatory molecules and antigen-presenting cells.

The atorvastatin, rosuvastatin, and pravastatin formulations are compatibles for co-administration with cART and all these drugs have showed efficacy in reducing levels of inflammatory markers (Chastain et al., 2017). Chloroquine and hydroxychloroquine have demonstrated to impair HIV replication by targeting gp120 envelope glycoprotein production reducing levels of intracellular iron that is required for viral replication. These immunomodulators also impact on Tat viral protein and integrase functionality (Piconi et al., 2011).

Another recent strategy aiming at reducing inflammation in HIV-infected patients is based on the use of a human monoclonal anti-IL-1β antibody (canakinumab) which was able to significantly reduce levels of some pro-inflammatory molecules after administration to HIV-infected patients.

More recent studies suggest that administering probiotics with cART may decrease the persistent status of activation/inflammation at mucosal level of GI, and microbial translocation by competing with pathogenic bacteria. The results showed that combination of two or four different probiotics resulted in moderately microbial translocation and immune activation in individuals on cART (Gonzalez-Hernandez et al., 2012; Schunter et al., 2012).

Conclusion

The innate immune response has a major role in the control of HIV infection and may have a long-term impact on the outcome of disease. However, HIV has evolved different mechanisms to evade host innate immune response. Moreover, the early establishment of long-lived viral reservoirs despite effective cART, residual immune activation, and chronic inflammation prevent the eradication of the virus. To address these challenges, novel strategies have recently been developed aimed at inactivating or eradicating viral reservoir and reducing the negative impact of chronic activation/inflammation on the immune system. Given the complexity of HIV infection, it is clear that for future studies it will be important to consider a comprehensive cure strategy that includes the combination of effective cART with molecular and immunomodulatory strategies, in order to achieve HIV functional cure.

References

Abdel-Mohsen, M., L. Kuri-Cervantes, J. Grau-Exposito, A. M. Spivak, R. A. Nell, C. Tomescu, S. K. Vadrevu, L. B. Giron, C. Serra-Peinado, M. Genesca, J. Castellvi, G. Wu, P. M. Del Rio Estrada, M. Gonzalez-Navarro, K. Lynn, C. T. King, S. Vemula, K. Cox, Y. Wan, Q. Li, K. Mounzer, J. Kostman, I. Frank, M. Paiardini, D. Hazuda, G. Reyes-Teran, D. Richman, B. Howell, P. Tebas, J. Martinez-Picado, V. Planelles, M. J. Buzon, M. R. Betts, and L. J. Montaner. 2018. "CD32 is expressed on cells with transcriptionally active HIV but does not enrich for HIV DNA in resting T cells." *Sci Transl Med* no. 10 (437). doi: 10.1126/scitranslmed.aar6759.

Abdulhaqq, S. A., C. Zorrilla, G. Kang, X. Yin, V. Tamayo, K. E. Seaton, J. Joseph, S. Garced, G. D. Tomaras, K. A. Linn, A. S. Foulkes, L. Azzoni, M. VerMilyea, C. Coutifaris, A. V. Kossenkov, L. Showe, E. N. Kraiselburd, Q. Li, and L. J. Montaner. 2016. "HIV-1-negative female sex workers sustain high cervical IFNvarepsilon, low immune activation, and low expression of HIV-1-required host genes." *Mucosal Immunol* no. 9 (4):1027-38. doi: 10.1038/ mi.2015.116.

Abner, E., and A. Jordan. 2019. "HIV "shock and kill" therapy: In need of revision." *Antiviral Res* no. 166:19-34. doi: 10.1016/j.antiviral. 2019.03.008.

Abreu, C. M., R. T. Veenhuis, C. R. Avalos, S. Graham, S. E. Queen, E. N. Shirk, B. T. Bullock, M. Li, K. A. Metcalf Pate, S. E. Beck, L. M. Mangus, J. L. Mankowski, J. E. Clements, and L. Gama. 2019. "Infectious Virus Persists in CD4(+) T Cells and Macrophages in Antiretroviral Therapy-Suppressed Simian Immunodeficiency Virus-Infected Macaques." *J Virol* no. 93 (15). doi: 10.1128/JVI.00065-19.

Ahlenstiel, C. L., G. Symonds, S. J. Kent, and A. D. Kelleher. 2020. "Block and Lock HIV Cure Strategies to Control the Latent Reservoir." *Front Cell Infect Microbiol* no. 10:424. doi: 10.3389/fcimb.2020.00424.

Alter, G., and M. Altfeld. 2009. "NK cells in HIV-1 infection: evidence for their role in the control of HIV-1 infection." *J Intern Med* no. 265 (1):29-42. doi: 10.1111/j.1365-2796.2008.02045.x.

Altfeld, M., L. Fadda, D. Frleta, and N. Bhardwaj. 2011. "DCs and NK cells: critical effectors in the immune response to HIV-1." *Nat Rev Immunol* no. 11 (3):176-86. doi: 10.1038/nri2935.

Aluri, J., M. A. Cooper, and L. G. Schuettpelz. 2021. "Toll-Like Receptor Signaling in the Establishment and Function of the Immune System." *Cells* no. 10 (6). doi: 10.3390/cells10061374.

Anand, P. K., R. K. Malireddi, and T. D. Kanneganti. 2011. "Role of the nlrp3 inflammasome in microbial infection." *Front Microbiol* no. 2:12. doi: 10.3389/fmicb.2011.00012.

Ancuta, P., K. J. Kunstman, P. Autissier, T. Zaman, D. Stone, S. M. Wolinsky, and D. Gabuzda. 2006. "CD16+ monocytes exposed to HIV promote highly efficient viral replication upon differentiation into macrophages and interaction with T cells." *Virology* no. 344 (2):267-76. doi: 10.1016/j.virol.2005.10.027.

Anderson, A. M., S. Kundu, B. Tang, F. Vaida, O. Okwuegbuna, D. McClernon, M. Cherner, R. Deutsch, D. Cookson, M. Crescini, I. Grant, H. Zetterberg, K. Blennow, M. Gisslen, R. J. Ellis, and S. L. Letendre. 2021. "Cerebrospinal fluid CXCL10 is associated with the presence of low level CSF HIV during suppressive antiretroviral therapy." *J Neuroimmunol* no. 353:577493. doi: 10.1016/j.jneuroim. 2021.577493.

Andre, P., C. Denis, C. Soulas, C. Bourbon-Caillet, J. Lopez, T. Arnoux, M. Blery, C. Bonnafous, L. Gauthier, A. Morel, B. Rossi, R. Remark, V. Breso, E. Bonnet, G. Habif, S. Guia, A. I. Lalanne, C. Hoffmann, O. Lantz, J. Fayette, A. Boyer-Chammard, R. Zerbib, P. Dodion, H. Ghadially, M. Jure-Kunkel, Y. Morel, R. Herbst, E. Narni-Mancinelli, R. B. Cohen, and E. Vivier. 2018. "Anti-NKG2A mAb Is a Checkpoint Inhibitor that Promotes Anti-tumor Immunity by Unleashing Both T and NK Cells." *Cell* no. 175 (7):1731-1743 e13. doi: 10.1016/j.cell.2018.10.014.

Archin, N. M., J. L. Kirchherr, J. A. Sung, G. Clutton, K. Sholtis, Y. Xu, B. Allard, E. Stuelke, A. D. Kashuba, J. D. Kuruc, J. Eron, C. L. Gay, N. Goonetilleke, and D. M. Margolis. 2017. "Interval dosing with the HDAC inhibitor vorinostat effectively reverses HIV latency." *J Clin Invest* no. 127 (8):3126-3135. doi: 10.1172/JCI92684.

Asahchop, E. L., O. Meziane, M. K. Mamik, W. F. Chan, W. G. Branton, L. Resch, M. J. Gill, E. Haddad, J. V. Guimond, M. A. Wainberg, G. B. Baker, E. A. Cohen, and C. Power. 2017. "Reduced antiretroviral drug efficacy and concentration in HIV-infected microglia contributes to viral persistence in brain." *Retrovirology* no. 14 (1):47. doi: 10.1186/s12977-017-0370-5.

Avalos, C. R., C. M. Abreu, S. E. Queen, M. Li, S. Price, E. N. Shirk, E. L. Engle, E. Forsyth, B. T. Bullock, F. Mac Gabhann, S. W. Wietgrefe, A. T. Haase, M. C. Zink, J. L. Mankowski, J. E. Clements, and L. Gama. 2017. "Brain Macrophages in Simian Immunodeficiency Virus-Infected, Antiretroviral-Suppressed Macaques: a Functional Latent Reservoir." *mBio* no. 8 (4). doi: 10.1128/mBio.01186-17.

Avalos, C. R., S. L. Price, E. R. Forsyth, J. N. Pin, E. N. Shirk, B. T. Bullock, S. E. Queen, M. Li, D. Gellerup, S. L. O'Connor, M. C. Zink, J. L. Mankowski, L. Gama, and J. E. Clements. 2016. "Quantitation of Productively Infected Monocytes and Macrophages of Simian Immunodeficiency Virus-Infected Macaques." *J Virol* no. 90 (12):5643-5656. doi: 10.1128/JVI.00290-16.

Avettand-Fenoel, V., L. Hocqueloux, J. Ghosn, A. Cheret, P. Frange, A. Melard, J. P. Viard, and C. Rouzioux. 2016. "Total HIV-1 DNA, a Marker of Viral Reservoir Dynamics with Clinical Implications." *Clin Microbiol Rev* no. 29 (4):859-80. doi: 10.1128/CMR.00015-16.

Azzoni, L., A. S. Foulkes, E. Papasavvas, A. M. Mexas, K. M. Lynn, K. Mounzer, P. Tebas, J. M. Jacobson, I. Frank, M. P. Busch, S. G. Deeks, M. Carrington, U. O'Doherty, J. Kostman, and L. J. Montaner. 2013. "Pegylated Interferon alfa-2a monotherapy results in suppression of HIV type 1 replication and decreased cell-associated HIV DNA integration." *J Infect Dis* no. 207 (2):213-22. doi: 10.1093/infdis/jis663.

Badley, A. D., A. Sainski, F. Wightman, and S. R. Lewin. 2013. "Altering cell death pathways as an approach to cure HIV infection." *Cell Death Dis* no. 4:e718. doi: 10.1038/cddis.2013.248.

Balcom, E. F., W. C. Roda, E. A. Cohen, M. Y. Li, and C. Power. 2019. "HIV-1 persistence in the central nervous system: viral and host determinants during antiretroviral therapy." *Curr Opin Virol* no. 38:54-62. doi: 10.1016/j. coviro.2019.06.004.

Bale, M. J., and M. F. Kearney. 2019. "Review: HIV-1 phylogeny during suppressive antiretroviral therapy." *Curr Opin HIV AIDS* no. 14 (3):188-193. doi: 10.1097/COH.0000000000000535.

Bandera, A., E. Colella, G. Rizzardini, A. Gori, and M. Clerici. 2017. "Strategies to limit immune-activation in HIV patients." *Expert Rev Anti Infect Ther* no. 15 (1):43-54. doi: 10.1080/14787210.2017. 1250624.

Banerjee, C., N. Archin, D. Michaels, A. C. Belkina, G. V. Denis, J. Bradner, P. Sebastiani, D. M. Margolis, and M. Montano. 2012. "BET bromodomain inhibition as a novel strategy for reactivation of HIV-1." *J Leukoc Biol* no. 92 (6):1147-54. doi: 10.1189/jlb.0312165.

Banga, R., F. A. Procopio, A. Noto, G. Pollakis, M. Cavassini, K. Ohmiti, J. M. Corpataux, L. de Leval, G. Pantaleo, and M. Perreau. 2016. "PD-1(+) and follicular helper T cells are responsible for persistent HIV-1 transcription in treated aviremic individuals." *Nat Med* no. 22 (7):754-61. doi: 10.1038/nm.4113.

Barblu, L., K. Machmach, C. Gras, J. F. Delfraissy, F. Boufassa, M. Leal, E. Ruiz-Mateos, O. Lambotte, J. P. Herbeuval, and Anrs Ep Hiv Controllers Study Group. 2012. "Plasmacytoid dendritic cells (pDCs) from HIV controllers produce interferon-alpha and differentiate into functional killer pDCs under HIV activation." *J Infect Dis* no. 206 (5):790-801. doi: 10.1093/infdis/jis384.

Bayigga, L., R. Nabatanzi, P. N. Sekiziyivu, H. Mayanja-Kizza, M. R. Kamya, A. Kambugu, J. Olobo, A. Kiragga, S. Kirimunda, M. Joloba, and D. Nakanjako. 2014. "High CD56++CD16- natural killer (NK) cells among suboptimal immune responders after four years of suppressive antiretroviral therapy in an

African adult HIV treatment cohort." *BMC Immunol* no. 15:2. doi: 10.1186/1471-2172-15-2.

Bednar, M. M., C. B. Sturdevant, L. A. Tompkins, K. T. Arrildt, E. Dukhovlinova, L. P. Kincer, and R. Swanstrom. 2015. "Compartmentalization, Viral Evolution, and Viral Latency of HIV in the CNS." *Curr HIV/AIDS Rep* no. 12 (2):262-71. doi: 10.1007/ s11904-015-0265-9.

Benlahrech, A., and S. Patterson. 2011. "HIV-1 infection and induction of interferon alpha in plasmacytoid dendritic cells." *Curr Opin HIV AIDS* no. 6 (5):373-8. doi: 10.1097/COH.0b013e328349592a.

Berenson, L. S., J. Yang, B. P. Sleckman, T. L. Murphy, and K. M. Murphy. 2006. "Selective requirement of p38alpha MAPK in cytokine-dependent, but not antigen receptor-dependent, Th1 responses." *J Immunol* no. 176 (8):4616-21. doi: 10.4049/jimmunol. 176.8.4616.

Bernard, M. A., X. Han, S. Inderbitzin, I. Agbim, H. Zhao, H. Koziel, and S. D. Tachado. 2014. "HIV-derived ssRNA binds to TLR8 to induce inflammation-driven macrophage foam cell formation." *PLoS One* no. 9 (8):e104039. doi: 10.1371/journal.pone.0104039.

Bertrand, L., H. J. Cho, and M. Toborek. 2019. "Blood-brain barrier pericytes as a target for HIV-1 infection." *Brain* no. 142 (3):502-511. doi: 10.1093/brain/awy339.

Beutler, B. 2004. "Innate immunity: an overview." *Mol Immunol* no. 40 (12):845-59. doi: 10.1016/j.molimm.2003.10.005.

Biswas, S. K., and E. Lopez-Collazo. 2009. "Endotoxin tolerance: new mechanisms, molecules and clinical significance." *Trends Immunol* no. 30 (10):475-87. doi: 10.1016/j.it.2009.07.009.

Blankson, J. N., D. Finzi, T. C. Pierson, B. P. Sabundayo, K. Chadwick, J. B. Margolick, T. C. Quinn, and R. F. Siliciano. 2000. "Biphasic decay of latently infected CD4+ T cells in acute human immunodeficiency virus type 1 infection." *J Infect Dis* no. 182 (6):1636-42. doi: 10.1086/317615.

Bolduc, J. F., M. Ouellet, L. Hany, and M. J. Tremblay. 2017. "Toll-Like Receptor 2 Ligation Enhances HIV-1 Replication in Activated CCR6+ CD4+ T Cells by Increasing Virus Entry and Establishing a More Permissive Environment to Infection." *J Virol* no. 91 (4). doi: 10.1128/JVI.01402-16.

Booiman, T., and N. A. Kootstra. 2014. "Polymorphism in IFI16 affects CD4(+) T-cell counts in HIV-1 infection." *Int J Immunogenet* no. 41 (6):518-20. doi: 10.1111/iji.12157.

Booiman, T., F. W. Wit, I. Maurer, D. De Francesco, C. A. Sabin, A. M. Harskamp, M. Prins, P. Garagnani, C. Pirazzini, C. Franceschi, D. Fuchs, M. Gisslen, A. Winston, P. Reiss, N. A. Kootstra, and Aids Collaboration Comorbidity in Relation to. 2017. "High Cellular Monocyte Activation in People Living With Human Immunodeficiency Virus on Combination

Antiretroviral Therapy and Lifestyle-Matched Controls Is Associated With Greater Inflammation in Cerebrospinal Fluid." *Open Forum Infect Dis* no. 4 (3):ofx108. doi: 10.1093/ofid/ofx108.

Borducchi, E. N., J. Liu, J. P. Nkolola, A. M. Cadena, W. H. Yu, S. Fischinger, T. Broge, P. Abbink, N. B. Mercado, A. Chandrashekar, D. Jetton, L. Peter, K. McMahan, E. T. Moseley, E. Bekerman, J. Hesselgesser, W. Li, M. G. Lewis, G. Alter, R. Geleziunas, and D. H. Barouch. 2018. "Antibody and TLR7 agonist delay viral rebound in SHIV-infected monkeys." *Nature* no. 563 (7731):360-364. doi: 10.1038/s41586-018-0600-6.

Bosinger, S. E., and N. S. Utay. 2015. "Type I interferon: understanding its role in HIV pathogenesis and therapy." *Curr HIV/AIDS Rep* no. 12 (1):41-53. doi: 10.1007/s11904-014-0244-6.

Bosisio, D., N. Polentarutti, M. Sironi, S. Bernasconi, K. Miyake, G. R. Webb, M. U. Martin, A. Mantovani, and M. Muzio. 2002. "Stimulation of toll-like receptor 4 expression in human mononuclear phagocytes by interferon-gamma: a molecular basis for priming and synergism with bacterial lipopolysaccharide." *Blood* no. 99 (9):3427-31. doi: 10.1182/blood.v99.9.3427.

Brenner, D., H. Blaser, and T. W. Mak. 2015. "Regulation of tumour necrosis factor signalling: live or let die." *Nat Rev Immunol* no. 15 (6):362-74. doi: 10.1038/nri3834.

Brites-Alves, C., E. Luz, E. M. Netto, T. Ferreira, R. S. Diaz, C. Pedroso, K. Page, and C. Brites. 2018. "Immune Activation, Proinflammatory Cytokines, and Conventional Risks for Cardiovascular Disease in HIV Patients: A Case-Control Study in Bahia, Brazil." *Front Immunol* no. 9:1469. doi: 10.3389/fimmu.2018.01469.

Bruel, T., and O. Schwartz. 2018. "Markers of the HIV-1 reservoir: facts and controversies." *Curr Opin HIV AIDS* no. 13 (5):383-388. doi: 10.1097/COH.0000000000000482.

Bruner, K. M., and L. B. Cohn. 2019. "HIV-1 reservoir dynamics in CD4+ T cells." *Curr Opin HIV AIDS* no. 14 (2):108-114. doi: 10.1097/COH.0000000000000521.

Burdo, T. H., J. Walker, and K. C. Williams. 2015. "Macrophage Polarization in AIDS: Dynamic Interface between Anti-Viral and Anti-Inflammatory Macrophages during Acute and Chronic Infection." *J Clin Cell Immunol* no. 6 (3).

Buzon, M. J., H. Sun, C. Li, A. Shaw, K. Seiss, Z. Ouyang, E. Martin-Gayo, J. Leng, T. J. Henrich, J. Z. Li, F. Pereyra, R. Zurakowski, B. D. Walker, E. S. Rosenberg, X. G. Yu, and M. Lichterfeld. 2014. "HIV-1 persistence in CD4+ T cells with stem cell-like properties." *Nat Med* no. 20 (2):139-42. doi: 10.1038/nm.3445.

Cai, J., M. Xu, X. Zhang, and H. Li. 2019. "Innate Immune Signaling in Nonalcoholic Fatty Liver Disease and Cardiovascular Diseases." *Annu Rev Pathol* no. 14:153-184. doi: 10.1146/annurev-pathmechdis-012418-013003.

Calantone, N., F. Wu, Z. Klase, C. Deleage, M. Perkins, K. Matsuda, E. A. Thompson, A. M. Ortiz, C. L. Vinton, I. Ourmanov, K. Lore, D. C. Douek, J. D. Estes, V. M. Hirsch, and J. M. Brenchley. 2014. "Tissue myeloid cells in SIV-infected primates acquire viral DNA through phagocytosis of infected T cells." *Immunity* no. 41 (3):493-502. doi: 10.1016/j.immuni.2014.08.014.

Cantero-Perez, J., J. Grau-Exposito, C. Serra-Peinado, D. A. Rosero, L. Luque-Ballesteros, A. Astorga-Gamaza, J. Castellvi, T. Sanhueza, G. Tapia, B. Lloveras, M. A. Fernandez, J. G. Prado, J. M. Sole-Sedeno, A. Tarrats, C. Lecumberri, L. Manalich-Barrachina, C. Centeno-Mediavilla, V. Falco, M. J. Buzon, and M. Genesca. 2019. "Resident memory T cells are a cellular reservoir for HIV in the cervical mucosa." *Nat Commun* no. 10 (1):4739. doi: 10.1038/s41467-019-12732-2.

Carbone, F. R. 2015. "Tissue-Resident Memory T Cells and Fixed Immune Surveillance in Nonlymphoid Organs." *J Immunol* no. 195 (1):17-22. doi: 10.4049/jimmunol.1500515.

Carranza, P., P. M. Del Rio Estrada, D. Diaz Rivera, Y. Ablanedo-Terrazas, and G. Reyes-Teran. 2016. "Lymph nodes from HIV-infected individuals harbor mature dendritic cells and increased numbers of PD-L1+ conventional dendritic cells." *Hum Immunol* no. 77 (7):584-93. doi: 10.1016/j.humimm.2016.05.019.

Carty, M., C. Guy, and A. G. Bowie. 2021. "Detection of Viral Infections by Innate Immunity." *Biochem Pharmacol* no. 183:114316. doi: 10.1016/j.bcp.2020.114316.

Castro-Gonzalez, S., M. Colomer-Lluch, and R. Serra-Moreno. 2018. "Barriers for HIV Cure: The Latent Reservoir." *AIDS Res Hum Retroviruses* no. 34 (9):739-759. doi: 10.1089/AID.2018.0118.

Clayton, K. L., D. R. Collins, J. Lengieza, M. Ghebremichael, F. Dotiwala, J. Lieberman, and B. D. Walker. 2018. "Resistance of HIV-infected macrophages to CD8(+) T lymphocyte-mediated killing drives activation of the immune system." *Nat Immunol* no. 19 (5):475-486. doi: 10.1038/s41590-018-0085-3.

Coelho, A. V., R. R. de Moura, A. J. Kamada, R. C. da Silva, R. L. Guimaraes, L. A. Brandao, L. C. de Alencar, and S. Crovella. 2016. "Dendritic Cell-Based Immunotherapies to Fight HIV: How Far from a Success Story? A Systematic Review and Meta-Analysis." *Int J Mol Sci* no. 17 (12). doi: 10.3390/ijms17121985.

Cohen, Y. Z., J. C. C. Lorenzi, L. Krassnig, J. P. Barton, L. Burke, J. Pai, C. L. Lu, P. Mendoza, T. Y. Oliveira, C. Sleckman, K. Millard, A. L. Butler, J. P.

Dizon, S. A. Belblidia, M. Witmer-Pack, I. Shimeliovich, R. M. Gulick, M. S. Seaman, M. Jankovic, M. Caskey, and M. C. Nussenzweig. 2018. "Relationship between latent and rebound viruses in a clinical trial of anti-HIV-1 antibody 3BNC117." *J Exp Med* no. 215 (9):2311-2324. doi: 10.1084/jem.20180936.

Cohn, L. B., N. Chomont, and S. G. Deeks. 2020. "The Biology of the HIV-1 Latent Reservoir and Implications for Cure Strategies." *Cell Host Microbe* no. 27 (4):519-530. doi: 10.1016/j.chom.2020.03.014.

Coiras, M., M. Bermejo, B. Descours, E. Mateos, J. Garcia-Perez, M. R. Lopez-Huertas, M. M. Lederman, M. Benkirane, and J. Alcami. 2016. "IL-7 Induces SAMHD1 Phosphorylation in CD4+ T Lymphocytes, Improving Early Steps of HIV-1 Life Cycle." *Cell Rep* no. 14 (9):2100-2107. doi: 10.1016/j.celrep.2016.02.022.

Colby, D. J., L. Trautmann, S. Pinyakorn, L. Leyre, A. Pagliuzza, E. Kroon, M. Rolland, H. Takata, S. Buranapraditkun, J. Intasan, N. Chomchey, R. Muir, E. K. Haddad, S. Tovanabutra, S. Ubolyam, D. L. Bolton, B. A. Fullmer, R. J. Gorelick, L. Fox, T. A. Crowell, R. Trichavaroj, R. O'Connell, N. Chomont, J. H. Kim, N. L. Michael, M. L. Robb, N. Phanuphak, J. Ananworanich, and R. V. study group. 2018. "Rapid HIV RNA rebound after antiretroviral treatment interruption in persons durably suppressed in Fiebig I acute HIV infection." *Nat Med* no. 24 (7):923-926. doi: 10.1038/s41591-018-0026-6.

Colonna, M., and M. Cella. 2007. "Crosspresentation: plasmacytoid dendritic cells are in the business." *Immunity* no. 27 (3):419-21. doi: 10.1016/j.immuni.2007.08.006.

Collin, M., and V. Bigley. 2018. "Human dendritic cell subsets: an update." *Immunology* no. 154 (1):3-20. doi: 10.1111/imm.12888.

Coomes, S. M., Y. Kannan, V. S. Pelly, L. J. Entwistle, R. Guidi, J. Perez-Lloret, N. Nikolov, W. Muller, and M. S. Wilson. 2017. "CD4(+) Th2 cells are directly regulated by IL-10 during allergic airway inflammation." *Mucosal Immunol* no. 10 (1):150-161. doi: 10.1038/mi.2016.47.

Cortes, F. H., H. H. S. de Paula, G. Bello, M. Ribeiro-Alves, S. S. D. de Azevedo, D. G. Caetano, S. L. M. Teixeira, B. Hoagland, B. Grinsztejn, V. G. Veloso, M. L. Guimaraes, and M. G. Morgado. 2018. "Plasmatic Levels of IL-18, IP-10, and Activated CD8(+) T Cells Are Potential Biomarkers to Identify HIV-1 Elite Controllers With a True Functional Cure Profile." *Front Immunol* no. 9:1576. doi: 10.3389/fimmu.2018.01576.

Couturier, J., J. W. Suliburk, J. M. Brown, D. J. Luke, N. Agarwal, X. Yu, C. Nguyen, D. Iyer, C. A. Kozinetz, P. A. Overbeek, M. L. Metzker, A. Balasubramanyam, and D. E. Lewis. 2015. "Human adipose tissue as a reservoir for memory CD4+ T cells and HIV." *AIDS* no. 29 (6):667-74. doi: 10.1097/QAD.0000000000000599.

Covino, D. A., M. G. Desimio, and M. Doria. 2021. "Combinations of Histone Deacetylase Inhibitors with Distinct Latency Reversing Agents Variably Affect HIV Reactivation and Susceptibility to NK Cell-Mediated Killing of T Cells That Exit Viral Latency." *Int J Mol Sci* no. 22 (13). doi: 10.3390/ijms22136654.

Cribbs, S. K., K. Crothers, and A. Morris. 2020. "Pathogenesis of HIV-Related Lung Disease: Immunity, Infection, and Inflammation." *Physiol Rev* no. 100 (2):603-632. doi: 10.1152/physrev.00039.2018.

Cribbs, S. K., J. Lennox, A. M. Caliendo, L. A. Brown, and D. M. Guidot. 2015. "Healthy HIV-1-infected individuals on highly active antiretroviral therapy harbor HIV-1 in their alveolar macrophages." *AIDS Res Hum Retroviruses* no. 31 (1):64-70. doi: 10.1089/AID. 2014.0133.

Crooks, A. M., R. Bateson, A. B. Cope, N. P. Dahl, M. K. Griggs, J. D. Kuruc, C. L. Gay, J. J. Eron, D. M. Margolis, R. J. Bosch, and N. M. Archin. 2015. "Precise Quantitation of the Latent HIV-1 Reservoir: Implications for Eradication Strategies." *J Infect Dis* no. 212 (9):1361-5. doi: 10.1093/infdis/jiv218.

Chaillon, A., S. Gianella, S. Dellicour, S. A. Rawlings, T. E. Schlub, M. F. De Oliveira, C. Ignacio, M. Porrachia, B. Vrancken, and D. M. Smith. 2020. "HIV persists throughout deep tissues with repopulation from multiple anatomical sources." *J Clin Invest* no. 130 (4):1699-1712. doi: 10.1172/JCI134815.

Chastain, D. B., K. R. Stover, and D. M. Riche. 2017. "Evidence-based review of statin use in patients with HIV on antiretroviral therapy." *J Clin Transl Endocrinol* no. 8:6-14. doi: 10.1016/j.jcte.2017.01.004.

Chatzidimitriou, D., E. Tsotridou, P. Grigoropoulos, and L. Skoura. 2020. "HIV-1: towards understanding the nature and quantifying the latent reservoir." *Acta Virol* no. 64 (1):3-9. doi: 10.4149/av_2020_101.

Chaudhry, A., R. M. Samstein, P. Treuting, Y. Liang, M. C. Pils, J. M. Heinrich, R. S. Jack, F. T. Wunderlich, J. C. Bruning, W. Muller, and A. Y. Rudensky. 2011. "Interleukin-10 signaling in regulatory T cells is required for suppression of Th17 cell-mediated inflammation." *Immunity* no. 34 (4):566-78. doi: 10.1016/j.immuni.2011.03.018.

Chavez, L., V. Calvanese, and E. Verdin. 2015. "HIV Latency Is Established Directly and Early in Both Resting and Activated Primary CD4 T Cells." *PLoS Pathog* no. 11 (6):e1004955. doi: 10.1371/journal.ppat.1004955.

Cheng, L., J. Ma, J. Li, D. Li, G. Li, F. Li, Q. Zhang, H. Yu, F. Yasui, C. Ye, L. C. Tsao, Z. Hu, L. Su, and L. Zhang. 2017. "Blocking type I interferon signaling enhances T cell recovery and reduces HIV-1 reservoirs." *J Clin Invest* no. 127 (1):269-279. doi: 10.1172/ JCI90745.

Chomont, N., M. El-Far, P. Ancuta, L. Trautmann, F. A. Procopio, B. Yassine-Diab, G. Boucher, M. R. Boulassel, G. Ghattas, J. M. Brenchley, T. W. Schacker, B. J. Hill, D. C. Douek, J. P. Routy, E. K. Haddad, and R. P. Sekaly. 2009. "HIV reservoir size and persistence are driven by T cell survival and homeostatic proliferation." *Nat Med* no. 15 (8):893-900. doi: 10.1038/nm.1972.

Dai, J., L. M. Agosto, C. Baytop, J. J. Yu, M. J. Pace, M. K. Liszewski, and U. O'Doherty. 2009. "Human immunodeficiency virus integrates directly into naive resting CD4+ T cells but enters naive cells less efficiently than memory cells." *J Virol* no. 83 (9):4528-37. doi: 10.1128/JVI.01910-08.

Damouche, A., T. Lazure, V. Avettand-Fenoel, N. Huot, N. Dejucq-Rainsford, A. P. Satie, A. Melard, L. David, C. Gommet, J. Ghosn, N. Noel, G. Pourcher, V. Martinez, S. Benoist, V. Bereziat, A. Cosma, B. Favier, B. Vaslin, C. Rouzioux, J. Capeau, M. Muller-Trutwin, N. Dereuddre-Bosquet, R. Le Grand, O. Lambotte, and C. Bourgeois. 2015. "Adipose Tissue Is a Neglected Viral Reservoir and an Inflammatory Site during Chronic HIV and SIV Infection." *PLoS Pathog* no. 11 (9):e1005153. doi: 10.1371/journal. ppat.1005153.

DeCarlo, C. A., A. Severini, L. Edler, N. G. Escott, P. F. Lambert, M. Ulanova, and I. Zehbe. 2010. "IFN-kappa, a novel type I IFN, is undetectable in HPV-positive human cervical keratinocytes." *Lab Invest* no. 90 (10):1482-91. doi: 10.1038/labinvest.2010.95.

Deeks, S. G., S. R. Lewin, and D. V. Havlir. 2013. "The end of AIDS: HIV infection as a chronic disease." *Lancet* no. 382 (9903):1525-33. doi: 10.1016/S0140-6736(13)61809-7.

Del Corno, M., A. Cappon, G. Donninelli, B. Varano, F. Marra, and S. Gessani. 2016. "HIV-1 gp120 signaling through TLR4 modulates innate immune activation in human macrophages and the biology of hepatic stellate cells." *J Leukoc Biol* no. 100 (3):599-606. doi: 10.1189/jlb.4A1215-534R.

Desimio, M. G., E. Giuliani, A. S. Ferraro, G. Adorno, and M. Doria. 2018. "In Vitro Exposure to Prostratin but Not Bryostatin-1 Improves Natural Killer Cell Functions Including Killing of CD4(+) T Cells Harboring Reactivated Human Immunodeficiency Virus." *Front Immunol* no. 9:1514. doi: 10.3389/fimmu.2018.01514.

Dhillon, N. K., F. Peng, R. M. Ransohoff, and S. Buch. 2007. "PDGF synergistically enhances IFN-gamma-induced expression of CXCL10 in blood-derived macrophages: implications for HIV dementia." *J Immunol* no. 179 (5):2722-30. doi: 10.4049/jimmunol.179.5.2722.

Dianzani, F., G. Rozera, I. Abbate, G. D'Offizi, A. Abdeddaim, C. Vlassi, G. Antonucci, P. Narciso, F. Martini, and M. R. Capobianchi. 2008. "Interferon may prevent HIV viral rebound after HAART interruption in HIV patients." *J Interferon Cytokine Res* no. 28 (1):1-3. doi: 10.1089/jir.2007.0076.

Dillon, S. M., K. B. Robertson, S. C. Pan, S. Mawhinney, A. L. Meditz, J. M. Folkvord, E. Connick, M. D. McCarter, and C. C. Wilson. 2008. "Plasmacytoid and myeloid dendritic cells with a partial activation phenotype accumulate in lymphoid tissue during asymptomatic chronic HIV-1 infection." *J Acquir Immune Defic Syndr* no. 48 (1):1-12. doi: 10.1097/QAI.0b013e3181664b60.

Dufour, C., P. Gantner, R. Fromentin, and N. Chomont. 2020. "The multifaceted nature of HIV latency." *J Clin Invest* no. 130 (7):3381-3390. doi: 10.1172/JCI136227.

Ellery, P. J., E. Tippett, Y. L. Chiu, G. Paukovics, P. U. Cameron, A. Solomon, S. R. Lewin, P. R. Gorry, A. Jaworowski, W. C. Greene, S. Sonza, and S. M. Crowe. 2007. "The CD16+ monocyte subset is more permissive to infection and preferentially harbors HIV-1 in vivo." *J Immunol* no. 178 (10):6581-9. doi: 10.4049/jimmunol.178. 10.6581.

Engelhardt, K. R., and B. Grimbacher. 2014. "IL-10 in humans: lessons from the gut, IL-10/IL-10 receptor deficiencies, and IL-10 polymorphisms." *Curr Top Microbiol Immunol* no. 380:1-18. doi: 10.1007/978-3-662-43492-5_1.

Engelman, A. N., and P. K. Singh. 2018. "Cellular and molecular mechanisms of HIV-1 integration targeting." *Cell Mol Life Sci* no. 75 (14):2491-2507. doi: 10.1007/s00018-018-2772-5.

Estes, J. D., C. Kityo, F. Ssali, L. Swainson, K. N. Makamdop, G. Q. Del Prete, S. G. Deeks, P. A. Luciw, J. G. Chipman, G. J. Beilman, T. Hoskuldsson, A. Khoruts, J. Anderson, C. Deleage, J. Jasurda, T. E. Schmidt, M. Hafertepe, S. P. Callisto, H. Pearson, T. Reimann, J. Schuster, J. Schoephoerster, P. Southern, K. Perkey, L. Shang, S. W. Wietgrefe, C. V. Fletcher, J. D. Lifson, D. C. Douek, J. M. McCune, A. T. Haase, and T. W. Schacker. 2017. "Defining total-body AIDS-virus burden with implications for curative strategies." *Nat Med* no. 23 (11):1271-1276. doi: 10.1038/nm.4411.

Farber, D. L., N. A. Yudanin, and N. P. Restifo. 2014. "Human memory T cells: generation, compartmentalization and homeostasis." *Nat Rev Immunol* no. 14 (1):24-35. doi: 10.1038/nri3567.

Fidler, S., and J. Fox. 2016. "Primary HIV infection: a medical and public health emergency requiring rapid specialist management." *Clin Med (Lond)* no. 16 (2):180-3. doi: 10.7861/clinmedicine.16-2-180.

Fiorentino, D. F., A. Zlotnik, T. R. Mosmann, M. Howard, and A. O'Garra. 1991a. "IL-10 inhibits cytokine production by activated macrophages." *J Immunol* no. 147 (11):3815-22.

Fiorentino, D. F., A. Zlotnik, P. Vieira, T. R. Mosmann, M. Howard, K. W. Moore, and A. O'Garra. 1991b. "IL-10 acts on the antigen-presenting cell to inhibit cytokine production by Th1 cells." *J Immunol* no. 146 (10):3444-51.

Fletcher, C. V., K. Staskus, S. W. Wietgrefe, M. Rothenberger, C. Reilly, J. G. Chipman, G. J. Beilman, A. Khoruts, A. Thorkelson, T. E. Schmidt, J. Anderson, K. Perkey, M. Stevenson, A. S. Perelson, D. C. Douek, A. T. Haase, and T. W. Schacker. 2014. "Persistent HIV-1 replication is associated with lower antiretroviral drug concentrations in lymphatic tissues." *Proc Natl Acad Sci U S A* no. 111 (6):2307-12. doi: 10.1073/pnas.1318249111.

Florez-Alvarez, L., J. C. Hernandez, and W. Zapata. 2018. "NK Cells in HIV-1 Infection: From Basic Science to Vaccine Strategies." *Front Immunol* no. 9:2290. doi: 10.3389/fimmu.2018.02290.

Fonteneau, J. F., M. Gilliet, M. Larsson, I. Dasilva, C. Munz, Y. J. Liu, and N. Bhardwaj. 2003. "Activation of influenza virus-specific CD4+ and CD8+ T cells: a new role for plasmacytoid dendritic cells in adaptive immunity." *Blood* no. 101 (9):3520-6. doi: 10.1182/blood-2002-10-3063.

Fromentin, R., W. Bakeman, M. B. Lawani, G. Khoury, W. Hartogensis, S. DaFonseca, M. Killian, L. Epling, R. Hoh, E. Sinclair, F. M. Hecht, P. Bacchetti, S. G. Deeks, S. R. Lewin, R. P. Sekaly, and N. Chomont. 2016. "CD4+ T Cells Expressing PD-1, TIGIT and LAG-3 Contribute to HIV Persistence during ART." *PLoS Pathog* no. 12 (7):e1005761. doi: 10.1371/journal.ppat.1005761.

Fukazawa, Y., R. Lum, A. A. Okoye, H. Park, K. Matsuda, J. Y. Bae, S. I. Hagen, R. Shoemaker, C. Deleage, C. Lucero, D. Morcock, T. Swanson, A. W. Legasse, M. K. Axthelm, J. Hesselgesser, R. Geleziunas, V. M. Hirsch, P. T. Edlefsen, M. Piatak, Jr., J. D. Estes, J. D. Lifson, and L. J. Picker. 2015. "B cell follicle sanctuary permits persistent productive simian immunodeficiency virus infection in elite controllers." *Nat Med* no. 21 (2):132-9. doi: 10.1038/nm.3781.

Galicia, G., and J. L. Gommerman. 2014. "Plasmacytoid dendritic cells and autoimmune inflammation." *Biol Chem* no. 395 (3):335-46. doi: 10.1515/hsz-2013-0213.

Gallin, J. I., J. M. Farber, S. M. Holland, and T. B. Nutman. 1995. "Interferon-gamma in the management of infectious diseases." *Ann Intern Med* no. 123 (3):216-24. doi: 10.7326/0003-4819-123-3-199508010-00009.

Gandhi, R. T., D. S. Kwon, E. A. Macklin, J. R. Shopis, A. P. McLean, N. McBrine, T. Flynn, L. Peter, A. Sbrolla, D. E. Kaufmann, F. Porichis, B. D. Walker, N. Bhardwaj, D. H. Barouch, and D. G. Kavanagh. 2016. "Immunization of HIV-1-Infected Persons With Autologous Dendritic Cells Transfected With mRNA Encoding HIV-1 Gag and Nef: Results of a

Randomized, Placebo-Controlled Clinical Trial." *J Acquir Immune Defic Syndr* no. 71 (3):246-53. doi: 10.1097/QAI.0000000000000852.

Ganor, Y., F. Real, A. Sennepin, C. A. Dutertre, L. Prevedel, L. Xu, D. Tudor, B. Charmeteau, A. Couedel-Courteille, S. Marion, A. R. Zenak, J. P. Jourdain, Z. Zhou, A. Schmitt, C. Capron, E. A. Eugenin, R. Cheynier, M. Revol, S. Cristofari, A. Hosmalin, and M. Bomsel. 2019. "HIV-1 reservoirs in urethral macrophages of patients under suppressive antiretroviral therapy." *Nat Microbiol* no. 4 (4):633-644. doi: 10.1038/s41564-018-0335-z.

Garcia-Basteiro, A. L., E. Mambuque, A. den Hertog, B. Saavedra, I. Cuamba, L. Oliveras, S. Blanco, H. Bulo, J. Brew, L. E. Cuevas, F. Cobelens, A. Nhabomba, and R. Anthony. 2017. "IP-10 Kinetics in the First Week of Therapy are Strongly Associated with Bacteriological Confirmation of Tuberculosis Diagnosis in HIV-Infected Patients." *Sci Rep* no. 7 (1):14302. doi: 10.1038/s41598-017-13785-3.

Garcia-Minambres, A., S. G. Eid, N. E. Mangan, C. Pade, S. S. Lim, A. Y. Matthews, N. A. de Weerd, P. J. Hertzog, and J. Mak. 2017. "Interferon epsilon promotes HIV restriction at multiple steps of viral replication." *Immunol Cell Biol* no. 95 (5):478-483. doi: 10.1038/icb.2016.123.

Garrido, C., M. Tolstrup, O. S. Sogaard, T. A. Rasmussen, B. Allard, N. Soriano-Sarabia, N. M. Archin, and D. M. Margolis. 2019. "In-vivo administration of histone deacetylase inhibitors does not impair natural killer cell function in HIV+ individuals." *AIDS* no. 33 (4):605-613. doi: 10.1097/QAD.0000000000002112.

Gavegnano, C., M. Detorio, C. Montero, A. Bosque, V. Planelles, and R. F. Schinazi. 2014. "Ruxolitinib and tofacitinib are potent and selective inhibitors of HIV-1 replication and virus reactivation in vitro." *Antimicrob Agents Chemother* no. 58 (4):1977-86. doi: 10.1128/AAC.02496-13.

Gay, N. J., M. F. Symmons, M. Gangloff, and C. E. Bryant. 2014. "Assembly and localization of Toll-like receptor signalling complexes." *Nat Rev Immunol* no. 14 (8):546-58. doi: 10.1038/ nri3713.

Geginat, J., M. Paroni, S. Maglie, J. S. Alfen, I. Kastirr, P. Gruarin, M. De Simone, M. Pagani, and S. Abrignani. 2014. "Plasticity of human CD4 T cell subsets." *Front Immunol* no. 5:630. doi: 10.3389/fimmu.2014.00630.

Geissmann, F., M. G. Manz, S. Jung, M. H. Sieweke, M. Merad, and K. Ley. 2010. "Development of monocytes, macrophages, and dendritic cells." *Science* no. 327 (5966):656-61. doi: 10.1126/science.1178331.

George, J., and J. J. Mattapallil. 2018. "Interferon-alpha Subtypes As an Adjunct Therapeutic Approach for Human Immunodeficiency Virus Functional Cure." *Front Immunol* no. 9:299. doi: 10.3389/fimmu. 2018.00299.

Gerosa, F., A. Gobbi, P. Zorzi, S. Burg, F. Briere, G. Carra, and G. Trinchieri. 2005. "The reciprocal interaction of NK cells with plasmacytoid or myeloid dendritic cells profoundly affects innate resistance functions." *J Immunol* no. 174 (2):727-34. doi: 10.4049/jimmunol.174.2.727.

Gianella, S., S. L. Kosakovsky Pond, M. F. Oliveira, K. Scheffler, M. C. Strain, A. De la Torre, S. Letendre, D. M. Smith, and R. J. Ellis. 2016. "Compartmentalized HIV rebound in the central nervous system after interruption of antiretroviral therapy." *Virus Evol* no. 2 (2):vew020. doi: 10.1093/ve/vew020.

Goicoechea, M., D. M. Smith, L. Liu, S. May, A. R. Tenorio, C. C. Ignacio, A. Landay, and R. Haubrich. 2006. "Determinants of CD4+ T cell recovery during suppressive antiretroviral therapy: association of immune activation, T cell maturation markers, and cellular HIV-1 DNA." *J Infect Dis* no. 194 (1):29-37. doi: 10.1086/504718.

Gonzalez-Hernandez, L. A., L. F. Jave-Suarez, M. Fafutis-Morris, K. E. Montes-Salcedo, L. G. Valle-Gutierrez, A. E. Campos-Loza, L. F. Enciso-Gomez, and J. F. Andrade-Villanueva. 2012. "Synbiotic therapy decreases microbial translocation and inflammation and improves immunological status in HIV-infected patients: a double-blind randomized controlled pilot trial." *Nutr J* no. 11:90. doi: 10.1186/1475-2891-11-90.

Gordon, S., and A. Pluddemann. 2017. "Tissue macrophages: hetero-geneity and functions." *BMC Biol* no. 15 (1):53. doi: 10.1186/s12915-017-0392-4.

Gosselin, A., P. Monteiro, N. Chomont, F. Diaz-Griffero, E. A. Said, S. Fonseca, V. Wacleche, M. El-Far, M. R. Boulassel, J. P. Routy, R. P. Sekaly, and P. Ancuta. 2010. "Peripheral blood CCR4+CCR6+ and CXCR3+CCR6+CD4+ T cells are highly permissive to HIV-1 infection." *J Immunol* no. 184 (3):1604-16. doi: 10.4049/jimmunol. 0903058.

Goujard, C., D. Emilie, C. Roussillon, V. Godot, C. Rouzioux, A. Venet, C. Colin, G. Pialoux, P. M. Girard, V. Boilet, M. L. Chaix, P. Galanaud, G. Chene, and Anrs-Interprim Study Group. 2012. "Continuous versus intermittent treatment strategies during primary HIV-1 infection: the randomized ANRS INTERPRIM Trial." *AIDS* no. 26 (15):1895-905. doi: 10.1097/QAD. 0b013e32835844d9.

Goujon, C., and M. H. Malim. 2010. "Characterization of the alpha interferon-induced postentry block to HIV-1 infection in primary human macrophages and T cells." *J Virol* no. 84 (18):9254-66. doi: 10.1128/JVI.00854-10.

Grau-Exposito, J., L. Luque-Ballesteros, J. Navarro, A. Curran, J. Burgos, E. Ribera, A. Torrella, B. Planas, R. Badia, M. Martin-Castillo, J. Fernandez-Sojo, M. Genesca, V. Falco, and M. J. Buzon. 2019. "Latency reversal agents affect differently the latent reservoir present in distinct CD4+ T subpopulations." *PLoS Pathog* no. 15 (8):e1007991. doi: 10.1371/journal. ppat.1007991.

Gubernatorova, E. O., E. A. Gorshkova, A. I. Polinova, and M. S. Drutskaya. 2020. "IL-6: Relevance for immunopathology of SARS-CoV-2." *Cytokine Growth Factor Rev* no. 53:13-24. doi: 10.1016/j.cytogfr.2020.05.009.

Gulick, R. M., and C. Flexner. 2019. "Long-Acting HIV Drugs for Treatment and Prevention." *Annu Rev Med* no. 70:137-150. doi: 10.1146/annurev-med-041217-013717.

Guo, H., J. Gao, D. J. Taxman, J. P. Ting, and L. Su. 2014. "HIV-1 infection induces interleukin-1 beta production via TLR8 protein-dependent and NLRP3 inflammasome mechanisms in human monocytes." *J Biol Chem* no. 289 (31):21716-26. doi: 10.1074/jbc. M114.566620.

Gupta, R. K., S. Abdul-Jawad, L. E. McCoy, H. P. Mok, D. Peppa, M. Salgado, J. Martinez-Picado, M. Nijhuis, A. M. J. Wensing, H. Lee, P. Grant, E. Nastouli, J. Lambert, M. Pace, F. Salasc, C. Monit, A. J. Innes, L. Muir, L. Waters, J. Frater, A. M. L. Lever, S. G. Edwards, I. H. Gabriel, and E. Olavarria. 2019. "HIV-1 remission following CCR5Delta32/Delta32 haematopoietic stem-cell transplantation." *Nature* no. 568 (7751):244-248. doi: 10.1038/s41586-019-1027-4.

Gurdasani, D., L. Iles, D. G. Dillon, E. H. Young, A. D. Olson, V. Naranbhai, S. Fidler, E. Gkrania-Klotsas, F. A. Post, P. Kellam, K. Porter, and M. S. Sandhu. 2014. "A systematic review of definitions of extreme phenotypes of HIV control and progression." *AIDS* no. 28 (2):149-62. doi: 10.1097/QAD.0000000000000049.

Hardy, G. A., S. Sieg, B. Rodriguez, D. Anthony, R. Asaad, W. Jiang, J. Mudd, T. Schacker, N. T. Funderburg, H. A. Pilch-Cooper, R. Debernardo, R. L. Rabin, M. M. Lederman, and C. V. Harding. 2013. "Interferon-alpha is the primary plasma type-I IFN in HIV-1 infection and correlates with immune activation and disease markers." *PLoS One* no. 8 (2):e56527. doi: 10.1371/journal. pone.0056527.

Harper, J., N. Huot, L. Micci, G. Tharp, C. King, P. Rascle, N. Shenvi, H. Wang, C. Galardi, A. A. Upadhyay, F. Villinger, J. Lifson, G. Silvestri, K. Easley, B. Jacquelin, S. Bosinger, M. Muller-Trutwin, and M. Paiardini. 2021. "IL-21 and IFNalpha therapy rescues terminally differentiated NK cells and limits SIV reservoir in ART-treated macaques." *Nat Commun* no. 12 (1):2866. doi: 10.1038/s41467-021-23189-7.

Harper, M. S., K. Guo, K. Gibbert, E. J. Lee, S. M. Dillon, B. S. Barrett, M. D. McCarter, K. J. Hasenkrug, U. Dittmer, C. C. Wilson, and M. L. Santiago. 2015. "Interferon-alpha Subtypes in an Ex Vivo Model of Acute HIV-1 Infection: Expression, Potency and Effector Mechanisms." *PLoS Pathog* no. 11 (11):e1005254. doi: 10.1371/journal.ppat.1005254.

Hearps, A. C., A. Maisa, W. J. Cheng, T. A. Angelovich, G. F. Lichtfuss, C. S. Palmer, A. L. Landay, A. Jaworowski, and S. M. Crowe. 2012. "HIV infection induces age-related changes to monocytes and innate immune activation in young men that persist despite combination antiretroviral therapy." *AIDS* no. 26 (7):843-53. doi: 10.1097/QAD.0b013e328351f756.

Heesters, B. A., M. Lindqvist, P. A. Vagefi, E. P. Scully, F. A. Schildberg, M. Altfeld, B. D. Walker, D. E. Kaufmann, and M. C. Carroll. 2015. "Follicular Dendritic Cells Retain Infectious HIV in Cycling Endosomes." *PLoS Pathog* no. 11 (12):e1005285. doi: 10.1371/journal.ppat.1005285.

Hemann, E. A., M. Gale, Jr., and R. Savan. 2017. "Interferon Lambda Genetics and Biology in Regulation of Viral Control." *Front Immunol* no. 8:1707. doi: 10.3389/fimmu.2017.01707.

Hemont, C., A. Neel, M. Heslan, C. Braudeau, and R. Josien. 2013. "Human blood mDC subsets exhibit distinct TLR repertoire and responsiveness." *J Leukoc Biol* no. 93 (4):599-609. doi: 10.1189/jlb.0912452.

Hendricks, C. M., T. Cordeiro, A. P. Gomes, and M. Stevenson. 2021. "The Interplay of HIV-1 and Macrophages in Viral Persistence." *Front Microbiol* no. 12:646447. doi: 10.3389/fmicb.2021.646447.

Henrick, B. M., X. D. Yao, K. L. Rosenthal, and Infant study team. 2015. "HIV-1 Structural Proteins Serve as PAMPs for TLR2 Heterodimers Significantly Increasing Infection and Innate Immune Activation." *Front Immunol* no. 6:426. doi: 10.3389/fimmu.2015.00426.

Herbein, G., L. J. Montaner, and S. Gordon. 1996. "Tumor necrosis factor alpha inhibits entry of human immunodeficiency virus type 1 into primary human macrophages: a selective role for the 75-kilodalton receptor." *J Virol* no. 70 (11):7388-97. doi: 10.1128/JVI.70.11.7388-7397.1996.

Heron, S. E., and S. Elahi. 2017. "HIV Infection and Compromised Mucosal Immunity: Oral Manifestations and Systemic Inflammation." *Front Immunol* no. 8:241. doi: 10.3389/fimmu.2017. 00241.

Hertoghs, N., A. M. van der Aar, L. C. Setiawan, N. A. Kootstra, S. I. Gringhuis, and T. B. Geijtenbeek. 2015. "SAMHD1 degradation enhances active suppression of dendritic cell maturation by HIV-1." *J Immunol* no. 194 (9):4431-7. doi: 10.4049/jimmunol.1403016.

Herzig, E., K. C. Kim, T. A. Packard, N. Vardi, R. Schwarzer, A. Gramatica, S. G. Deeks, S. R. Williams, K. Landgraf, N. Killeen, D. W. Martin, L. S. Weinberger, and W. C. Greene. 2019. "Attacking Latent HIV with convertible CAR-T Cells, a Highly Adaptable Killing Platform." *Cell* no. 179 (4):880-894 e10. doi: 10.1016/j.cell. 2019.10.002.

Hiener, B., B. A. Horsburgh, J. S. Eden, K. Barton, T. E. Schlub, E. Lee, S. von Stockenstrom, L. Odevall, J. M. Milush, T. Liegler, E. Sinclair, R. Hoh, E. A. Boritz, D. Douek, R. Fromentin, N. Chomont, S. G. Deeks, F. M. Hecht, and S. Palmer. 2017. "Identification of Genetically Intact HIV-1 Proviruses in

Specific CD4(+) T Cells from Effectively Treated Participants." *Cell Rep* no. 21 (3):813-822. doi: 10.1016/j.celrep.2017.09.081.

Hilldorfer, B. B., A. R. Cillo, G. J. Besson, M. A. Bedison, and J. W. Mellors. 2012. "New tools for quantifying HIV-1 reservoirs: plasma RNA single copy assays and beyond." *Curr HIV/AIDS Rep* no. 9 (1):91-100. doi: 10.1007/s11904-011-0104-6.

Hoeffel, G., and F. Ginhoux. 2018. "Fetal monocytes and the origins of tissue-resident macrophages." *Cell Immunol* no. 330:5-15. doi: 10.1016/j.cellimm.2018.01.001.

Hogan, L. E., J. Vasquez, K. S. Hobbs, E. Hanhauser, B. Aguilar-Rodriguez, R. Hussien, C. Thanh, E. A. Gibson, A. B. Carvidi, L. C. B. Smith, S. Khan, M. Trapecar, S. Sanjabi, M. Somsouk, C. A. Stoddart, D. R. Kuritzkes, S. G. Deeks, and T. J. Henrich. 2018. "Increased HIV-1 transcriptional activity and infectious burden in peripheral blood and gut-associated CD4+ T cells expressing CD30." *PLoS Pathog* no. 14 (2):e1006856. doi: 10.1371/journal.ppat. 1006856.

Honeycutt, J. B., A. Wahl, N. Archin, S. Choudhary, D. Margolis, and J. V. Garcia. 2013. "HIV-1 infection, response to treatment and establishment of viral latency in a novel humanized T cell-only mouse (TOM) model." *Retrovirology* no. 10:121. doi: 10.1186/1742-4690-10-121.

Horton, R. E., P. J. McLaren, K. Fowke, J. Kimani, and T. B. Ball. 2010. "Cohorts for the study of HIV-1-exposed but uninfected individuals: benefits and limitations." *J Infect Dis* no. 202 Suppl 3:S377-81. doi: 10.1086/655971.

Hua, S., S. Vigano, S. Tse, O. Zhengyu, S. Harrington, J. Negron, P. Garcia-Broncano, G. Marchetti, M. Genebat, M. Leal, S. Resino, E. Ruiz-Mateos, M. Lichterfeld, and X. G. Yu. 2018. "Pegylated Interferon-alpha-Induced Natural Killer Cell Activation Is Associated With Human Immunodeficiency Virus-1 DNA Decline in Antiretroviral Therapy-Treated HIV-1/Hepatitis C Virus-Coinfected Patients." *Clin Infect Dis* no. 66 (12):1910-1917. doi: 10.1093/cid/cix1111.

Igarashi, T., C. R. Brown, Y. Endo, A. Buckler-White, R. Plishka, N. Bischofberger, V. Hirsch, and M. A. Martin. 2001. "Macrophage are the principal reservoir and sustain high virus loads in rhesus macaques after the depletion of CD4+ T cells by a highly pathogenic simian immunodeficiency virus/HIV type 1 chimera (SHIV): Implications for HIV-1 infections of humans." *Proc Natl Acad Sci U S A* no. 98 (2):658-63. doi: 10.1073/pnas.021551798.

Imamichi, H., M. Smith, J. W. Adelsberger, T. Izumi, F. Scrimieri, B. T. Sherman, C. A. Rehm, T. Imamichi, A. Pau, M. Catalfamo, A. S. Fauci, and H. C. Lane. 2020. "Defective HIV-1 proviruses produce viral proteins." *Proc Natl Acad Sci U S A* no. 117 (7):3704-3710. doi: 10.1073/pnas.1917876117.

Iyer, S. S., F. Bibollet-Ruche, S. Sherrill-Mix, G. H. Learn, L. Plenderleith, A. G. Smith, H. J. Barbian, R. M. Russell, M. V. Gondim, C. Y. Bahari, C. M. Shaw, Y. Li, T. Decker, B. F. Haynes, G. M. Shaw, P. M. Sharp, P. Borrow, and B. H. Hahn. 2017. "Resistance to type 1 interferons is a major determinant of HIV-1 transmission fitness." *Proc Natl Acad Sci U S A* no. 114 (4):E590-E599. doi: 10.1073/pnas.1620144114.

Jacobson, J. M., J. P. Routy, S. Welles, M. DeBenedette, I. Tcherepanova, J. B. Angel, D. M. Asmuth, D. K. Stein, J. G. Baril, M. McKellar, D. M. Margolis, B. Trottier, K. Wood, and C. Nicolette. 2016. "Dendritic Cell Immunotherapy for HIV-1 Infection Using Autologous HIV-1 RNA: A Randomized, Double-Blind, Placebo-Controlled Clinical Trial." *J Acquir Immune Defic Syndr* no. 72 (1):31-8. doi: 10.1097/QAI.0000000000000926.

Jacquelin, B., V. Mayau, B. Targat, A. S. Liovat, D. Kunkel, G. Petitjean, M. A. Dillies, P. Roques, C. Butor, G. Silvestri, L. D. Giavedoni, P. Lebon, F. Barre-Sinoussi, A. Benecke, and M. C. Muller-Trutwin. 2009. "Nonpathogenic SIV infection of African green monkeys induces a strong but rapidly controlled type I IFN response." *J Clin Invest* no. 119 (12):3544-55. doi: 10.1172/JCI40093.

Jakobsen, M. R., R. O. Bak, A. Andersen, R. K. Berg, S. B. Jensen, J. Tengchuan, A. Laustsen, K. Hansen, L. Ostergaard, K. A. Fitzgerald, T. S. Xiao, J. G. Mikkelsen, T. H. Mogensen, and S. R. Paludan. 2013. "IFI16 senses DNA forms of the lentiviral replication cycle and controls HIV-1 replication." *Proc Natl Acad Sci U S A* no. 110 (48):E4571-80. doi: 10.1073/pnas.1311669110.

Jakobsen, M. R., D. Olagnier, and J. Hiscott. 2015. "Innate immune sensing of HIV-1 infection." *Curr Opin HIV AIDS* no. 10 (2):96-102. doi: 10.1097/COH.0000000000000129.

Jambo, K. C., D. H. Banda, A. M. Kankwatira, N. Sukumar, T. J. Allain, R. S. Heyderman, D. G. Russell, and H. C. Mwandumba. 2014. "Small alveolar macrophages are infected preferentially by HIV and exhibit impaired phagocytic function." *Mucosal Immunol* no. 7 (5):1116-26. doi: 10.1038/mi.2013.127.

Jaworowski, A., A. C. Hearps, T. A. Angelovich, and J. F. Hoy. 2019. "How Monocytes Contribute to Increased Risk of Atherosclerosis in Virologically-Suppressed HIV-Positive Individuals Receiving Combination Antiretroviral Therapy." *Front Immunol* no. 10:1378. doi: 10.3389/fimmu.2019.01378.

Jego, G., A. K. Palucka, J. P. Blanck, C. Chalouni, V. Pascual, and J. Banchereau. 2003. "Plasmacytoid dendritic cells induce plasma cell differentiation through type I interferon and interleukin 6." *Immunity* no. 19 (2):225-34. doi: 10.1016/s1074-7613(03)00208-5.

Jiang, C., X. Lian, C. Gao, X. Sun, K. B. Einkauf, J. M. Chevalier, S. M. Y. Chen, S. Hua, B. Rhee, K. Chang, J. E. Blackmer, M. Osborn, M. J. Peluso, R. Hoh,

M. Somsouk, J. Milush, L. N. Bertagnolli, S. E. Sweet, J. A. Varriale, P. D. Burbelo, T. W. Chun, G. M. Laird, E. Serrao, A. N. Engelman, M. Carrington, R. F. Siliciano, J. M. Siliciano, S. G. Deeks, B. D. Walker, M. Lichterfeld, and X. G. Yu. 2020. "Distinct viral reservoirs in individuals with spontaneous control of HIV-1." *Nature* no. 585 (7824):261-267. doi: 10.1038/s41586-020-2651-8.

Jiao, Y., T. Zhang, R. Wang, H. Zhang, X. Huang, J. Yin, L. Zhang, X. Xu, and H. Wu. 2012. "Plasma IP-10 is associated with rapid disease progression in early HIV-1 infection." *Viral Immunol* no. 25 (4):333-7. doi: 10.1089/vim.2012.0011.

Josefowicz, S. Z., L. F. Lu, and A. Y. Rudensky. 2012. "Regulatory T cells: mechanisms of differentiation and function." *Annu Rev Immunol* no. 30:531-64. doi: 10.1146/annurev.immunol.25.022106. 141623.

Josefsson, L., S. von Stockenstrom, N. R. Faria, E. Sinclair, P. Bacchetti, M. Killian, L. Epling, A. Tan, T. Ho, P. Lemey, W. Shao, P. W. Hunt, M. Somsouk, W. Wylie, D. C. Douek, L. Loeb, J. Custer, R. Hoh, L. Poole, S. G. Deeks, F. Hecht, and S. Palmer. 2013. "The HIV-1 reservoir in eight patients on long-term suppressive antiretroviral therapy is stable with few genetic changes over time." *Proc Natl Acad Sci U S A* no. 110 (51):E4987-96. doi: 10.1073/pnas. 1308313110.

Kalinski, P., H. Edington, H. J. Zeh, H. Okada, L. H. Butterfield, J. M. Kirkwood, and D. L. Bartlett. 2011. "Dendritic cells in cancer immunotherapy: vaccines or autologous transplants?" *Immunol Res* no. 50 (2-3):235-47. doi: 10.1007/s12026-011-8224-z.

Kalyan, S., and D. Kabelitz. 2013. "Defining the nature of human gammadelta T cells: a biographical sketch of the highly empathetic." *Cell Mol Immunol* no. 10 (1):21-9. doi: 10.1038/cmi.2012.44.

Kamat, A., V. Misra, E. Cassol, P. Ancuta, Z. Yan, C. Li, S. Morgello, and D. Gabuzda. 2012. "A plasma biomarker signature of immune activation in HIV patients on antiretroviral therapy." *PLoS One* no. 7 (2):e30881. doi: 10.1371/journal.pone.0030881.

Kandathil, A. J., S. Sugawara, and A. Balagopal. 2017. "Erratum to: Are T cells the only HIV-1 reservoir?" *Retrovirology* no. 14 (1):11. doi: 10.1186/s12977-017-0333-x.

Kapellos, T. S., L. Bonaguro, I. Gemund, N. Reusch, A. Saglam, E. R. Hinkley, and J. L. Schultze. 2019. "Human Monocyte Subsets and Phenotypes in Major Chronic Inflammatory Diseases." *Front Immunol* no. 10:2035. doi: 10.3389/fimmu.2019.02035.

Katlama, C., S. G. Deeks, B. Autran, J. Martinez-Picado, J. van Lunzen, C. Rouzioux, M. Miller, S. Vella, J. E. Schmitz, J. Ahlers, D. D. Richman, and R. P. Sekaly. 2013. "Barriers to a cure for HIV: new ways to target and eradicate HIV-1 reservoirs." *Lancet* no. 381 (9883):2109-17. doi: 10.1016/S0140-6736(13)60104-X.

Keane, J., S. Gershon, R. P. Wise, E. Mirabile-Levens, J. Kasznica, W. D. Schwieterman, J. N. Siegel, and M. M. Braun. 2001. "Tuberculosis associated with infliximab, a tumor necrosis factor alpha-neutralizing agent." *N Engl J Med* no. 345 (15):1098-104. doi: 10.1056/NEJMoa011110.

Kedzierska, K., and S. M. Crowe. 2001. "Cytokines and HIV-1: interactions and clinical implications." *Antivir Chem Chemother* no. 12 (3):133-50. doi: 10.1177/095632020101200301.

Kelly, J., M. H. Beddall, D. Yu, S. R. Iyer, J. W. Marsh, and Y. Wu. 2008. "Human macrophages support persistent transcription from unintegrated HIV-1 DNA." *Virology* no. 372 (2):300-12. doi: 10.1016/j.virol.2007.11.007.

Khoury, G., R. Fromentin, A. Solomon, W. Hartogensis, M. Killian, R. Hoh, M. Somsouk, P. W. Hunt, V. Girling, E. Sinclair, P. Bacchetti, J. L. Anderson, F. M. Hecht, S. G. Deeks, P. U. Cameron, N. Chomont, and S. R. Lewin. 2017. "Human Immunodeficiency Virus Persistence and T-Cell Activation in Blood, Rectal, and Lymph Node Tissue in Human Immunodeficiency Virus-Infected Individuals Receiving Suppressive Antiretroviral Therapy." *J Infect Dis* no. 215 (6):911-919. doi: 10.1093/infdis/jix039.

Kilzer, J. M., T. Stracker, B. Beitzel, K. Meek, M. Weitzman, and F. D. Bushman. 2003. "Roles of host cell factors in circularization of retroviral dna." *Virology* no. 314 (1):460-7. doi: 10.1016/s0042-6822(03)00455-0.

Kim, M., and R. F. Siliciano. 2016. "Reservoir expansion by T-cell proliferation may be another barrier to curing HIV infection." *Proc Natl Acad Sci U S A* no. 113 (7):1692-4. doi: 10.1073/pnas. 1600097113.

Ko, A., G. Kang, J. B. Hattler, H. I. Galadima, J. Zhang, Q. Li, and W. K. Kim. 2019. "Macrophages but not Astrocytes Harbor HIV DNA in the Brains of HIV-1-Infected Aviremic Individuals on Suppressive Antiretroviral Therapy." *J Neuroimmune Pharmacol* no. 14 (1):110-119. doi: 10.1007/s11481-018-9809-2.

Kodama, A., R. Tanaka, L. F. Zhang, T. Adachi, M. Saito, A. A. Ansari, and Y. Tanaka. 2010. "Impairment of in vitro generation of monocyte-derived human dendritic cells by inactivated human immunodeficiency virus-1: Involvement of type I interferon produced from plasmacytoid dendritc cells." *Hum Immunol* no. 71 (6):541-50. doi: 10.1016/j.humimm.2010.02.020.

Krapp, C., K. Jonsson, and M. R. Jakobsen. 2018. "STING dependent sensing - Does HIV actually care?" *Cytokine Growth Factor Rev* no. 40:68-76. doi: 10.1016/j.cytogfr.2018.03.002.

Krarup, A. R., M. Abdel-Mohsen, M. H. Schleimann, L. Vibholm, P. A. Engen, A. Dige, B. Wittig, M. Schmidt, S. J. Green, A. Naqib, A. Keshavarzian, X. Deng, R. Olesen, A. M. Petersen, T. Benfield, L. Ostergaard, T. A.

Rasmussen, J. Agnholt, J. R. Nyengaard, A. Landay, O. S. Sogaard, S. K. Pillai, M. Tolstrup, and P. W. Denton. 2018. "The TLR9 agonist MGN1703 triggers a potent type I interferon response in the sigmoid colon." *Mucosal Immunol* no. 11 (2):449-461. doi: 10.1038/mi.2017.59.

Kristoff, J., M. L. Palma, T. M. Garcia-Bates, C. Shen, N. Sluis-Cremer, P. Gupta, C. R. Rinaldo, and R. B. Mailliard. 2019. "Type 1-programmed dendritic cells drive antigen-specific latency reversal and immune elimination of persistent HIV-1." *EBioMedicine* no. 43:295-306. doi: 10.1016/j.ebiom.2019.03.077.

Kristoff, J., C. R. Rinaldo, and R. B. Mailliard. 2019. "Role of Dendritic Cells in Exposing Latent HIV-1 for the Kill." *Viruses* no. 12 (1). doi: 10.3390/v12010037.

Kruize, Z., and N. A. Kootstra. 2019. "The Role of Macrophages in HIV-1 Persistence and Pathogenesis." *Front Microbiol* no. 10:2828. doi: 10.3389/fmicb.2019.02828.

Kumar, N. A., K. Cheong, D. R. Powell, C. da Fonseca Pereira, J. Anderson, V. A. Evans, S. R. Lewin, and P. U. Cameron. 2015. "The role of antigen presenting cells in the induction of HIV-1 latency in resting CD4(+) T-cells." *Retrovirology* no. 12:76. doi: 10.1186/ s12977-015-0204-2.

Kwon, K. J., A. E. Timmons, S. Sengupta, F. R. Simonetti, H. Zhang, R. Hoh, S. G. Deeks, J. D. Siliciano, and R. F. Siliciano. 2020. "Different human resting memory CD4(+) T cell subsets show similar low inducibility of latent HIV-1 proviruses." *Sci Transl Med* no. 12 (528). doi: 10.1126/scitranslmed.aax6795.

Laplana, M., M. J. Bravo, M. Fernandez-Fuertes, C. Ruiz-Garcia, E. Alarcon-Martin, J. D. Colmenero, A. Caruz, J. Fibla, L. M. Real, and J. L. Royo. 2020. "Toll-Like Receptor 2 Promoter -196 to -174 Deletion Affects CD4 Levels Along Human Immunodeficiency Virus Infection Progression." *J Infect Dis* no. 222 (12):2007-2011. doi: 10.1093/infdis/jiaa327.

Lavender, K. J., K. Gibbert, K. E. Peterson, E. Van Dis, S. Francois, T. Woods, R. J. Messer, A. Gawanbacht, J. A. Muller, J. Munch, K. Phillips, B. Race, M. S. Harper, K. Guo, E. J. Lee, M. Trilling, H. Hengel, J. Piehler, J. Verheyen, C. C. Wilson, M. L. Santiago, K. J. Hasenkrug, and U. Dittmer. 2016. "Interferon Alpha Subtype-Specific Suppression of HIV-1 Infection In Vivo." *J Virol* no. 90 (13):6001-6013. doi: 10.1128/JVI.00451-16.

Lazear, H. M., J. W. Schoggins, and M. S. Diamond. 2019. "Shared and Distinct Functions of Type I and Type III Interferons." *Immunity* no. 50 (4):907-923. doi: 10.1016/j.immuni.2019.03.025.

Lee, G. Q., and M. Lichterfeld. 2016. "Diversity of HIV-1 reservoirs in CD4+ T-cell subpopulations." *Curr Opin HIV AIDS* no. 11 (4):383-7. doi: 10.1097/COH.0000000000000281.

Lehmann, C., M. Lafferty, A. Garzino-Demo, N. Jung, P. Hartmann, G. Fatkenheuer, J. S. Wolf, J. van Lunzen, and F. Romerio. 2010. "Plasmacytoid dendritic cells accumulate and secrete interferon alpha in lymph nodes of HIV-1 patients." *PLoS One* no. 5 (6):e11110. doi: 10.1371/journal.pone.0011110.

Lewis, D. E., Couturier, J. P. . 2019. *Chapter 6- Chronic Inflammation in HIV Pathogenesis: Effects on Immune Cells, Organ Systems, and Systemic Consequences., Translational Inflammation.*

Li, G. H., L. Henderson, and A. Nath. 2016. "Astrocytes as an HIV Reservoir: Mechanism of HIV Infection." *Curr HIV Res* no. 14 (5):373-381. doi: 10.2174/1570162x14666161006121455.

Li, J. Z., B. Etemad, H. Ahmed, E. Aga, R. J. Bosch, J. W. Mellors, D. R. Kuritzkes, M. M. Lederman, M. Para, and R. T. Gandhi. 2016. "The size of the expressed HIV reservoir predicts timing of viral rebound after treatment interruption." *AIDS* no. 30 (3):343-53. doi: 10.1097/QAD.00000 00000000953.

Li, S. F., F. R. Zhao, J. J. Shao, Y. L. Xie, H. Y. Chang, and Y. G. Zhang. 2017. "Interferon-omega: Current status in clinical applications." *Int Immunopharmacol* no. 52:253-260. doi: 10.1016/j.intimp.2017.08.028.

Li Vecchi, V., P. Maggi, M. Rizzo, and G. Montalto. 2014. "The metabolic syndrome and HIV infection." *Curr Pharm Des* no. 20 (31):4975-5003. doi: 10.2174/1381612819666131206104209.

Liang, S. C., X. Y. Tan, D. P. Luxenberg, R. Karim, K. Dunussi-Joannopoulos, M. Collins, and L. A. Fouser. 2006. "Interleukin (IL)-22 and IL-17 are coexpressed by Th17 cells and cooperatively enhance expression of antimicrobial peptides." *J Exp Med* no. 203 (10):2271-9. doi: 10.1084/jem.20061308.

Lim, S. Y., C. E. Osuna, P. T. Hraber, J. Hesselgesser, J. M. Gerold, T. L. Barnes, S. Sanisetty, M. S. Seaman, M. G. Lewis, R. Geleziunas, M. D. Miller, T. Cihlar, W. A. Lee, A. L. Hill, and J. B. Whitney. 2018. "TLR7 agonists induce transient viremia and reduce the viral reservoir in SIV-infected rhesus macaques on antiretroviral therapy." *Sci Transl Med* no. 10 (439). doi: 10.1126/scitranslmed.aao4521.

Liou, L. Y., C. H. Herrmann, and A. P. Rice. 2002. "Transient induction of cyclin T1 during human macrophage differentiation regulates human immunodeficiency virus type 1 Tat transactivation function." *J Virol* no. 76 (21):10579-87. doi: 10.1128/jvi.76.21.10579-10587.2002.

Liu, E., D. Marin, P. Banerjee, H. A. Macapinlac, P. Thompson, R. Basar, L. Nassif Kerbauy, B. Overman, P. Thall, M. Kaplan, V. Nandivada, I. Kaur, A. Nunez Cortes, K. Cao, M. Daher, C. Hosing, E. N. Cohen, P. Kebriaei, R. Mehta, S. Neelapu, Y. Nieto, M. Wang, W. Wierda, M. Keating, R. Champlin, E. J. Shpall, and K. Rezvani. 2020. "Use of CAR-Transduced Natural Killer

Cells in CD19-Positive Lymphoid Tumors." *N Engl J Med* no. 382 (6):545-553. doi: 10.1056/NEJMoa1910607.

Liu, E., Y. Tong, G. Dotti, H. Shaim, B. Savoldo, M. Mukherjee, J. Orange, X. Wan, X. Lu, A. Reynolds, M. Gagea, P. Banerjee, R. Cai, M. H. Bdaiwi, R. Basar, M. Muftuoglu, L. Li, D. Marin, W. Wierda, M. Keating, R. Champlin, E. Shpall, and K. Rezvani. 2018. "Cord blood NK cells engineered to express IL-15 and a CD19-targeted CAR show long-term persistence and potent antitumor activity." *Leukemia* no. 32 (2):520-531. doi: 10.1038/leu.2017.226.

Liu, S., X. Cai, J. Wu, Q. Cong, X. Chen, T. Li, F. Du, J. Ren, Y. T. Wu, N. V. Grishin, and Z. J. Chen. 2015. "Phosphorylation of innate immune adaptor proteins MAVS, STING, and TRIF induces IRF3 activation." *Science* no. 347 (6227):aaa2630. doi: 10.1126/science. aaa2630.

Lopez-Galindez, C., M. Pernas, C. Casado, I. Olivares, and R. Lorenzo-Redondo. 2019. "Elite controllers and lessons learned for HIV-1 cure." *Curr Opin Virol* no. 38:31-36. doi: 10.1016/j.coviro. 2019.05.010.

Lopez-Huertas, M. R., E. Mateos, M. Sanchez Del Cojo, F. Gomez-Esquer, G. Diaz-Gil, S. Rodriguez-Mora, J. A. Lopez, E. Calvo, G. Lopez-Campos, J. Alcami, and M. Coiras. 2013. "The presence of HIV-1 Tat protein second exon delays fas protein-mediated apoptosis in CD4+ T lymphocytes: a potential mechanism for persistent viral production." *J Biol Chem* no. 288 (11):7626-7644. doi: 10.1074/jbc.M112.408294.

Lorenzo-Redondo, R., H. R. Fryer, T. Bedford, E. Y. Kim, J. Archer, S. L. K. Pond, Y. S. Chung, S. Penugonda, J. Chipman, C. V. Fletcher, T. W. Schacker, M. H. Malim, A. Rambaut, A. T. Haase, A. R. McLean, and S. M. Wolinsky. 2016. "Persistent HIV-1 replication maintains the tissue reservoir during therapy." *Nature* no. 530 (7588):51-56. doi: 10.1038/nature16933.

Luo, X., and J. J. He. 2015. "Cell-cell contact viral transfer contributes to HIV infection and persistence in astrocytes." *J Neurovirol* no. 21 (1):66-80. doi: 10.1007/s13365-014-0304-0.

Macedo, A. B., C. L. Novis, and A. Bosque. 2019. "Targeting Cellular and Tissue HIV Reservoirs With Toll-Like Receptor Agonists." *Front Immunol* no. 10:2450. doi: 10.3389/fimmu.2019.02450.

Manel, N., B. Hogstad, Y. Wang, D. E. Levy, D. Unutmaz, and D. R. Littman. 2010. "A cryptic sensor for HIV-1 activates antiviral innate immunity in dendritic cells." *Nature* no. 467 (7312):214-7. doi: 10.1038/nature09337.

Manjati, T., B. Nkambule, and H. Ipp. 2016. "Immune activation is associated with decreased thymic function in asymptomatic, untreated HIV-infected individuals." *South Afr J HIV Med* no. 17 (1):445. doi: 10.4102/sajhivmed.v17i1.445.

Marchetti, G., C. Tincati, and G. Silvestri. 2013. "Microbial translocation in the pathogenesis of HIV infection and AIDS." *Clin Microbiol Rev* no. 26 (1):2-18. doi: 10.1128/CMR.00050-12.

Margolis, D. M., N. M. Archin, M. S. Cohen, J. J. Eron, G. Ferrari, J. V. Garcia, C. L. Gay, N. Goonetilleke, S. B. Joseph, R. Swanstrom, A. W. Turner, and A. Wahl. 2020. "Curing HIV: Seeking to Target and Clear Persistent Infection." *Cell* no. 181 (1):189-206. doi: 10.1016/j.cell.2020.03.005.

Marini, A., J. M. Harper, and F. Romerio. 2008. "An in vitro system to model the establishment and reactivation of HIV-1 latency." *J Immunol* no. 181 (11):7713-20. doi: 10.4049/jimmunol.181.11.7713.

Martin-Moreno, A., and M. A. Munoz-Fernandez. 2019. "Dendritic Cells, the Double Agent in the War Against HIV-1." *Front Immunol* no. 10:2485. doi: 10.3389/fimmu.2019.02485.

Martinsen, J. T., J. D. Gunst, J. F. Hojen, M. Tolstrup, and O. S. Sogaard. 2020. "The Use of Toll-Like Receptor Agonists in HIV-1 Cure Strategies." *Front Immunol* no. 11:1112. doi: 10.3389/fimmu.2020.01112.

McElrath, M. J., K. Smythe, J. Randolph-Habecker, K. R. Melton, T. A. Goodpaster, S. M. Hughes, M. Mack, A. Sato, G. Diaz, G. Steinbach, R. M. Novak, M. E. Curlin, J. D. Lord, J. Maenza, A. Duerr, N. Frahm, F. Hladik, and Niaid Hiv Vaccine Trials Network. 2013. "Comprehensive assessment of HIV target cells in the distal human gut suggests increasing HIV susceptibility toward the anus." *J Acquir Immune Defic Syndr* no. 63 (3):263-71. doi: 10.1097/QAI. 0b013e3182898392.

Mhandire, K., T. Mlambo, L. S. Zijenah, K. Duri, K. Mateveke, M. Tshabalala, D. Z. Mhandire, C. Musarurwa, P. T. Wekare, L. R. Mazengera, H. T. Matarira, and B. Stray-Pedersen. 2017. "Plasma IP-10 Concentrations Correlate Positively with Viraemia and Inversely with CD4 Counts in Untreated HIV Infection." *Open AIDS J* no. 11:24-31. doi: 10.2174/1874613601711010024.

Milush, J. M., S. Lopez-Verges, V. A. York, S. G. Deeks, J. N. Martin, F. M. Hecht, L. L. Lanier, and D. F. Nixon. 2013. "CD56negCD16(+) NK cells are activated mature NK cells with impaired effector function during HIV-1 infection." *Retrovirology* no. 10:158. doi: 10.1186/1742-4690-10-158.

Mintz, M. A., and J. G. Cyster. 2020. "T follicular helper cells in germinal center B cell selection and lymphomagenesis." *Immunol Rev* no. 296 (1):48-61. doi: 10.1111/imr.12860.

Misse, D., H. Yssel, D. Trabattoni, C. Oblet, S. Lo Caputo, F. Mazzotta, J. Pene, J. P. Gonzalez, M. Clerici, and F. Veas. 2007. "IL-22 participates in an innate anti-HIV-1 host-resistance network through acute-phase protein induction." *J Immunol* no. 178 (1):407-15. doi: 10.4049/jimmunol.178.1.407.

Mohamed, H., V. Miller, S. R. Jennings, B. Wigdahl, and F. C. Krebs. 2020. "The Evolution of Dendritic Cell Immunotherapy against HIV-1 Infection: Improvements and Outlook." *J Immunol Res* no. 2020:9470102. doi: 10.1155/2020/9470102.

Monroe, K. M., Z. Yang, J. R. Johnson, X. Geng, G. Doitsh, N. J. Krogan, and W. C. Greene. 2014. "IFI16 DNA sensor is required for death of lymphoid CD4 T cells abortively infected with HIV." *Science* no. 343 (6169):428-32. doi: 10.1126/science.1243640.

Moradi, M., A. Tabibzadeh, D. Javanmard, S. Ghorbani, F. Bokharaei-Salim, H. Keivani, M. Khazeni, and S. H. R. Monavari. 2020. "Assessment of Key Elements in the Innate Immunity System Among Patients with HIV, HCV, and Coinfections of HIV/HCV." *Curr HIV Res* no. 18 (3):194-200. doi: 10.2174/1570162X18999 200325162533.

Mosoian, A., L. Zhang, F. Hong, F. Cunyat, A. Rahman, R. Bhalla, A. Panchal, Y. Saiman, M. I. Fiel, S. Florman, S. Roayaie, M. Schwartz, A. Branch, M. Stevenson, and M. B. Bansal. 2017. "Frontline Science: HIV infection of Kupffer cells results in an amplified proinflammatory response to LPS." *J Leukoc Biol* no. 101 (5):1083-1090. doi: 10.1189/jlb.3HI0516-242R.

Mu, W., M. A. Carrillo, and S. G. Kitchen. 2020. "Engineering CAR T Cells to Target the HIV Reservoir." *Front Cell Infect Microbiol* no. 10:410. doi: 10.3389/fcimb.2020.00410.

Mueller, S. N., T. Gebhardt, F. R. Carbone, and W. R. Heath. 2013. "Memory T cell subsets, migration patterns, and tissue residence." *Annu Rev Immunol* no. 31:137-61. doi: 10.1146/annurev-immunol-032712-095954.

Muller, F., P. Aukrust, E. Lien, C. J. Haug, and S. S. Froland. 1998. "Enhanced interleukin-10 production in response to Mycobacterium avium products in mononuclear cells from patients with human immunodeficiency virus infection." *J Infect Dis* no. 177 (3):586-94. doi: 10.1086/514222.

Muranski, P., and N. P. Restifo. 2013. "Essentials of Th17 cell commitment and plasticity." *Blood* no. 121 (13):2402-14. doi: 10.1182/blood-2012-09-378653.

Murray, A. J., K. J. Kwon, D. L. Farber, and R. F. Siliciano. 2016. "The Latent Reservoir for HIV-1: How Immunologic Memory and Clonal Expansion Contribute to HIV-1 Persistence." *J Immunol* no. 197 (2):407-17. doi: 10.4049/jimmunol.1600343.

Murray, P. J. 2007. "The JAK-STAT signaling pathway: input and output integration." *J Immunol* no. 178 (5):2623-9. doi: 10.4049/jimmunol. 178.5.2623.

Murray, P. J. 2017. "Macrophage Polarization." *Annu Rev Physiol* no. 79:541-566. doi: 10.1146/annurev-physiol-022516-034339.

Musinguzi, K., A. Obuku, N. Nakasujja, H. Birabwa, J. Nakku, J. Levin, and E. Kinyanda. 2018. "Association between major depressive disorder and pro-inflammatory cytokines and acute phase proteins among HIV-1 positive patients in Uganda." *BMC Immunol* no. 19 (1):1. doi: 10.1186/s12865-017-0239-3.

Nagai, T., O. Devergne, G. A. van Seventer, and J. M. van Seventer. 2007. "Interferon-beta mediates opposing effects on interferon-gamma-dependent Interleukin-12 p70 secretion by human monocyte-derived dendritic cells." *Scand J Immunol* no. 65 (2):107-17. doi: 10.1111/j.1365-3083.2006.01880.x.

Narazaki, M., and T. Kishimoto. 2018. "The Two-Faced Cytokine IL-6 in Host Defense and Diseases." *Int J Mol Sci* no. 19 (11). doi: 10.3390/ijms19113528.

Nazli, A., J. K. Kafka, V. H. Ferreira, V. Anipindi, K. Mueller, B. J. Osborne, S. Dizzell, S. Chauvin, M. F. Mian, M. Ouellet, M. J. Tremblay, K. L. Mossman, A. A. Ashkar, C. Kovacs, D. M. Bowdish, D. P. Snider, R. Kaul, and C. Kaushic. 2013. "HIV-1 gp120 induces TLR2- and TLR4-mediated innate immune activation in human female genital epithelium." *J Immunol* no. 191 (8):4246-58. doi: 10.4049/jimmunol.1301482.

Nchioua, R., M. Bosso, D. Kmiec, and F. Kirchhoff. 2020. "Cellular Factors Targeting HIV-1 Transcription and Viral RNA Transcripts." *Viruses* no. 12 (5). doi: 10.3390/v12050495.

Neuhaus, J., D. R. Jacobs, Jr., J. V. Baker, A. Calmy, D. Duprez, A. La Rosa, L. H. Kuller, S. L. Pett, M. Ristola, M. J. Ross, M. G. Shlipak, R. Tracy, and J. D. Neaton. 2010. "Markers of inflammation, coagulation, and renal function are elevated in adults with HIV infection." *J Infect Dis* no. 201 (12):1788-95. doi: 10.1086/652749.

Nganou-Makamdop, K., J. M. Billingsley, Z. Yaffe, G. O'Connor, G. K. Tharp, A. Ransier, F. Laboune, R. Matus-Nicodemos, A. Lerner, L. Gharu, J. M. Robertson, M. L. Ford, M. Schlapschy, N. Kuhn, A. Lensch, J. Lifson, M. Nason, A. Skerra, G. Schreiber, S. E. Bosinger, and D. C. Douek. 2018. "Type I IFN signaling blockade by a PASylated antagonist during chronic SIV infection suppresses specific inflammatory pathways but does not alter T cell activation or virus replication." *PLoS Pathog* no. 14 (8):e1007246. doi: 10.1371/ journal.ppat.1007246.

Nixon, D. E., and A. L. Landay. 2010. "Biomarkers of immune dysfunction in HIV." *Curr Opin HIV AIDS* no. 5 (6):498-503. doi: 10.1097/COH. 0b013e32833ed6f4.

Noel, N., B. Jacquelin, N. Huot, C. Goujard, O. Lambotte, and M. Muller-Trutwin. 2018. "Interferon-associated therapies toward HIV control: The back and forth." *Cytokine Growth Factor Rev* no. 40:99-112. doi: 10.1016/j.cytogfr .2018.03.004.

Novis, C. L., N. M. Archin, M. J. Buzon, E. Verdin, J. L. Round, M. Lichterfeld, D. M. Margolis, V. Planelles, and A. Bosque. 2013. "Reactivation of latent HIV-1 in central memory CD4(+) T cells through TLR-1/2 stimulation." *Retrovirology* no. 10:119. doi: 10.1186/1742-4690-10-119.

O'Neil, T. R., K. Hu, N. R. Truong, S. Arshad, B. L. Shacklett, A. L. Cunningham, and N. Nasr. 2021. "The Role of Tissue Resident Memory CD4 T Cells in Herpes Simplex Viral and HIV Infection." *Viruses* no. 13 (3). doi: 10.3390/v13030359.

O'Shea, J. J., M. Gadina, and R. D. Schreiber. 2002. "Cytokine signaling in 2002: new surprises in the Jak/Stat pathway." *Cell* no. 109 Suppl:S121-31. doi: 10.1016/s0092-8674(02)00701-8.

Oldstone, M. B. 2015. "A Jekyll and Hyde Profile: Type 1 Interferon Signaling Plays a Prominent Role in the Initiation and Maintenance of a Persistent Virus Infection." *J Infect Dis* no. 212 Suppl 1:S31-6. doi: 10.1093/infdis/jiu501.

Olesen, R., S. Vigano, T. A. Rasmussen, O. S. Sogaard, Z. Ouyang, M. Buzon, A. Bashirova, M. Carrington, S. Palmer, C. R. Brinkmann, X. G. Yu, L. Ostergaard, M. Tolstrup, and M. Lichterfeld. 2015. "Innate Immune Activity Correlates with CD4 T Cell-Associated HIV-1 DNA Decline during Latency-Reversing Treatment with Panobinostat." *J Virol* no. 89 (20):10176-89. doi: 10.1128/JVI. 01484-15.

Orenstein, J. M., C. Fox, and S. M. Wahl. 1997. "Macrophages as a source of HIV during opportunistic infections." *Science* no. 276 (5320):1857-61. doi: 10.1126/science.276.5320.1857.

Pace, M. J., L. Agosto, E. H. Graf, and U. O'Doherty. 2011. "HIV reservoirs and latency models." *Virology* no. 411 (2):344-54. doi: 10.1016/j.virol. 2010.12.041.

Paiardini, M., and M. Muller-Trutwin. 2013. "HIV-associated chronic immune activation." *Immunol Rev* no. 254 (1):78-101. doi: 10.1111/imr.12079.

Pardons, M., A. E. Baxter, M. Massanella, A. Pagliuzza, R. Fromentin, C. Dufour, L. Leyre, J. P. Routy, D. E. Kaufmann, and N. Chomont. 2019. "Single-cell characterization and quantification of translation-competent viral reservoirs in treated and untreated HIV infection." *PLoS Pathog* no. 15 (2):e1007619. doi: 10.1371/journal.ppat. 1007619.

Pardons, M., R. Fromentin, A. Pagliuzza, J. P. Routy, and N. Chomont. 2019. "Latency-Reversing Agents Induce Differential Responses in Distinct Memory CD4 T Cell Subsets in Individuals on Antiretroviral Therapy." *Cell Rep* no. 29 (9):2783-2795 e5. doi: 10.1016/j.celrep. 2019.10.101.

Pasternak, A. O., V. V. Lukashov, and B. Berkhout. 2013. "Cell-associated HIV RNA: a dynamic biomarker of viral persistence." *Retrovirology* no. 10:41. doi: 10.1186/1742-4690-10-41.

Pastor, L., A. Casellas, J. Carrillo, S. Alonso, E. Parker, L. Fuente-Soro, C. Jairoce, I. Mandomando, J. Blanco, and D. Naniche. 2017. "IP-10 Levels as an Accurate Screening Tool to Detect Acute HIV Infection in Resource-Limited Settings." *Sci Rep* no. 7 (1):8104. doi: 10.1038/s41598-017-08218-0.

Pedro, K. D., A. J. Henderson, and L. M. Agosto. 2019. "Mechanisms of HIV-1 cell-to-cell transmission and the establishment of the latent reservoir." *Virus Res* no. 265:115-121. doi: 10.1016/j.virusres.2019. 03.014.

Perdiguero, E. G., and F. Geissmann. 2016. "The development and maintenance of resident macrophages." *Nat Immunol* no. 17 (1):2-8. doi: 10.1038/ni.3341.

Perreau, M., A. L. Savoye, E. De Crignis, J. M. Corpataux, R. Cubas, E. K. Haddad, L. De Leval, C. Graziosi, and G. Pantaleo. 2013. "Follicular helper T cells serve as the major CD4 T cell compartment for HIV-1 infection, replication, and production." *J Exp Med* no. 210 (1):143-56. doi: 10.1084/jem.20121932.

Pertel, T., S. Hausmann, D. Morger, S. Zuger, J. Guerra, J. Lascano, C. Reinhard, F. A. Santoni, P. D. Uchil, L. Chatel, A. Bisiaux, M. L. Albert, C. Strambio-De-Castillia, W. Mothes, M. Pizzato, M. G. Grutter, and J. Luban. 2011. "TRIM5 is an innate immune sensor for the retrovirus capsid lattice." *Nature* no. 472 (7343):361-5. doi: 10.1038/nature09976.

Piconi, S., S. Parisotto, G. Rizzardini, S. Passerini, R. Terzi, B. Argenteri, P. Meraviglia, A. Capetti, M. Biasin, D. Trabattoni, and M. Clerici. 2011. "Hydroxychloroquine drastically reduces immune activation in HIV-infected, antiretroviral therapy-treated immunologic nonresponders." *Blood* no. 118 (12):3263-72. doi: 10.1182/blood-2011-01-329060.

Pillai, S. K., S. L. Pond, Y. Liu, B. M. Good, M. C. Strain, R. J. Ellis, S. Letendre, D. M. Smith, H. F. Gunthard, I. Grant, T. D. Marcotte, J. A. McCutchan, D. D. Richman, and J. K. Wong. 2006. "Genetic attributes of cerebrospinal fluid-derived HIV-1 env." *Brain* no. 129 (Pt 7):1872-83. doi: 10.1093/brain/awl136.

Procopio, F. A., R. Fromentin, D. A. Kulpa, J. H. Brehm, A. G. Bebin, M. C. Strain, D. D. Richman, U. O'Doherty, S. Palmer, F. M. Hecht, R. Hoh, R. J. Barnard, M. D. Miller, D. J. Hazuda, S. G. Deeks, R. P. Sekaly, and N. Chomont. 2015. "A Novel Assay to Measure the Magnitude of the Inducible Viral Reservoir in HIV-infected Individuals." *EBioMedicine* no. 2 (8):874-83. doi: 10.1016/j.ebiom. 2015.06.019.

Ramirez, L. A., T. A. Arango, E. Thompson, M. Naji, P. Tebas, and J. D. Boyer. 2014. "High IP-10 levels decrease T cell function in HIV-1-infected individuals on ART." *J Leukoc Biol* no. 96 (6):1055-63. doi: 10.1189/jlb.3A0414-232RR.

Rausch, J. W., and S. F. J. Le Grice. 2020. "Characterizing the Latent HIV-1 Reservoir in Patients with Viremia Suppressed on cART: Progress, Challenges, and Opportunities." *Curr HIV Res* no. 18 (2):99-113. doi: 10.2174/1570162X18666191231105438.

Reizis, B. 2019. "Plasmacytoid Dendritic Cells: Development, Regulation, and Function." *Immunity* no. 50 (1):37-50. doi: 10.1016/ j.immuni.2018.12.027.

Ren, X. X., L. Ma, W. W. Sun, W. D. Kuang, T. S. Li, X. Jin, and J. H. Wang. 2017. "Dendritic cells maturated by co-culturing with HIV-1 latently infected Jurkat T cells or stimulating with AIDS-associated pathogens secrete TNF-alpha to reactivate HIV-1 from latency." *Virulence* no. 8 (8):1732-1743. doi: 10.1080/21505594.2017.1356535.

Reuter, M. A., P. M. Del Rio Estrada, M. Buggert, C. Petrovas, S. Ferrando-Martinez, S. Nguyen, A. Sada Japp, Y. Ablanedo-Terrazas, A. Rivero-Arrieta, L. Kuri-Cervantes, H. M. Gunzelman, E. Gostick, D. A. Price, R. A. Koup, A. Naji, D. H. Canaday, G. Reyes-Teran, and M. R. Betts. 2017. "HIV-Specific CD8(+) T Cells Exhibit Reduced and Differentially Regulated Cytolytic Activity in Lymphoid Tissue." *Cell Rep* no. 21 (12):3458-3470. doi: 10.1016/j. celrep.2017.11.075.

Reuven, E. M., M. Ali, E. Rotem, R. Schwarzer, A. Gramatica, A. H. Futerman, and Y. Shai. 2014. "The HIV-1 envelope transmembrane domain binds TLR2 through a distinct dimerization motif and inhibits TLR2-mediated responses." *PLoS Pathog* no. 10 (8):e1004248. doi: 10.1371/journal.ppat.1004248.

Rezaei, S. D., H. K. Lu, J. J. Chang, A. Rhodes, S. R. Lewin, and P. U. Cameron. 2018. "The Pathway To Establishing HIV Latency Is Critical to How Latency Is Maintained and Reversed." *J Virol* no. 92 (13). doi: 10.1128/JVI.02225-17.

Robb, M. L., and J. Ananworanich. 2016. "Lessons from acute HIV infection." *Curr Opin HIV AIDS* no. 11 (6):555-560. doi: 10.1097/ COH.0000000 000000316.

Roche, M., C. Tumpach, J. Symons, M. Gartner, J. L. Anderson, G. Khoury, K. Cashin, P. U. Cameron, M. J. Churchill, S. G. Deeks, P. R. Gorry, and S. R. Lewin. 2020. "CXCR4-Using HIV Strains Predominate in Naive and Central Memory CD4(+) T Cells in People Living with HIV on Antiretroviral Therapy: Implications for How Latency Is Established and Maintained." *J Virol* no. 94 (6). doi: 10.1128/JVI.01736-19.

Rodger, A. J., Z. Fox, J. D. Lundgren, L. H. Kuller, C. Boesecke, D. Gey, A. Skoutelis, M. B. Goetz, A. N. Phillips, and Insight Strategies for Management of Antiretroviral Therapy Study Group. 2009. "Activation and coagulation biomarkers are independent predictors of the development of opportunistic disease in patients with HIV infection." *J Infect Dis* no. 200 (6):973-83. doi: 10.1086/605447.

Rogez-Kreuz, C., B. Maneglier, M. Martin, N. Dereuddre-Bosquet, J. Martal, D. Dormont, and P. Clayette. 2005. "Involvement of IL-6 in the anti-human immunodeficiency virus activity of IFN-tau in human macrophages." *Int Immunol* no. 17 (8):1047-57. doi: 10.1093/intimm/ dxh285.

Rothenberger, M. K., B. F. Keele, S. W. Wietgrefe, C. V. Fletcher, G. J. Beilman, J. G. Chipman, A. Khoruts, J. D. Estes, J. Anderson, S. P. Callisto, T. E. Schmidt, A. Thorkelson, C. Reilly, K. Perkey, T. G. Reimann, N. S. Utay, K. Nganou Makamdop, M. Stevenson, D. C. Douek, A. T. Haase, and T. W. Schacker. 2015. "Large number of rebounding/founder HIV variants emerge from multifocal infection in lymphatic tissues after treatment interruption." *Proc Natl Acad Sci U S A* no. 112 (10):E1126-34. doi: 10.1073/pnas. 1414926112.

Routy, J. P., J. B. Angel, M. Patel, C. Kanagaratham, D. Radzioch, I. Kema, N. Gilmore, P. Ancuta, J. Singer, and M. A. Jenabian. 2015. "Assessment of chloroquine as a modulator of immune activation to improve CD4 recovery in immune nonresponding HIV-infected patients receiving antiretroviral therapy." *HIV Med* no. 16 (1):48-56. doi: 10.1111/hiv.12171.

Rouzioux, C., and V. Avettand-Fenoel. 2018. "Total HIV DNA: a global marker of HIV persistence." *Retrovirology* no. 15 (1):30. doi: 10.1186/s12977-018-0412-7.

Russell, R. A., J. Chojnacki, D. M. Jones, E. Johnson, T. Do, C. Eggeling, S. Padilla-Parra, and Q. J. Sattentau. 2017. "Astrocytes Resist HIV-1 Fusion but Engulf Infected Macrophage Material." *Cell Rep* no. 18 (6):1473-1483. doi: 10.1016/j.celrep.2017.01.027.

Rutz, S., X. Wang, and W. Ouyang. 2014. "The IL-20 subfamily of cytokines--from host defence to tissue homeostasis." *Nat Rev Immunol* no. 14 (12):783-95. doi: 10.1038/nri3766.

Ryoo, J., J. Choi, C. Oh, S. Kim, M. Seo, S. Y. Kim, D. Seo, J. Kim, T. E. White, A. Brandariz-Nunez, F. Diaz-Griffero, C. H. Yun, J. A. Hollenbaugh, B. Kim, D. Baek, and K. Ahn. 2014. "The ribonuclease activity of SAMHD1 is required for HIV-1 restriction." *Nat Med* no. 20 (8):936-41. doi: 10.1038/nm.3626.

Sabado, R. L., M. O'Brien, A. Subedi, L. Qin, N. Hu, E. Taylor, O. Dibben, A. Stacey, J. Fellay, K. V. Shianna, F. Siegal, M. Shodell, K. Shah, M. Larsson, J. Lifson, A. Nadas, M. Marmor, R. Hutt, D. Margolis, D. Garmon, M. Markowitz, F. Valentine, P. Borrow, and N. Bhardwaj. 2010. "Evidence of dysregulation of dendritic cells in primary HIV infection." *Blood* no. 116 (19):3839-52. doi: 10.1182/blood-2010-03-273763.

Saez-Cirion, A., C. Bacchus, L. Hocqueloux, V. Avettand-Fenoel, I. Girault, C. Lecuroux, V. Potard, P. Versmisse, A. Melard, T. Prazuck, B. Descours, J. Guergnon, J. P. Viard, F. Boufassa, O. Lambotte, C. Goujard, L. Meyer, D.

Costagliola, A. Venet, G. Pancino, B. Autran, C. Rouzioux, and Anrs Visconti Study Group. 2013. "Post-treatment HIV-1 controllers with a long-term virological remission after the interruption of early initiated antiretroviral therapy ANRS VISCONTI Study." *PLoS Pathog* no. 9 (3):e1003211. doi: 10.1371/journal.ppat.1003211.

Saha, P., and F. Geissmann. 2011. "Toward a functional characterization of blood monocytes." *Immunol Cell Biol* no. 89 (1):2-4. doi: 10.1038/icb.2010.130.

Sandler, N. G., S. E. Bosinger, J. D. Estes, R. T. Zhu, G. K. Tharp, E. Boritz, D. Levin, S. Wijeyesinghe, K. N. Makamdop, G. Q. del Prete, B. J. Hill, J. K. Timmer, E. Reiss, G. Yarden, S. Darko, E. Contijoch, J. P. Todd, G. Silvestri, M. Nason, R. B. Norgren, Jr., B. F. Keele, S. Rao, J. A. Langer, J. D. Lifson, G. Schreiber, and D. C. Douek. 2014. "Type I interferon responses in rhesus macaques prevent SIV infection and slow disease progression." *Nature* no. 511 (7511):601-5. doi: 10.1038/nature13554.

Sandler, N. G., H. Wand, A. Roque, M. Law, M. C. Nason, D. E. Nixon, C. Pedersen, K. Ruxrungtham, S. R. Lewin, S. Emery, J. D. Neaton, J. M. Brenchley, S. G. Deeks, I. Sereti, D. C. Douek, and Insight Smart Study Group. 2011. "Plasma levels of soluble CD14 independently predict mortality in HIV infection." *J Infect Dis* no. 203 (6):780-90. doi: 10.1093/infdis/jiq118.

Sandstrom, T. S., N. Ranganath, and J. B. Angel. 2017. "Impairment of the type I interferon response by HIV-1: Potential targets for HIV eradication." *Cytokine Growth Factor Rev* no. 37:1-16. doi: 10.1016/j.cytogfr.2017.04.004.

Saravia, J., N. M. Chapman, and H. Chi. 2019. "Helper T cell differentiation." *Cell Mol Immunol* no. 16 (7):634-643. doi: 10.1038/s41423-019-0220-6.

Saxena, M., R. L. Sabado, M. La Mar, H. Mohri, A. M. Salazar, H. Dong, J. Correa Da Rosa, M. Markowitz, N. Bhardwaj, and E. Miller. 2019. "Poly-ICLC, a TLR3 Agonist, Induces Transient Innate Immune Responses in Patients With Treated HIV-Infection: A Randomized Double-Blinded Placebo Controlled Trial." *Front Immunol* no. 10:725. doi: 10.3389/fimmu.2019.00725.

Schechter, M. E., B. B. Andrade, T. He, G. H. Richter, K. W. Tosh, B. B. Policicchio, A. Singh, K. D. Raehtz, V. Sheikh, D. Ma, E. Brocca-Cofano, C. Apetrei, R. Tracy, R. M. Ribeiro, A. Sher, I. M. B. Francischetti, I. Pandrea, and I. Sereti. 2017. "Inflammatory monocytes expressing tissue factor drive SIV and HIV coagulopathy." *Sci Transl Med* no. 9 (405). doi: 10.1126/scitranslmed.aam5441.

Schlitzer, A., N. McGovern, and F. Ginhoux. 2015. "Dendritic cells and monocyte-derived cells: Two complementary and integrated functional systems." *Semin Cell Dev Biol* no. 41:9-22. doi: 10.1016/j.semcdb.2015.03.011.

Schmid, M. A., H. Takizawa, D. R. Baumjohann, Y. Saito, and M. G. Manz. 2011. "Bone marrow dendritic cell progenitors sense pathogens via Toll-like receptors and subsequently migrate to inflamed lymph nodes." *Blood* no. 118 (18):4829-40. doi: 10.1182/blood-2011-03-344960.

Schreiber, G., and J. Piehler. 2015. "The molecular basis for functional plasticity in type I interferon signaling." *Trends Immunol* no. 36 (3):139-49. doi: 10.1016/j.it.2015.01.002.

Schroder, A. R., P. Shinn, H. Chen, C. Berry, J. R. Ecker, and F. Bushman. 2002. "HIV-1 integration in the human genome favors active genes and local hotspots." *Cell* no. 110 (4):521-9. doi: 10.1016/s0092-8674(02)00864-4.

Schunter, M., H. Chu, T. L. Hayes, D. McConnell, S. S. Crawford, P. A. Luciw, S. Bengmark, D. M. Asmuth, J. Brown, C. L. Bevins, B. L. Shacklett, and J. W. Critchfield. 2012. "Randomized pilot trial of a synbiotic dietary supplement in chronic HIV-1 infection." *BMC Complement Altern Med* no. 12:84. doi: 10.1186/1472-6882-12-84.

Shacklett, B. L., A. L. Ferre, and B. E. Kiniry. 2020. "Defining T Cell Tissue Residency in Humans: Implications for HIV Pathogenesis and Vaccine Design." *Curr HIV/AIDS Rep* no. 17 (2):109-117. doi: 10.1007/s11904-020-00481-7.

Shan, L., K. Deng, H. Gao, S. Xing, A. A. Capoferri, C. M. Durand, S. A. Rabi, G. M. Laird, M. Kim, N. N. Hosmane, H. C. Yang, H. Zhang, J. B. Margolick, L. Li, W. Cai, R. Ke, R. A. Flavell, J. D. Siliciano, and R. F. Siliciano. 2017. "Transcriptional Reprogramming during Effector-to-Memory Transition Renders CD4(+) T Cells Permissive for Latent HIV-1 Infection." *Immunity* no. 47 (4):766-775 e3. doi: 10.1016/j.immuni.2017.09.014.

Shukla, A., N. P. Ramirez, and I. D'Orso. 2020. "HIV-1 Proviral Transcription and Latency in the New Era." *Viruses* no. 12 (5). doi: 10.3390/v12050555.

Siegal, F. P., N. Kadowaki, M. Shodell, P. A. Fitzgerald-Bocarsly, K. Shah, S. Ho, S. Antonenko, and Y. J. Liu. 1999. "The nature of the principal type 1 interferon-producing cells in human blood." *Science* no. 284 (5421):1835-7. doi: 10.1126/science.284.5421.1835.

Sittig, S. P., G. Bakdash, J. Weiden, A. E. Skold, J. Tel, C. G. Figdor, I. J. de Vries, and G. Schreibelt. 2016. "A Comparative Study of the T Cell Stimulatory and Polarizing Capacity of Human Primary Blood Dendritic Cell Subsets." *Mediators Inflamm* no. 2016:3605643. doi: 10.1155/2016/3605643.

So-Armah K, Benjamin LA, Bloomfield GS, Feinstein MJ, Hsue P, Njuguna B, Freiberg MS. 2020. *HIV and cardiovascular disease*. edited by Lancet HIV.

Sonnenberg, G. F., L. A. Fouser, and D. Artis. 2011. "Border patrol: regulation of immunity, inflammation and tissue homeostasis at barrier surfaces by IL-22." *Nat Immunol* no. 12 (5):383-90. doi: 10.1038/ni.2025.

Soriano-Sarabia, N., N. M. Archin, R. Bateson, N. P. Dahl, A. M. Crooks, J. D. Kuruc, C. Garrido, and D. M. Margolis. 2015. "Peripheral V gamma 9 V delta 2 T Cells Are a Novel Reservoir of Latent HIV Infection." *PLoS Pathog* no. 11 (10):e1005201. doi: 10.1371/journal.ppat.1005201.

Soriano-Sarabia, N., R. E. Bateson, N. P. Dahl, A. M. Crooks, J. D. Kuruc, D. M. Margolis, and N. M. Archin. 2014. "Quantitation of replication-competent HIV-1 in populations of resting CD4+ T cells." *J Virol* no. 88 (24):14070-7. doi: 10.1128/JVI.01900-14.

Spivak, A. M., E. T. Larragoite, M. L. Coletti, A. B. Macedo, L. J. Martins, A. Bosque, and V. Planelles. 2016. "Janus kinase inhibition suppresses PKC-induced cytokine release without affecting HIV-1 latency reversal ex vivo." *Retrovirology* no. 13 (1):88. doi: 10.1186/s12977-016-0319-0.

Stacey, A. R., P. J. Norris, L. Qin, E. A. Haygreen, E. Taylor, J. Heitman, M. Lebedeva, A. DeCamp, D. Li, D. Grove, S. G. Self, and P. Borrow. 2009. "Induction of a striking systemic cytokine cascade prior to peak viremia in acute human immunodeficiency virus type 1 infection, in contrast to more modest and delayed responses in acute hepatitis B and C virus infections." *J Virol* no. 83 (8):3719-33. doi: 10.1128/JVI.01844-08.

Stackaruk, M. L., A. J. Lee, and A. A. Ashkar. 2013. "Type I interferon regulation of natural killer cell function in primary and secondary infections." *Expert Rev Vaccines* no. 12 (8):875-84. doi: 10.1586/14760584.2013.814871.

Stary, G., I. Klein, S. Kohlhofer, F. Koszik, T. Scherzer, L. Mullauer, H. Quendler, N. Kohrgruber, and G. Stingl. 2009. "Plasmacytoid dendritic cells express TRAIL and induce CD4+ T-cell apoptosis in HIV-1 viremic patients." *Blood* no. 114 (18):3854-63. doi: 10.1182/blood-2009-04-217927.

Strategies for Management of Antiretroviral Therapy Study, Group, W. M. El-Sadr, J. Lundgren, J. D. Neaton, F. Gordin, D. Abrams, R. C. Arduino, A. Babiker, W. Burman, N. Clumeck, C. J. Cohen, D. Cohn, D. Cooper, J. Darbyshire, S. Emery, G. Fatkenheuer, B. Gazzard, B. Grund, J. Hoy, K. Klingman, M. Losso, N. Markowitz, J. Neuhaus, A. Phillips, and C. Rappoport. 2006. "CD4+ count-guided interruption of antiretroviral treatment." *N Engl J Med* no. 355 (22):2283-96. doi: 10.1056/NEJMoa062360.

Sturdevant, C. B., S. B. Joseph, G. Schnell, R. W. Price, R. Swanstrom, and S. Spudich. 2015. "Compartmentalized replication of R5 T cell-tropic HIV-1 in the central nervous system early in the course of infection." *PLoS Pathog* no. 11 (3):e1004720. doi: 10.1371/journal. ppat.1004720.

Stylianou, E., P. Aukrust, D. Kvale, F. Muller, and S. S. Froland. 1999. "IL-10 in HIV infection: increasing serum IL-10 levels with disease progression--down-regulatory effect of potent anti-retroviral therapy." *Clin Exp Immunol* no. 116 (1):115-20. doi: 10.1046/j.1365-2249.1999.00865.x.

Sun, H., D. Kim, X. Li, M. Kiselinova, Z. Ouyang, L. Vandekerckhove, H. Shang, E. S. Rosenberg, X. G. Yu, and M. Lichterfeld. 2015. "Thl/17 Polarization of CD4 T Cells Supports HIV-1 Persistence during Antiretroviral Therapy." *J Virol* no. 89 (22):11284-93. doi: 10.1128/JVI.01595-15.

Sung, J. M., and D. M. Margolis. 2018. "Correction to: HIV Persistence on Antiretroviral Therapy and Barriers to a Cure." *Adv Exp Med Biol* no. 1075:E1. doi: 10.1007/978-981-13-0484-2_13.

Svicher, V., F. Ceccherini-Silberstein, A. Antinori, S. Aquaro, and C. F. Perno. 2014. "Understanding HIV compartments and reservoirs." *Curr HIV/AIDS Rep* no. 11 (2):186-94. doi: 10.1007/s11904-014-0207-y.

Tamalet, C., C. Devaux, G. Dubourg, and P. Colson. 2021. "Resistance to human immunodeficiency virus infection: a rare but neglected state." *Ann N Y Acad Sci* no. 1485 (1):22-42. doi: 10.1111/nyas.14452.

Tanaka, Y., and Z. J. Chen. 2012. "STING specifies IRF3 phospho-rylation by TBK1 in the cytosolic DNA signaling pathway." *Sci Signal* no. 5 (214):ra20. doi: 10.1126/scisignal.2002521.

Thompson, C. G., C. L. Gay, and A. D. M. Kashuba. 2017. "HIV Persistence in Gut-Associated Lymphoid Tissues: Pharmacological Challenges and Opportunities." *AIDS Res Hum Retroviruses* no. 33 (6):513-523. doi: 10.1089/AID.2016.0253.

Thompson, K. A., C. L. Cherry, J. E. Bell, and C. A. McLean. 2011. "Brain cell reservoirs of latent virus in presymptomatic HIV-infected individuals." *Am J Pathol* no. 179 (4):1623-9. doi: 10.1016/j. ajpath.2011.06.039.

Tomaras, G. D., G. Ferrari, X. Shen, S. M. Alam, H. X. Liao, J. Pollara, M. Bonsignori, M. A. Moody, Y. Fong, X. Chen, B. Poling, C. O. Nicholson, R. Zhang, X. Lu, R. Parks, J. Kaewkungwal, S. Nitayaphan, P. Pitisuttithum, S. Rerks-Ngarm, P. B. Gilbert, J. H. Kim, N. L. Michael, D. C. Montefiori, and B. F. Haynes. 2013. "Vaccine-induced plasma IgA specific for the C1 region of the HIV-1 envelope blocks binding and effector function of IgG." *Proc Natl Acad Sci U S A* no. 110 (22):9019-24. doi: 10.1073/pnas. 1301456110.

Tran, T. A., M. G. de Goer de Herve, H. Hendel-Chavez, B. Dembele, E. Le Nevot, K. Abbed, C. Pallier, C. Goujard, J. Gasnault, J. F. Delfraissy, A. M. Balazuc, and Y. Taoufik. 2008. "Resting regulatory CD4 T cells: a site of HIV persistence in patients on long-term effective antiretroviral therapy." *PLoS One* no. 3 (10):e3305. doi: 10.1371/journal.pone.0003305.

Tsai, A., A. Irrinki, J. Kaur, T. Cihlar, G. Kukolj, D. D. Sloan, and J. P. Murry. 2017. "Toll-Like Receptor 7 Agonist GS-9620 Induces HIV Expression and HIV-Specific Immunity in Cells from HIV-Infected Individuals on Suppressive Antiretroviral Therapy." *J Virol* no. 91 (8). doi: 10.1128/JVI.02166-16.

Utay, N. S., and D. C. Douek. 2016. "Interferons and HIV Infection: The Good, the Bad, and the Ugly." *Pathog Immun* no. 1 (1):107-116. doi: 10.20411/pai.v1i1.125.

Vaidya, S. A., C. Korner, M. N. Sirignano, M. Amero, S. Bazner, J. Rychert, T. M. Allen, E. S. Rosenberg, R. J. Bosch, and M. Altfeld. 2014. "Tumor necrosis factor alpha is associated with viral control and early disease progression in patients with HIV type 1 infection." *J Infect Dis* no. 210 (7):1042-6. doi: 10.1093/infdis/jiu206.

Valcour, V. G., B. T. Shiramizu, and C. M. Shikuma. 2010. "HIV DNA in circulating monocytes as a mechanism to dementia and other HIV complications." *J Leukoc Biol* no. 87 (4):621-6. doi: 10.1189/jlb. 0809571.

Valverde-Villegas, J. M., R. M. de Medeiros, J. H. Ellwanger, B. R. Santos, M. G. Melo, S. E. M. Almeida, and J. A. B. Chies. 2018. "High CXCL10/IP-10 levels are a hallmark in the clinical evolution of the HIV infection." *Infect Genet Evol* no. 57:51-58. doi: 10.1016/j.meegid.2017.11.002.

van der Sluis, R. M., T. M. van Capel, D. Speijer, R. W. Sanders, B. Berkhout, E. C. de Jong, R. E. Jeeninga, and T. van Montfort. 2015. "Dendritic cell type-specific HIV-1 activation in effector T cells: implications for latent HIV-1 reservoir establishment." *AIDS* no. 29 (9):1003-14. doi: 10.1097/ QAD.0000000000000637.

Van Gulck, E., E. Vlieghe, M. Vekemans, V. F. Van Tendeloo, A. Van De Velde, E. Smits, S. Anguille, N. Cools, H. Goossens, L. Mertens, W. De Haes, J. Wong, E. Florence, G. Vanham, and Z. N. Berneman. 2012. "mRNA-based dendritic cell vaccination induces potent antiviral T-cell responses in HIV-1-infected patients." *AIDS* no. 26 (4):F1-12. doi: 10.1097/QAD. 0b013e32834f33e8.

Vandergeeten, C., R. Fromentin, S. DaFonseca, M. B. Lawani, I. Sereti, M. M. Lederman, M. Ramgopal, J. P. Routy, R. P. Sekaly, and N. Chomont. 2013. "Interleukin-7 promotes HIV persistence during antiretroviral therapy." *Blood* no. 121 (21):4321-9. doi: 10.1182/blood-2012-11-465625.

Vatakis, D. N., S. Kim, N. Kim, S. A. Chow, and J. A. Zack. 2009. "Human immunodeficiency virus integration efficiency and site selection in quiescent CD4+ T cells." *J Virol* no. 83 (12):6222-33. doi: 10.1128/JVI.00356-09.

Venanzi Rullo, E., L. Cannon, M. R. Pinzone, M. Ceccarelli, G. Nunnari, and U. O'Doherty. 2019. "Genetic Evidence That Naive T Cells Can Contribute Significantly to the Human Immunodeficiency Virus Intact Reservoir: Time to Re-evaluate Their Role." *Clin Infect Dis* no. 69 (12):2236-2237. doi: 10.1093/cid/ciz378.

Ventura, J. D. 2020. "Human Immunodeficiency Virus 1 (HIV-1): Viral Latency, the Reservoir, and the Cure." *Yale J Biol Med* no. 93 (4):549-560.

Vergara-Mendoza, M., L. E. Gomez-Quiroz, R. U. Miranda-Labra, L. L. Fuentes-Romero, D. P. Romero-Rodriguez, J. Gonzalez-Ruiz, S. Hernandez-Rizo, and M. Viveros-Rogel. 2020. "Regulation of Cas9 by viral proteins Tat and Rev for HIV-1 inactivation." *Antiviral Res* no. 180:104856. doi: 10.1016/j.antiviral.2020.104856.

Vibholm, L. K., C. V. Konrad, M. H. Schleimann, G. Frattari, A. Winckelmann, V. Klastrup, N. M. Jensen, S. S. Jensen, M. Schmidt, B. Wittig, K. Zuwala, K. Mack, R. Olesen, S. Hua, M. Lichterfeld, L. Ostergaard, P. W. Denton, M. Tolstrup, and O. S. Sogaard. 2019. "Effects of 24-week Toll-like receptor 9 agonist treatment in HIV type 1+ individuals." *AIDS* no. 33 (8):1315-1325. doi: 10.1097/QAD.0000000000002213.

Vibholm, L., M. H. Schleimann, J. F. Hojen, T. Benfield, R. Offersen, K. Rasmussen, R. Olesen, A. Dige, J. Agnholt, J. Grau, M. Buzon, B. Wittig, M. Lichterfeld, A. M. Petersen, X. Deng, M. Abdel-Mohsen, S. K. Pillai, S. Rutsaert, W. Trypsteen, W. De Spiegelaere, L. Vandekerchove, L. Ostergaard, T. A. Rasmussen, P. W. Denton, M. Tolstrup, and O. S. Sogaard. 2017. "Short-Course Toll-Like Receptor 9 Agonist Treatment Impacts Innate Immunity and Plasma Viremia in Individuals With Human Immunodeficiency Virus Infection." *Clin Infect Dis* no. 64 (12):1686-1695. doi: 10.1093/cid/cix201.

Vidyant, S., A. Chatterjee, V. Agarwal, and T. N. Dhole. 2017. "Susceptibility to HIV-1 infection is influenced by toll like receptor-2 (-196 to -174) polymorphism in a north Indian population." *J Gene Med* no. 19 (8). doi: 10.1002/jgm.2971.

Villani, A. C., R. Satija, G. Reynolds, S. Sarkizova, K. Shekhar, J. Fletcher, M. Griesbeck, A. Butler, S. Zheng, S. Lazo, L. Jardine, D. Dixon, E. Stephenson, E. Nilsson, I. Grundberg, D. McDonald, A. Filby, W. Li, P. L. De Jager, O. Rozenblatt-Rosen, A. A. Lane, M. Haniffa, A. Regev, and N. Hacohen. 2017. "Single-cell RNA-seq reveals new types of human blood dendritic cells, monocytes, and progenitors." *Science* no. 356 (6335). doi: 10.1126/science.aah4573.

Wacleche, V. S., J. P. Goulet, A. Gosselin, P. Monteiro, H. Soudeyns, R. Fromentin, M. A. Jenabian, S. Vartanian, S. G. Deeks, N. Chomont, J. P. Routy, and P. Ancuta. 2016. "New insights into the heterogeneity of Th17 subsets contributing to HIV-1 persistence during antiretroviral therapy." *Retrovirology* no. 13 (1):59. doi: 10.1186/s12977-016-0293-6.

Wallet, C., M. De Rovere, J. Van Assche, F. Daouad, S. De Wit, V. Gautier, P. W. G. Mallon, A. Marcello, C. Van Lint, O. Rohr, and C. Schwartz. 2019. "Microglial Cells: The Main HIV-1 Reservoir in the Brain." *Front Cell Infect Microbiol* no. 9:362. doi: 10.3389/ fcimb.2019.00362.

Wang, G., N. Zhao, B. Berkhout, and A. T. Das. 2016. "CRISPR-Cas9 Can Inhibit HIV-1 Replication but NHEJ Repair Facilitates Virus Escape." *Mol Ther* no. 24 (3):522-6. doi: 10.1038/mt.2016.24.

Wang, H., S. Hu, X. Chen, H. Shi, C. Chen, L. Sun, and Z. J. Chen. 2017. "cGAS is essential for the antitumor effect of immune checkpoint blockade." *Proc Natl Acad Sci U S A* no. 114 (7):1637-1642. doi: 10.1073/pnas.1621363114.

Wang, N. D., and T. S. Li. 2017. "Factors Associated with the Size of HIV DNA Reservoir." *Chin Med J (Engl)* no. 130 (2):224-230. doi: 10.4103/0366-6999.198009.

Wang, Y., L. Lifshitz, K. Gellatly, C. L. Vinton, K. Busman-Sahay, S. McCauley, P. Vangala, K. Kim, A. Derr, S. Jaiswal, A. Kucukural, P. McDonel, P. W. Hunt, T. Greenough, J. Houghton, M. Somsouk, J. D. Estes, J. M. Brenchley, M. Garber, S. G. Deeks, and J. Luban. 2020. "HIV-1-induced cytokines deplete homeostatic innate lymphoid cells and expand TCF7-dependent memory NK cells." *Nat Immunol* no. 21 (3):274-286. doi: 10.1038/s41590-020-0593-9.

Wang, Z., F. R. Simonetti, R. F. Siliciano, and G. M. Laird. 2018. "Measuring replication competent HIV-1: advances and challenges in defining the latent reservoir." *Retrovirology* no. 15 (1):21. doi: 10.1186/s12977-018-0404-7.

Westhorpe, C. L., A. Maisa, T. Spelman, J. F. Hoy, E. M. Dewar, S. Karapanagiotidis, A. C. Hearps, W. J. Cheng, J. Trevillyan, S. R. Lewin, D. Sviridov, J. H. Elliott, A. Jaworowski, A. M. Dart, and S. M. Crowe. 2014. "Associations between surface markers on blood monocytes and carotid atherosclerosis in HIV-positive individuals." *Immunol Cell Biol* no. 92 (2):133-8. doi: 10.1038/icb.2013.84.

Wiegand, A., J. Spindler, F. F. Hong, W. Shao, J. C. Cyktor, A. R. Cillo, E. K. Halvas, J. M. Coffin, J. W. Mellors, and M. F. Kearney. 2017. "Single-cell analysis of HIV-1 transcriptional activity reveals expression of proviruses in expanded clones during ART." *Proc Natl Acad Sci U S A* no. 114 (18):E3659-E3668. doi: 10.1073/pnas. 1617961114.

Williams, D. W., M. Veenstra, P. J. Gaskill, S. Morgello, T. M. Calderon, and J. W. Berman. 2014. "Monocytes mediate HIV neuropathoge-nesis: mechanisms that contribute to HIV associated neurocognitive disorders." *Curr HIV Res* no. 12 (2):85-96. doi: 10.2174/1570162x12666140526114526.

Williams, R., H. Yao, N. K. Dhillon, and S. J. Buch. 2009. "HIV-1 Tat co-operates with IFN-gamma and TNF-alpha to increase CXCL10 in human astrocytes." *PLoS One* no. 4 (5):e5709. doi: 10.1371/journal. pone.0005709.

Wong, J. K., and S. A. Yukl. 2016. "Tissue reservoirs of HIV." *Curr Opin HIV AIDS* no. 11 (4):362-70. doi: 10.1097/COH. 0000000000000293.

Wong, M. E., A. Jaworowski, and A. C. Hearps. 2019. "The HIV Reservoir in Monocytes and Macrophages." *Front Immunol* no. 10:1435. doi: 10.3389/fimmu.2019.01435.

Wu, J., L. Sun, X. Chen, F. Du, H. Shi, C. Chen, and Z. J. Chen. 2013. "Cyclic GMP-AMP is an endogenous second messenger in innate immune signaling by cytosolic DNA." *Science* no. 339 (6121):826-30. doi: 10.1126/science.1229963.

Wu, L., and V. N. KewalRamani. 2006. "Dendritic-cell interactions with HIV: infection and viral dissemination." *Nat Rev Immunol* no. 6 (11):859-68. doi: 10.1038/nri1960.

Wu, X., M. A. Newbold, Z. Gao, and C. L. Haynes. 2017. "A versatile microfluidic platform for the study of cellular interactions between endothelial cells and neutrophils." *Biochim Biophys Acta Gen Subj* no. 1861 (5 Pt A):1122-1130. doi: 10.1016/j.bbagen.2017.02.012.

Xia, P., Y. Wu, S. Lian, L. Yan, X. Meng, Q. Duan, and G. Zhu. 2021. "Research progress on Toll-like receptor signal transduction and its roles in antimicrobial immune responses." *Appl Microbiol Biotechnol* no. 105 (13):5341-5355. doi: 10.1007/s00253-021-11406-8.

Xiao, Q., D. Guo, and S. Chen. 2019. "Application of CRISPR/Cas9-Based Gene Editing in HIV-1/AIDS Therapy." *Front Cell Infect Microbiol* no. 9:69. doi: 10.3389/fcimb.2019.00069.

Yin, Z., J. Dai, J. Deng, F. Sheikh, M. Natalia, T. Shih, A. Lewis-Antes, S. B. Amrute, U. Garrigues, S. Doyle, R. P. Donnelly, S. V. Kotenko, and P. Fitzgerald-Bocarsly. 2012. "Type III IFNs are produced by and stimulate human plasmacytoid dendritic cells." *J Immunol* no. 189 (6):2735-45. doi: 10.4049/jimmunol.1102038.

Yoder, K. E., and R. Bundschuh. 2016. "Host Double Strand Break Repair Generates HIV-1 Strains Resistant to CRISPR/Cas9." *Sci Rep* no. 6:29530. doi: 10.1038/srep29530.

Zack, J. A., S. J. Arrigo, S. R. Weitsman, A. S. Go, A. Haislip, and I. S. Chen. 1990. "HIV-1 entry into quiescent primary lymphocytes: molecular analysis reveals a labile, latent viral structure." *Cell* no. 61 (2):213-22. doi: 10.1016/0092-8674(90)90802-l.

Zenere, G., O. A. Olwenyi, S. N. Byrareddy, and S. E. Braun. 2019. "Optimizing intracellular signaling domains for CAR NK cells in HIV immunotherapy: a comprehensive review." *Drug Discov Today* no. 24 (4):983-991. doi: 10.1016/j.drudis.2019.02.002.

Zeng, M., A. J. Smith, S. W. Wietgrefe, P. J. Southern, T. W. Schacker, C. S. Reilly, J. D. Estes, G. F. Burton, G. Silvestri, J. D. Lifson, J. V. Carlis, and A. T. Haase. 2011. "Cumulative mechanisms of lymphoid tissue fibrosis and

T cell depletion in HIV-1 and SIV infections." *J Clin Invest* no. 121 (3):998-1008. doi: 10.1172/JCI45157.

Zhang, Y., and J. Zheng. 2020. "Functions of Immune Checkpoint Molecules Beyond Immune Evasion." *Adv Exp Med Biol* no. 1248:201-226. doi: 10.1007/978-981-15-3266-5_9.

Zhen, A., V. Rezek, C. Youn, B. Lam, N. Chang, J. Rick, M. Carrillo, H. Martin, S. Kasparian, P. Syed, N. Rice, D. G. Brooks, and S. G. Kitchen. 2017. "Targeting type I interferon-mediated activation restores immune function in chronic HIV infection." *J Clin Invest* no. 127 (1):260-268. doi: 10.1172/JCI89488.

Zhou, H., S. Chen, M. Wang, and A. Cheng. 2014. "Interferons and their receptors in birds: a comparison of gene structure, phylogenetic analysis, and cross modulation." *Int J Mol Sci* no. 15 (11):21045-68. doi: 10.3390/ijms151121045.

Zicari, S., L. Sessa, N. Cotugno, A. Ruggiero, E. Morrocchi, C. Concato, S. Rocca, P. Zangari, E. C. Manno, and P. Palma. 2019. "Immune Activation, Inflammation, and Non-AIDS Co-Morbidities in HIV-Infected Patients under Long-Term ART." *Viruses* no. 11 (3). doi: 10.3390/v11030200.

Ziegler-Heitbrock, L., P. Ancuta, S. Crowe, M. Dalod, V. Grau, D. N. Hart, P. J. Leenen, Y. J. Liu, G. MacPherson, G. J. Randolph, J. Scherberich, J. Schmitz, K. Shortman, S. Sozzani, H. Strobl, M. Zembala, J. M. Austyn, and M. B. Lutz. 2010. "Nomenclature of monocytes and dendritic cells in blood." *Blood* no. 116 (16):e74-80. doi: 10.1182/blood-2010-02-258558.

Zulu, M. Z., K. K. Naidoo, Z. Mncube, M. Jaggernath, P. J. R. Goulder, T. Ndung'u, M. Altfeld, and C. F. Thobakgale. 2017. "Reduced Expression of Siglec-7, NKG2A, and CD57 on Terminally Differentiated CD56(-)CD16(+) Natural Killer Cell Subset Is Associated with Natural Killer Cell Dysfunction in Chronic HIV-1 Clade C Infection." *AIDS Res Hum Retroviruses* no. 33 (12):1205-1213. doi: 10.1089/AID.2017.0095.

Chapter 2

Vitamin D as a Modulator of the Immune Response during Viral Infections

Georgina I. López-Cortés, Alejandro Magallanes-Puebla and Enrique Ortega*

Departamento de Inmunología, Instituto de Investigaciones Biomédicas, Universidad Nacional Autónoma de México, México City, Mexico

Abstract

The active form of Vitamin D is 1,25-dihydroxyvitamin D3 (1,25$(OH)_2$D3), a metabolite that is involved in the absorption of minerals such as calcium, magnesium, and phosphate. Vitamin D can be obtained either from the diet or by synthesis induced by UVB radiation in the lower layers of the epidermis. Binding of 1,25 $(OH)_2$D3 to its nuclear receptor enhances the expression of proteins required for calcium absorption in the intestine and for maintaining its homeostatic levels in other tissues. Also, the active form of vitamin D3 has been shown to have immunomodulatory properties *in vitro* and *in vivo*. Several cells of the immune system can synthesize the active form of Vitamin D from its precursor, and express the Vitamin D receptor. Thus, vitamin D or its precursor can influence the immune response by up-regulating the production of anti-microbial and other proteins in cells of the innate immune system, while it can also modulate the activity of cells of the adaptive immune response. Adequate levels of Vitamin D are important in viral infections such as that produced by SARS-CoV-2. Low serum

* Corresponding Author's E-mail: ortsoto@biomedicas.unam.mx.

In: The Innate Immune System in Health and Disease
Editor: Jorge Morales-Montor
ISBN: 978-1-68507-507-1

levels of Vitamin D and its metabolites have been found to be associated with increased risk of infection by SARS-CoV-2, as well as severe presentation and mortality in COVID-19. This has led to an interest in the potential prophylactic and/or therapeutic use of Vitamin D in this disease. Here, we summarize the known actions of Vitamin D on cells of the immune system, with emphasis on its effects in viral diseases, and review several studies on the effect of supplementation with Vitamin D in patients with COVID-19.

Keywords: Vitamin D, 1,25-dihydroxyvitamin D3, immunomodulation, viruses, SARS-CoV-2, COVID-19, rinovirus, influenza

Introduction

Vitamin D

Vitamin D (VitD) is a fat soluble vitamin of the group of sercosteroids, which are steroids with a broken ring. They are lipophilic molecules that serve as hormones mediating the homeostasis of calcium (Ca^{2+}), magnesium (Mg^{2+}) and phosphate [$(PO_4)^{3-}$] in the body. Specifically, VitD promotes absorption of Ca^{2+} by the intestine and induces bone resorption to maintain accurate levels of calcium and phosphate in the bloodstream. VitD activation is induced by the parathyroid hormone, which is then regulated by these same ions' levels. The most studied function of VitD is that related to the metabolism of calcium. However, VitD has pleiotropic effects that impact on diverse functions, including proliferation, differentiation, migration, apoptosis, and immune functions [1].

In humans, one of the VitD precursors can be synthesized in the lower layers of the skin by a photochemical reaction induced by UVB radiation in the epidermis, where the 7-dehydrocholesterol is converted into cholecalciferol. UV radiation induces the reaction in a variety of animals and fungi. That is why, apart from the endogenous sources, other VitD precursors can be obtained from the diet (i.e., exogenous ergocalciferol or cholecalciferol). Any one of these precursors must be further hydroxylated in the body to become active. First, an enzyme vitamin D 25-hydroxylase in the hepatocytes converts either cholecalciferol into calcifediol (25-hydroxycholecalciferol), or ergocalciferol into 25-hydroxyergocalciferol. These two hydroxylated metabolites are considered as 25-hydroxyvitamin D (25(OH)D) which can be measured in the blood because it is secreted into the circulation, where it binds to the vitamin D-binding protein. Finally, 25(OH)D is transported to the proximal tubules of the kidney where the 1-α hydroxylase adds another hydroxyl group in the carbon at position 1, to generate calcitriol also known as 1, 25 dihydroxy vitamin D (1, 25 $(OH)_2D$) [1], which is the active form of VitD (Figure 1). Apart from the kidney, this last hydroxylation may occur in different cells such as in those of the placenta and colon, keratinocytes, osteoblasts, and some cells of the immune system. The 1,25 $(OH)_2D$ is then transported in the bloodstream associated with the vitamin D-binding protein to reach the intestine, bone, kidneys, and immune cells. The activity of the 1-α hydroxylase is induced by the parathyroid hormone-insulin-like growth factor type I and calcitonin, while it is negatively regulated by the serum levels of 25(OH)D, calcium, and phosphorus, the inhibition of the parathyroid hormone, and the fibroblast growth factor 23 [1]. Besides that, both calcifediol and calcitriol are susceptible to another hydroxylation at Carbon 24 by the vitamin D3 24-hydroxylase. The resulting molecules, secalciferol and calcitetrol respectively, have reduced activity.

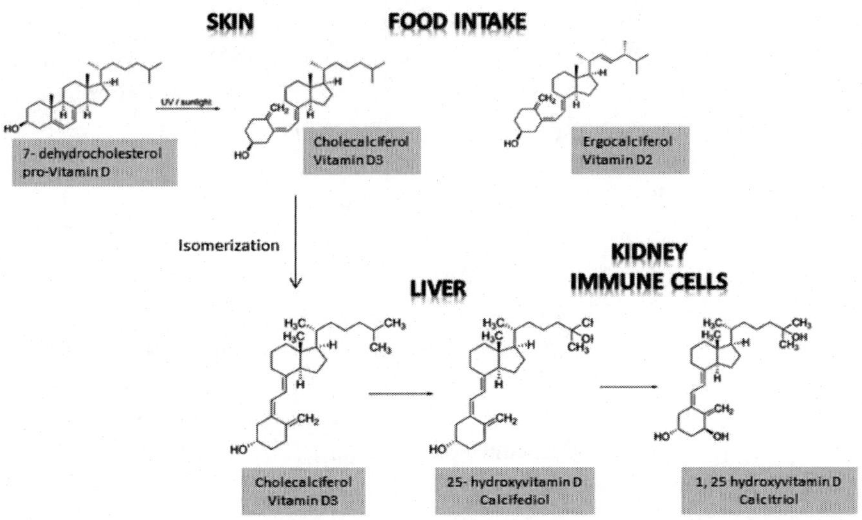

Figure 1. Synthesis of the active form of vitamin D. In the skin, UV light catalyzes the reaction that produces cholecalciferol or vitamin D3. This is then hydroxylated in the liver, and the product cholecalciferol is again hydroxylated by a different enzyme producing calcitriol, the active form of Vitamin D. The last reaction may occur in kidney cells and also in some immune cells. Ergocalciferol from the diet may be converted to cholecalciferol and enter this pathway; thus the synthesis of the active form of Vitamin D may start from endogenous cholecalciferol or by the intake of it or of ergocalciferol.

The active form of VitD, calcitriol or 1,25 $(OH)_2D$, regulates gene expression by ligation with its nuclear receptor, Vitamin D Receptor (VDR). After ligation, VDR forms a heterodimer with the Retinoid X Receptor (RXR). Next, the heterodimer binds to the Vitamin D Response Elements (VDRE) which causes the recruitment of histone acetylation enzymes that enable transcription and gene expression. Among the up-regulated proteins, TRPV6 and calbindin are involved in calcium absorption in the intestine. In osteoclasts, VDR induces proliferation to accomplish bone resorption. The circulating levels of calcium and phosphorous are modulated by VitD which in turn is regulated by the parathyroid hormone. Another VitD target gene is a tyrosine hydrolase which in the adrenal medullar cells modulates the synthesis of neurotrophic factors, nitric oxide synthase, and glutathione.

Blood levels of 25(OH) D fluctuate in a person depending on the diet, sun exposure, and age. The lowest levels of VitD can be detected in a person during winter because of the lower exposure to solar UVB radiation. Also, the decrease of 25(OH) D blood levels in elderly people is related to diminished intake of the VitD precursors through the diet and to lower exposure to the sun. According to the National Institutes of Health, the sufficient level of 25(OH)D is above 40-50 ng/mL and approximately 1,000 to 500 times lower for 1, 25 $(OH)_2D$ [2]. VitD deficiency causes low levels of calcium in serum, formation of calcium-phosphates complexes, nephron damage, and resistance to the parathyroid hormone.

Vitamin D in the Immune System

VitD effects on the immune system are wide because almost all immune cells express the VDR. The expression levels of VDR are very dynamic, for example, T cells increase the receptor's expression for up to 48 hours after activation, while monocytes diminish the expression while differentiating to macrophages or dendritic cells [3]. The active 1, 25 $(OH)_2D$ may be synthesized in monocytes and macrophages [4], and after being secreted it can act like a cytokine that modulates other immune cells. The activity of the 1α-hydroxylase is stimulated by inflammatory cytokines (i.e., IFN-γ, IL-1, IL-2, and TNF-α) [5] and by the activation of inflammatory pathways (i.e., TLRs, COX2, and STAT-1α) [3]. Interestingly, in turn, 1, 25 $(OH)_2D$ negatively regulates the TLRs'-induced inflammatory pathways by stimulating the Suppressor of Cytokine Signaling 1 (SOCS-1) in monocytes [6]. VitD also induces the release of antimicrobial peptides such as cathelicidin, β defensin

2, hCAP18, and LL37, and up-regulates the expression of glutathione reductase and glutamate-cysteine ligase, which are necessary for the oxidant stress [3] in monocytes, Natural Killer cells, and neutrophils. In dendritic cells, 1, 25 $(OH)_2D$ decreases expression of Major Histocompatibility Complex II (MHC II), the co-stimulatory molecules CD80 and CD86, and the adhesion molecule CD54; and increases expression of CCR5, DEC205, CD40 [7], and Programmed Death Ligand- 1 (PD-L1) [8]. VitD also alters the polarization of T cells, because it inhibits the production of IL-12 and IL-23 while it induces CTLA-4, IL-10 and TNF-α, favoring the polarization of activated T CD4 cells to a Th2 or Treg profile. VDR ligation in B lymphocytes promotes apoptosis and decreases their differentiation to plasma cells, and increases the production of IL-10 [3].

A research report shows that VitD treatment down-regulates production of pro-inflammatory cytokines after LPS stimulation of human monocytes and murine bone marrow-derived macrophages (BMDM). The mechanism involved is that VitD induces the expression of MAPK-1 phosphatase 1 (MKP-1), which dephosphorylates MAPK p38. As a consequence, NF-KB is not activated and the synthesis of pro-inflammatory cytokines (i.e., Il-6 and TNF-α) diminishes. The authors demonstrated *in vitro* that inhibition of pro-inflammatory cytokines is achieved at sufficient levels of serum VitD (30 ng/mL of $25(OH)D_3$) and the maximum inhibition was achieved with 50 ng/mL of $25(OH)D_3$ or with 0.4 ng/mL of 1, 25 $(OH)_2D_3$ [4].

Role of Vitamin D in the Immune Response during Selected Viral Infections of the Respiratory Tract

As viral particles express pathogen-associated molecular patterns, they can be recognized by innate response cell's pattern recognition receptors, like Retinoic acid-Inducible Gene-I protein (RIG-I) which can recognize viral ssRNA, or Toll-like receptors (TLRs) where TLR-3, TLR-7, and TLR-8 recognize viral products inside the cell. Innate response cells like macrophages and dendritic cells, can recognize viral PAMPs and produce a pro-inflammatory response mediated by type I and type III IFNs as a result of this recognition [9]. In viral infections in the lower respiratory tract, and as a consequence of cell death induced by the virus and by the immune response to PAMPs or DAMPs, lung tissue is damaged due to the inflammatory response which can produce the acute respiratory distress syndrome, in which excessive lung tissue damage impairs lung function [10].

One of the most important points to consider when the infection has been established is the immune response against a virus, as this response causes inflammation, which is beneficial to control the infection; however, the inflammatory response can also cause damage to respiratory tissue which results in respiratory complications for the infected person [11]. For this reason, support treatments must be directed to control an excessive inflammatory response. To tackle this issue, the use of immunoregulatory molecules as support treatment has been proposed.

As mentioned before, VitD has an important role in modulating the immune response [12]. Specifically during respiratory infections, VitD has been shown to be important for maintaining tight junctions between cells, inducing antimicrobial peptides as cathelicidin and defensins, and down-regulating the synthesis of pro-inflammatory cytokines. Many reports have highlighted the observation of an increased susceptibility to respiratory infections when VitD levels are insufficient. Unfortunately, VitD deficiency is common around the world, and the rate increases among people with dark skin and of advanced age. Here we present some information gathered from reports and analysis regarding respiratory infections and VitD serum levels.

Influenza

One of the respiratory diseases that constitute a public health problem is that caused by Influenza virus. Because of its high mutation rates, control of this disease has become a challenge for health authorities, and the potential risk of a pandemic caused by these virus is still present [13]. Furthermore, this virus was the cause of the most deadly pandemics registered in the past century [14]. Being part of the family of *Orthomyxoviridae*, Influenza virus is an enveloped negative-sense single-stranded RNA virus. The cause of its high variability is due to co-infection of a cell with multiple virus strains that allows gene re-assortment, resulting in the generation of novel subtypes that could give rise to new outbreaks [15]. Viral RNA contains information for the synthesis of the envelope glycoproteins haemagglutinin (HG) and neuraminidase (NA). Located at the virus's surface, HG and NA are the most antigenically variable proteins [14].

Regarding the VitD role in Influenza infection, since the last century it was proposed that some "seasonal stimulus" protected people from the disease [16]. This stimulus was correlated with seasonal sun exposure which is important for VitD synthesis, however, back then the knowledge of the role of

VitD was scarce and the VitD influence on immune response wasn't clear [17]. Nevertheless, studies demonstrated that there was a correlation between deaths caused by influenza and seasonal variations of VitD deficiency [18]. Based on new knowledge about VitD mechanisms on immune system cells, it was proposed that VitD improves the response of innate immune cells, which can then eliminate the virus efficiently.

Clinical studies showed mixed results. Zhou et al, reported that seasonal influenza was prevented in children who had received a high dose of VitD, as symptoms were less severe, and viral loads were lower in this group compared with children who received a low dose of VitD [19]. Similar results were observed in another study conducted in Japan, where VitD supplementation prevented seasonal influenza A in schoolchildren measuring the same parameters [20]. As children are the most important transmission vector of seasonal influenza, these studies suggest that VitD supplementation may be a way to ameliorate transmission of the disease [21]. However, studies performed with healthy children show that the immune response doesn't improve significantly, as after VitD supplementation only the production of IL-6 decreased in comparison with the control group [22]. Another study reported that eventhough VitD supplementation helped in the reduction of respiratory infections in healthy children, the viral load in the children that got infected was the same in the supplemented and the placebo group [23]. Children possess an active immune system and their normal activities include constant sun exposure, for this reason, VitD deficiency was rare in those trials. Other studies focused on groups where VitD levels could be lower. In the case of older people, VitD deficiency can be associated with various geriatric syndromes [24]. Studies show that VitD supplementation induced the production of TGF-β, an anti-inflammatory cytokine, which can induce a tolerogenic response [25]. This result can be correlated with another study, where it was demonstrated that in VitD-deficient elderly adults, production of pro-inflammatory cytokines like IL-6, was increased [26]. It's important to consider that excessive inflammation can induce severe damage in tissues and in this case, a low VitD concentration is associated with a poor prognosis in elder people [27].

At the light of these mixed results, we can conclude that VitD supplementation can help in maintaining healthy levels of pro- and anti-inflammatory response that help to control the disease, although the mechanisms involved are far from clear. VitD supplementation doesn't seem to improve the effectiveness of vaccination. Certainly, in the case of influenza,

VitD supplementation is important to cover VitD deficiencies which can be seen mostly in elder people.

Rhinovirus

Human rhinoviruses (HRV) are members of the family *Picornaviridae,* whose genome is a positive-sense, single-stranded RNA. HRV are one of the most common causes of cold-like diseases; hence, it represents a serious public health problem that costs billions of dollars annually in medical visits. They were discovered in the 1950s in an effort to isolate the etiologic agent of the common flu [28]. Commonly related to upper respiratory tract illness like the common flu, acute otitis media, or rhinosinusitis, HRV is linked with the exacerbation of various diseases including chronic pulmonary disease, asthma development, or cystic fibrosis [29]. In HRV infection, the cell surface receptor Intercellular Adhesion Molecule-1 (ICAM-1) plays an important role in virus infection since it serves as the receptor mediating entry of the virus into epithelial cells [30, 31]. The virus can be recognized via TLR-2 and intrecellularly via TLR-7 and TLR-8, inducing the expression of pro-inflammatory cytokines like RANTES, IP-10, IL-6, and IL-8 [32]. Because of the existence of more than 100 HRV serotypes with high sequence variability in the antigenic sites, vaccine development is not an optimal option and the treatment is primarily symptomatic. Nutrient supplementation has been considered to aid in the treatment, such as the use of Vitamin C, which effectiveness hasn´t been conclusively demostrated [33].

In recent years, VitD supplementation has also been considered as a supportive treatment to prevent infection, based on results from *in vitro* studies. Telcian et al, described that VitD decreased HRV replication and release in bronchial epithelial cells, and also increased rhinovirus-induced interferon genes and cathelicidin [34]. Another study showed that physiological concentrations of VitD induced resistance to infection against the virus, as the level of viral RNA inside the cells was reduced significantly; another interesting finding is that the cell surface receptor (ICAM-1) expression was attenuated in cells treated with VitD. As mentioned before, this receptor mediates virus internalization [35]. Another study performed in primary bronchial cells obtained from children with cystic fibrosis, shows that VitD supplementation represses viral replication in a dose-dependent manner. Interestingly, this study reports that VitD supplementation doesn't reduce proinflammatory cytokine secretion, which is a characteristic of VitD treatments [36].

In vitro studies show that VitD supplementation can enhance the immune response against rhinovirus; however, more clinical trials are needed to confirm if VitD can prevent the infection.

Respiratory Syncytial Virus

Respiratory syncytial virus (RSV) is a single-stranded RNA virus that belongs to the family *Paramyxoviridae* and is the most frequent cause of acute lower respiratory tract infections in infants [37]. One particular characteristic is its capacity to form syncytia derived from the fusion of infected cells, which led to the formation of multinucleated cells. For its treatment, symptomatic care remains the main therapy, including gentle suctioning of nasal secretions or oxygen and respiratory support in severe cases [38]. Humoral immunity plays an important role in infection control as trans-placental transfer of maternal antibodies into newborn infants takes an important role in protection; after that, antibodies are produced in response to natural infection [39]. Also, cellular immunity participates in control of the infection via cytotoxic cells, as there is an increase of $CD8^+$ T cells; the cytotoxic response is accompanied by an inflammatory reaction that helps in controlling the infection but can also produce tissue damage [40].

Regarding VitD role in RVS infection, *in vitro* studies shows that VitD can inhibit NF-kB translocation into the nucleus of airway epithelial cells infected with RSV, which in turn diminishes the inflammatory response without an increase in viral replication. The authors propose that the induction of antimicrobial peptides, like cathelicidin, induced by VitD could control the infection [41]. In clinical studies, it has been demonstrated that there is a correlation between VitD deficiency and the increased risk to develop an RSV infection. A cohort study that measured serum VitD levels in neonates, revealed that infants whose VitD concentration was lower than 50 ng/mL had a six-fold increased risk to develop lower respiratory tract infection caused by RSV. Normal VitD concentrations in neonates were associated with VitD supplementation in pregnancy and, for this reason, authors suggested VitD supplementation in pregnancy to diminish the incidence of this disease in infants [42].

Interestingly, another cohort study indicated that VitD levels are unrelated to RSV disease severity since there was no difference in measured parameters like duration of hospitalization, oxygen saturation, and bronchiolitis severity score in VitD sufficient vs deficient children [43]. However, another recent study shows that children with severe deficiency of VitD (serum concentrations ≤29 ng/mL) can develop a life-threatening disease [44].

Role of Vitamin D in the Immune Response during SARS-CoV-2 Infection

As mentioned before, VitD plays an important role in modulating the inflammatory response towards viral infections. That is why since the beginning of the pandemic of COVID-19, there was great interest in determining if there was an association between the severity of the disease caused by infection with SARS-CoV-2 and the serum levels of VitD, specially because during the first wave of the virus spread, in the northern hemisphere there was a seasonal deficiency of VitD [45]. The first observational studies correlating the serum levels of VitD with the severity of the symptoms and the outcome of the patients, found a positive correlation between sufficient levels and better outcome. VitD supplementation has been used widely in some countries as there is evidence supporting the benefits of VitD in viral infections (e.g., rotavirus, influenza, dengue, HIV). Particularly, the Centers for Disease Control and Prevention of the U.S.A. proposed the use of VitD as a supplement since March 2020 [46]. Several clinical trials have been running to confirm the advantage of incorporating VitD as part of COVID-19 treatment. It has been reported that having serum levels of 25(OH)D over 50 nmol/L is important in preventing the severe symptoms of COVID-19 [2].

At the beginning of the pandemic, physicians started to measure serum VitD since there is a seasonal VitD deficiency in North East England associated with the winter. They noticed that 66.4% of the hospitalized patients had VitD insufficiency (25- 50 nmol/L), 37.3% were deficient (< 25 nmol/L) and 21.6% were severely deficient (< 15 nmol/L). Besides, 81% of the patients at the Intensive Care Unit (ICU) had insufficiency levels. Another retrospective study was focused on determining if serum levels of VitD influenced COVID-19 progress. The VitD serum levels of twenty COVID-19 patients were analyzed, among them, 13 were admitted to ICU, and 11 of them (84.6%) were found to have VitD deficiency. Of the 7 patients that remained in floor, 4 (57.1%) had VitD insufficiency [47]. From this data, it was concluded that the worse symptoms of COVID-19 occurred in people that presented VitD deficiency levels previous to the infection. Another trial reported the relationships between COVID-19 mortality and other variables, including VitD serum levels. Here the correlation between these variables was again confirmed; in the deficient patients, sufficient serum levels were recovered by supplementing with cholecalciferol and, as a consequence, the mortality rate was reduced, even when patients had high baseline levels of C-reactive protein and creatinine [48].

A pilot randomized study looked for optimizing the treatment for hospitalized COVID-19 patients by administrating calcifediol. All patients were treated with the standard care therapy (hydroxychloroquine and azithromycin) but additionally, some of them had calcifediol doses until discharge or ICU admission. The results showed that only 2% of the calcifediol group required ICU, while 50% of the non-calcifediol group required it, and the calcifediol group had cero deaths against 7.6% of mortality rate in the control group [49]. Similarly, another study of COVID-19 patients older than 50 years in the General Hospital of Singapore concluded that administration of a combination of Vitamin D, Magnesium, and Vitamin B12 prevented patients to require oxygen therapy and, most importantly, ingress into the ICU. During an average of 5 days, patients in a group received a single daily dose of VitD 1,000 IU, magnesium 150 mg, and Vitamin B12 500μg. Only 17.6% of the patients who received the treatment required oxygen therapy while 61.5% of the patients of the control group needed it [50]. After these preliminary results, the General Hospital of Singapore continued to orally supply all patients with these three components. Both reports showed that supplementing with VitD reduced the admission to ICU and mortality for hospitalized patients receiving standard treatment.

The high frequency of VitD deficiency found in hospitalized patients raised the question whether patients that showed a less severe form of the disease had better serum levels of the vitamin, and if supplementation during the course of the disease could be beneficial. To answer this, serum levels of asymptomatic and mildly symptomatic patients of COVID-19 were tested during the course of infection. They were placed in two groups: one which received cholecalciferol and a control group who did not. All of them were VitD deficient at the beginning of the study and ingested or not 60,000 IU daily doses of cholecalciferol until the serum levels rose to 50 ng/mL. Both the clearance of SARS-CoV-2 and the inflammatory markers ceased by day 21 for 62.5% of the VitD group compared to 20.8% of the placebo group [51]. The conclusion was that sufficient VitD availability could accelerate the clearance of virus in the body even in less severe cases of COVID-19.

The COVIT-TRIAL is a multicenter, randomized study aimed to determine whether giving a single high dose (2x 200,000 IU) or a standard dose (50,000 IU) of supplementary VitD could ameliorate the outcome of COVID-19 patients over 65 years old. The design of the trial would monitor the levels of the vitamin, creatinine, calcium, albumin, and the World Health Organization Ordinal Scale for Clinical Improvement (OSCI) scale for COVID-19 at the day on inclusion, and at day 14 and day 28 [52]. The trial pretends to be more conclusive by correlating the VitD and the OSCI scale level to describe the patients' progress. This trial is currently underway. However, another multicenter and randomized trial conducted in two hospitals in Sao Paulo, Brazil found that a single high dose (200,000 IU) of VitD is not sufficient to reduce the hospitalization days of COVID19 patients. Although this dose raised the serum levels of VitD, the treatment did not resulted in a better patients' outcome and the mortality was not significantly different [53]. The possible explanation is that high doses of VitD suppress the proliferation of myeloid precursors, shutting down an efficient innate immune response. Thus, it seems that to improve patient's outcome it is important not only to increase the availability of circulating VitD by a single high dose, but maintaining it by frequent administration of the vitamin.

Despite differences in design, observational studies and clinical trials have been useful for documenting the beneficial effect of VitD supplementation in acute respiratory infections. One meta-analysis included 11 controlled randomized trials for preventing respiratory tract infections with VitD supplementation. This analysis showed that VitD supplementation resulted in a 36% reduction in the risk to get a respiratory infection as compared to not supplementing [54]. Another group performed an individual participant data meta-analysis from 25 eligible randomized controlled trials. They showed that supplementing with VitD analogue reduced 12% the risk of having an acute respiratory infection. Interestingly, eventhough all patients ended the studies with sufficient levels of VitD, the ones with baselines VitD deficiency benefited more than the ones that had sufficient serum levels at the beginning. Also, it was confirmed that the better prophylaxis treatment is the daily dose, rather than the high bolus ingestion spaced by months [55]. Another meta-analysis done in 2012 included trials giving VitD as monoprofilaxis to prevent respiratory tract infections in children and adults. Out of 5 trials that fulfilled the inclusive criteria, it was concluded that VitD supplementation reduced 8.4% the risk of developing an acute respiratory infection [56]. An interesting review of a large number of meta-analysis of studies on non-skeletal effects of vitamin D supplementation found that beneficial effects of VitD supplementation were reported in 3 out of 7 meta-analyses for respiratory tract infections and 8 out of 12 meta-analyses related to mortality [57].

The question remains whether VitD supplementation is recommended as a prophylactic treatment to reduce the risk of severe symptoms when contracting a viral infection. Some studies have shown that VitD supplementation could lead to a better immune response towards viral infections, but other studies showed contradictory results. At present, many clinicians think that there is a need for more evidence before a definitive conclusion can be made.

Conclusion

Several decades of basic research and clinical studies have strongly supported the idea that maintaining sufficient levels of VitD is necessary for the correct functioning of the immune system. VitD has pleiotropic effects on the immune function, inducing the expression of antimicrobial peptides and anti-inflammatory cytokines, as well as inhibiting inflammatory cytokines resulting in the reduction of the inflammation. VitD can enhance cellular immunity by inducing proliferation and differentiation of T cells, but reduce inflammation by inducing anti-inflammatory cytokines. Therefore, VitD supplementation is proposed as a complementary treatment for multiple conditions where inflammation could result in worsening of the disease. According to the results of different correlational studies and clinical trials, supplementation with VitD *per se* does not reduce significantly the risk of contracting respiratory infections. Nevertheless, it seems clear that VitD deficiency is a risk factor for worse outcomes in viral infections of the respiratory tract, mainly by favoring exacerbated inflammatory responses. Further research is necessary to dissect more precisely the mechanisms involved, which might also help in understanding the contradictory results found in some studies. Also, it would help to better define the conditions and circumstances (relating to clinical characteristics of the individual patient and even possibly to its genetic background, the specific pathogen involved, etc) where VitD supplementation would have optimal effects. Eventhough VitD supplementation should not be used as the only treatment for combating viral diseases, there is enough data to consider that at least sufficient levels are key for the optimal functioning of the immune system and therefore for a better outcome for patients of acute respiratory diseases, including COVID-19.

Acknowledgments

Georgina I. Lopez-Cortes is a doctoral student from the Program Maestría y Doctorado en Ciencias Bioquímicas, Universidad Autónoma de México (UNAM), and gratefully thanks the fellowship (699886) received from CONACyT to pursue her PhD. Work in the authors´ laboratory is supported by grants from CONACYT (Grant A1-S-9488) and Direccion General de Asuntos del Personal Académico, (DGAPA-PAPIIT), UNAM (Grant IN208320).

References

[1] Olmos-Ortiz, A., Avila, E., Durand-Carbajal, M. & Díaz, L. Regulation of calcitriol biosynthesis and activity: Focus on gestational vitamin D deficiency and adverse pregnancy outcomes. *Nutrients* 7, 443–480 (2015).

[2] Calder, P., Carr, A., Gombart, A. & Eggersdorfer, M. Optimal Nutritional Status for a Well-Functioning Immune System Is an Important Factor to Protect against Viral Infections. *Nutrients* 12, 1181 (2020).

[3] Martens, P. J., Gysemans, C., Verstuyf, A. & Mathieu, C. Vitamin d's effect on immune function. *Nutrients* vol. 12 (2020).

[4] Zhang, Y. et al. Vitamin D Inhibits Monocyte/Macrophage Proinflammatory Cytokine Production by Targeting MAPK Phosphatase-1. *J. Immunol.* 188, 2127–2135 (2012).

[5] Gyetko, M. R., Chen Hsaing Hsu, Wilkinson, C. C., Patel, S. & Young, E. Monocyte 1α-hydroxylase regulation: Induction by inflammatory cytokines and suppression by dexamethasone and uremia toxin. *J. Leukoc. Biol.* 54, 17–22 (1993).

[6] Chen, Y. et al. 1,25-Dihydroxyvitamin D Promotes Negative Feedback Regulation of TLR Signaling via Targeting MicroRNA-155–SOCS1 in Macrophages. *J. Immunol.* 190, 3687–3695 (2013).

[7] Ferreira, G. B. et al. 1,25-Dihydroxyvitamin D 3 alters murine dendritic cell behaviour in vitro and in vivo. *Diabetes. Metab. Res. Rev.* 27, 933–941 (2011).

[8] Unger, W. W. J., Laban, S., Kleijwegt, F. S., Van Der Slik, A. R. & Roep, B. O. Induction of Treg by monocyte-derived DC modulated by vitamin D3 or dexamethasone: Differential role for PD-L1. *Eur. J. Immunol.* 39, 3147–3159 (2009).

[9] Chen, X. et al. Host immune response to influenza A virus infection. *Front. Immunol.* 320 (2018) doi:10.3389/fimmu.2018.00320.

[10] Kalil, A. C. & Thomas, P. G. Influenza virus-related critical illness: Pathophysiology and epidemiology. *Crit. Care* 23, 1–7 (2019).

[11] Simmons, C. & Farrar, J. Insights into Inflammation and Influenza. *N. Engl. J. Med.* 359, 1621–1623 (2008).

[12] Hui, D. S., Lee, N., Chan, P. K. & Beigel, J. H. The role of adjuvant immunomodulatory agents for treatment of severe influenza. *Antiviral Research* vol. 150 202–216 (2018).

[13] Smith, B. L. & Wilke, C. O. A new twist in measuring mutation rates. *Elife* 6, (2017).

[14] Hutchinson, E. C. Influenza Virus. *Trends in microbiology* vol. 26 809–810 (2018).

[15] Krammer, F. et al. Influenza. *Nature Reviews Disease Primers* vol. 4 1–21 (2018).

[16] Hope-Simpson, R. E. The method of transmission of epidemic influenza: Further evidence from archival mortality data. *J. Hyg. (Lond).* 96, 353–375 (1986).

[17] Cannell, J. J. et al. Epidemic influenza and vitamin D. *Epidemiol. Infect.* 134, 1129–1140 (2006).

[18] Moan, J. E., Dahlback, A., Ma, L. & Juzeniene, A. Influenza, solar radiation and vitamin D. *Dermatoendocrinol.* 1, 308–310 (2009).

[19] Zhou, J. et al. Preventive effects of Vitamin D on seasonal influenza a in infants: A multicenter, randomized, open, controlled clinical trial. *Pediatr. Infect. Dis. J.* 37, 749–754 (2018).

[20] Urashima, M. et al. Randomized trial of vitamin D supplementation to prevent seasonal influenza A in schoolchildren. *Am. J. Clin. Nutr.* 91, 1255–1260 (2010).

[21] Endo, A., Uchida, M., Kucharski, A. J. & Funk, S. Fine-scale family structure shapes influenza transmission risk in households: Insights from primary schools in Matsumoto city, 2014/15. *PLoS Comput. Biol.* 15, e1007589 (2019).

[22] Brett, N. R. et al. Vitamin D status and immune health outcomes in a cross-sectional study and a randomized trial of healthy young children. *Nutrients* 10, (2018).

[23] Loeb, M. et al. Effect of Vitamin D supplementation to reduce respiratory infections in children and adolescents in Vietnam: A randomized controlled trial. *Influenza Other Respi. Viruses* 13, 176–183 (2019).

[24] Gutiérrez-Robledo, L. M. et al. Association of low serum 25-hydroxyvitamin D levels with the frailty syndrome in Mexican community-dwelling elderly. *Aging Male* 19, 58–63 (2016).

[25] Goncalves-Mendes, N. et al. Impact of Vitamin D supplementation on influenza vaccine response and immune functions in deficient elderly

persons: A randomized placebo-controlled trial. *Front. Immunol.* 10, (2019).

[26] Laird, E. et al. Vitamin D deficiency is associated with inflammation in older irish adults. *J. Clin. Endocrinol. Metab.* 99, 1807–1815 (2014).

[27] Lang, P. O. & Samaras, D. Aging adults and seasonal influenza: Does the vitamin D status (H)arm the body? *Journal of Aging Research* vol. 2012 (2012).

[28] Bertino, J. S. Cost burden of viral respiratory infections: Issues for formulary decision makers. *Am. J. Med.* 112, 42–49 (2002).

[29] Jacobs, S. E., Lamson, D. M., Kirsten, S. & Walsh, T. J. Human rhinoviruses. *Clin. Microbiol. Rev.* 26, 135–162 (2013).

[30] Greve, J. M. et al. The major human rhinovirus receptor is ICAM-1. *Cell* 56, 839–847 (1989).

[31] Traub, S. et al. An Anti-Human ICAM-1 Antibody Inhibits Rhinovirus-Induced Exacerbations of Lung Inflammation. *PLoS Pathog.* 9, (2013).

[32] Triantafilou, K. et al. Human rhinovirus recognition in non-immune cells is mediated by toll-like receptors and MDA-5, which trigger a synergetic pro-inflammatory immune response. *Virulence* 2, 22–29 (2011).

[33] Hemilä, H. & Chalker, E. Vitamin C for preventing and treating the common cold. *Cochrane Database Syst. Rev.* 2013, (2013).

[34] Telcian, A. G. et al. Vitamin D increases the antiviral activity of bronchial epithelial cells in vitro. *Antiviral Res.* 137, 93–101 (2017).

[35] Greiller, C. L. et al. Vitamin D attenuates rhinovirus-induced expression of intercellular adhesion molecule-1 (ICAM-1) and platelet-activating factor receptor (PAFR) in respiratory epithelial cells. *J. Steroid Biochem. Mol. Biol.* 187, 152–159 (2019).

[36] Schögler, A. et al. Vitamin D represses rhinovirus replication in cystic fibrosis cells by inducing LL-37. *Eur. Respir. J.* 47, 520–530 (2016).

[37] Kimura, D. et al. Pulmonary hypertension during respiratory syncytial virus bronchiolitis: A risk factor for severity of illness. *Cardiol. Young* 29, 615–619 (2019).

[38] Mammas, I. N. et al. Update on current views and advances on RSV infection (Review). *International Journal of Molecular Medicine* vol. 46 509–520 (2020).

[39] McIntosh, K., Masters, H. B., Orr, I., Kwa Chao, R. & Barkin, R. M. The immunologic response to infection with respiratory syncytial virus in infants. *J. Infect. Dis.* 138, 24–32 (1978).

[40] Russell, C. D., Unger, S. A., Walton, M. & Schwarze, J. The human immune response to respiratory syncytial virus infection. *Clin. Microbiol. Rev.* 30, 481–502 (2017).

[41] Hansdottir, S. et al. Vitamin D Decreases Respiratory Syncytial Virus Induction of NF-κB–Linked Chemokines and Cytokines in Airway Epithelium While Maintaining the Antiviral State. *J. Immunol.* 184, 965–974 (2010).

[42] Belderbos, M. E. et al. Cord blood vitamin D deficiency is associated with respiratory syncytial virus bronchiolitis. *Pediatrics* 127, (2011).

[43] Beigelman, A. et al. Vitamin D levels are unrelated to the severity of respiratory syncytial virus bronchiolitis among hospitalized infants. *J. Pediatric Infect. Dis. Soc.* 4, 182–188 (2015).

[44] Martin Ferolla, F. et al. Low Serum Vitamin D Levels Are Related to Life-Threatening Respiratory Syncytial Virus Infection in Previously Healthy Infants. *Open Forum Infect. Dis.* 6, S801–S801 (2019).

[45] Panagiotou, G. et al. Low serum 25-hydroxyvitamin D (25[OH]D) levels in patients hospitalized with COVID-19 are associated with greater disease severity. *Clinical Endocrinology* vol. 93 508–511 (2020).

[46] Grant, W. B. et al. Evidence that vitamin d supplementation could reduce risk of influenza and covid-19 infections and deaths. *Nutrients* vol. 12 988 (2020).

[47] Lau, F. H. et al. Vitamin D insufficiency is prevalent in severe COVID-19. *medRxiv* (2020) doi:10.1101/2020.04.24.20075838.

[48] Ling, S. F. et al. High-Dose Cholecalciferol Booster Therapy is Associated with a Reduced Risk of Mortality in Patients with COVID-19: A Cross-Sectional Multi-Centre Observational Study. *Nutrients* 12, 3799 (2020).

[49] Entrenas, M., Manuel, L., Costa, E., Bouillon, R. & Francisco, J. Effect of calcifediol treatment and best available therapy versus best available therapy on intensive care unit admission and mortality among patients hospitalized for COVID-19 : A pilot randomized clinical study. *J. Steroid Biochem. Mol. Biol.* (2020) doi:10.1016/j.jsbmb.2020.105751.

[50] Tan, C. W. et al. A cohort study to evaluate the effect of combination Vitamin D, Magnesium and Vitamin B12 (DMB) on progression to severe outcome in older COVID-19 patients. *medRxiv* 12, (2020).

[51] Rastogi, A. et al. Short term, high-dose vitamin D supplementation for COVID-19 disease: A randomised, placebo-controlled, study (SHADE study). *Postgrad. Med. J.* 1–4 (2020) doi:10.1136/postgradmedj-2020-139065.

[52] Annweiler, C. et al. COvid-19 and high-dose VITamin D supplementation TRIAL in high-risk older patients (COVIT-TRIAL): study protocol for a randomized controlled trial. *Trials* 21, 1–10 (2020).

[53] Murai, I. H. et al. *Effect of a Single High Dose of Vitamin D3 on Hospital Length of Stay in Patients With Moderate to Severe COVID-19: A Randomized Clinical Trial.* 903, 1–8 (2021).

[54] Bergman, P., Lindh, Å. U., Björkhem-Bergman, L. & Lindh, J. D. Vitamin D and Respiratory Tract Infections: A Systematic Review and Meta-Analysis of Randomized Controlled Trials. *PLoS One* 8, (2013).
[55] Martineau, A. R. et al. Vitamin D supplementation to prevent acute respiratory tract infections: Systematic review and meta-analysis of individual participant data. *BMJ* 356, (2017).
[56] Charan, J., Goyal, J. P., Saxena, D. & Yadav, P. Vitamin D for prevention of respiratory tract infections: A systematic review and meta-analysis. *J. Pharmacol. Pharmacother.* 3, 300–303 (2012).
[57] Rejnmark, L. et al. Non-skeletal health effects of Vitamin D supplementation: A systematic review on findings from meta-Analyses summarizing trial data. *PLoS ONE* vol. 12 (2017).

Chapter 3

Immunometabolic Interactions in Type 2 Diabetes and SARS-CoV-2 Infection (COVID-19)

Rebeca Viurcos-Sanabria[1,2] **and Galileo Escobedo**[1,*]
[1]Laboratory of Immunometabolism, Research Division,
General Hospital of Mexico "Dr. Eduardo Liceaga," Mexico City, Mexico
[2]PECEM, Facultad de Medicina, Universidad Nacional Autónoma de México,
Coyoacán, Mexico City, Mexico

Abstract

Coronavirus disease 2019 (COVID-19) is caused by the severe acute respiratory syndrome coronavirus-2 (SARS-CoV-2) that has affected the lives of millions of people worldwide. Type 2 diabetes (T2D) acts in synergy with SARS-CoV-2 infection to increase the progression, severity, and mortality of COVID-19. For this reason, there is an urgent need to understand the immunometabolic bases of the main etiological factors of T2D that affect the severity of COVID-19. Herein, we revise how hyperglycemia contributes to the cytokine storm commonly associated with COVID-19 by stimulating monocytes and macrophages to produce interleukin (IL-) 1 beta, IL-6, and tumor necrosis factor alpha (TNF-alpha) in the airway epithelium. The main mechanisms through which hyperglycemia promotes reactive oxygen species (ROS) release, inhibition of T cell activation, and neutrophil extracellular traps (NETs) in the lungs of patients with severe SARS-CoV-2 infection are also

* Corresponding Author's E-mail: gescobedog@msn.com.

In: The Innate Immune System in Health and Disease
Editor: Jorge Morales-Montor
ISBN: 978-1-68507-507-1

discussed. We further study the molecular mechanisms by which proinflammatory cytokines induce insulin resistance, and their deleterious effects on pancreatic β-cell exhaustion in T2D patients critically ill with COVID-19. We revise the effect of excess glucose on advanced glycation end product (AGE) formation and the role of AGEs in perpetuating pneumonia and acute respiratory distress syndrome (ARDS). We examine the contribution of preexisting endothelial dysfunction secondary to diabetes in the development of neutrophil trafficking, vascular leaking, and thrombotic events in patients with severe SARS-CoV-2 infection. Finally, we discuss the role of other mechanisms potentially involved in COVID-19 evolution such as hyperglucagonemia, adiposity, dyslipidemia, and glomerular damage. As we have outlined here, T2D acts in synergy with SARS-CoV-2 infection to increase the progression, severity, and mortality of COVID-19. We think a better understanding of the immunometabolic factors involved in T2D that contribute to the severity of COVID-19 will improve our ability to recognize patients at much higher mortality risk.

Keywords: COVID-19, SARS-CoV-2, type 2 diabetes, inflammation, hyperglycemia, prothrombotic state

Introduction

The severe acute respiratory syndrome coronavirus-2 (SARS-CoV-2) was identified in the city of Wuhan, China in December 2019 [1]. SARS-CoV-2 is highly transmissible to humans, with case fatality rates ranging from 3 to 11% [2]. A study conducted in 1099 Chinese patients showed that incubation of SARS-CoV-2 takes a median of 4 days on average [3], while in the United States, the Centers for Disease Control and Prevention (CDC) has estimated that symptoms can appear within 2-14 days after exposure [4]. The main modes of SARS-CoV-2 transmission are respiratory droplets produced by infected individuals, aerosols, and direct contact with contaminated surfaces or objects [5]. Transmission of SARS-CoV-2 is more likely to occur in the early stage of infection; by day 10 after the onset of symptoms, 90% of patients with mild disease show a negative ribonucleic acid (RNA) test [6,7]. SARS-CoV-2 is the causal agent of the Coronavirus disease 2019 (COVID-19), an ongoing global pandemic that is affecting the lives of millions of people worldwide [8].

Even though more than 85% of patients with COVID-19 experience a self-limiting illness with symptoms such as fever, headache, myalgia, and diarrhea, some patients develop the most severe forms of the disease including pneumonia, acute respiratory distress syndrome (ARDS), sepsis, multiple organ failure, and death [9] . For this reason, there is a deep need to understand the variety of factors that underlie increased severity of COVID-19.

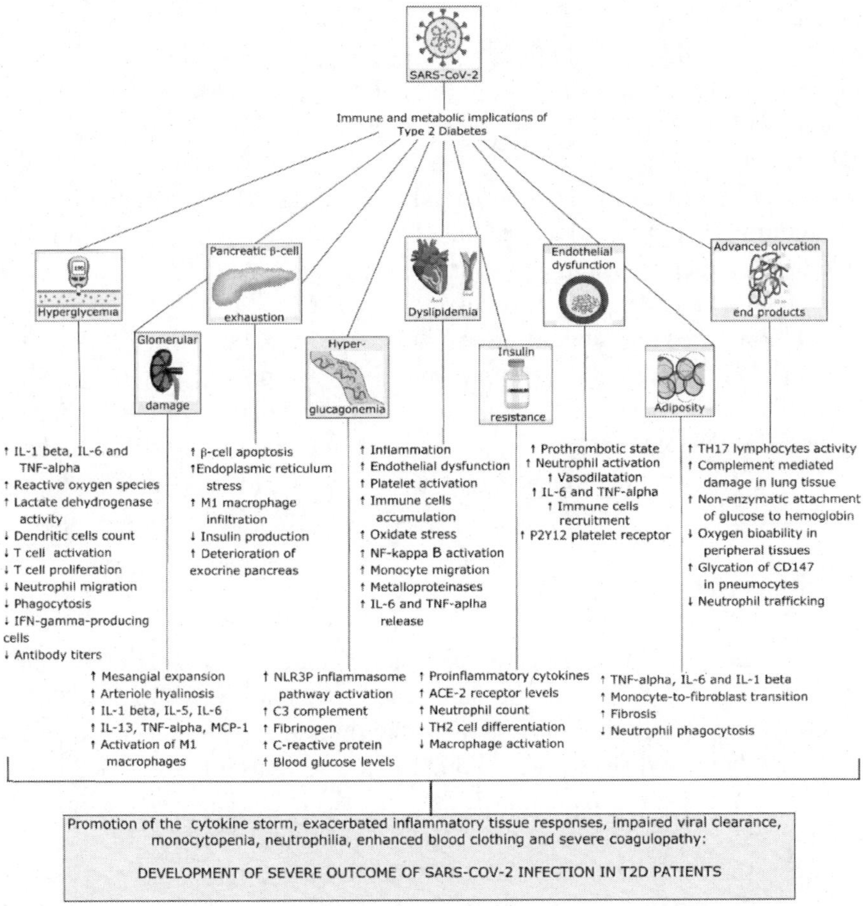

Figure 1. Immunometabolic mechanisms of the main etiological factors associated with type 2 diabetes and their implications in the development of severe SARS-CoV-2 infection.

The most severe outcomes of COVID-19 have been reported to occur in patients with preexisting comorbidities such as cancer, hypertension, and Diabetes Mellitus [10, 12]. In fact, Diabetes Mellitus is one of the most prevalent comorbidities in patients critically ill with COVID-19 [12]. Diabetes

Mellitus is a complex disorder characterized by abnormally high blood glucose levels that affect blood vessels and nerves, resulting in chronic damage to skin, feet, immune system, eyes, kidneys, brain, and heart [13]. People living with diabetes do not exhibit increased susceptibility to SARS-CoV-2 infection as compared with non-diabetic individuals [12]. However, patients with diabetes and COVID-19 are at much higher risk for adverse outcomes including admission to intensive care units, need for invasive ventilatory support, hospital-acquired infections, and death [14, 15]. Diabetic patients with COVID-19 have a mortality rate of 7.3% with respect to non-diabetic subjects, among whom a mortality of 2.3% has been reported [14].

Type 2 Diabetes (T2D), formerly known as adult-onset diabetes, is the most common form of diabetes and is characterized by multiple disorders including hyperglycemia, insulin resistance, systemic inflammation, hyperinsulinemia, β-cell exhaustion, vascular damage, endothelial dysfunction, nerve degeneration, increased platelet reactivity and prothrombotic state, and dyslipidemia, among others [16, 17]. These conditions affect the immune response to pathogens through different mechanisms that are not yet fully understood and might influence the severity of COVID-19, as will be further reviewed (Figure 1) [16, 18–23].

Hyperglycemia

Hyperglycemia is characterized by persistently high blood sugar levels and is the most common metabolic alteration associated with T2D [24]. Hyperglycemia has numerous effects on the immune response and *vice versa*, which has been shown to directly impact the severity of the SARS-CoV-2 infection. It is well known that high glucose levels can activate multiple immune cell types, leading to enhanced production of proinflammatory cytokines such as interleukin (IL-) 1 beta, IL-6, and tumor necrosis factor-alpha (TNF-alpha), among others [25]. This phenomenon is especially relevant to SARS-CoV-2 infection, because patients with T2D and COVID-19 exhibit increased serum levels of proinflammatory cytokines including IL-1 beta, IL-6, TNF-alpha, IL-8, and interferon (IFN-) gamma [26–30]. In patients with SARS-CoV-2 infection, this so-called "cytokine storm," accompanied by other inflammation-related markers such as C-reactive protein, D-dimer, and ferritin, has been linked to the severity of COVID-19 [31, 32]. Exacerbation of the inflammatory response causes a more severe acute infection that leads to ARDS and multiple organ failure [31].

Particularly, IL-6 has been shown to increase in both the airway epithelium and blood stream of patients with COVID-19 that develop severe ARDS [33].

Accumulating evidence confirms that hyperglycemia is a negative predictor in COVID-19 due to increased release of inflammatory mediators, endothelial dysfunction, thrombosis, and production of reactive oxygen species (ROS) [34 -36]. Accordingly, T2D patients critically ill with COVID-19 that show acute hyperglycemic peaks at hospital admission have a worse prognosis than patients with glycated hemoglobin (HbA1c) levels near to 6.5% [37, 38]. A high blood glucose value at hospital admission is therefore a risk factor for mortality in patients with severe SARS-CoV-2 infection that is independent of whether or not the patient has T2D [39, 40].

Hyperglycemia has been shown to promote mitochondrial oxidative stress and ROS production [41] that in turn lead to blood vessel damage and pancreatic beta cell dysfunction, which has deleterious effects on insulin secretion [42, 43]. Hyperglycemia-induced mitochondrial dysfunction triggers release of intracellular signaling molecules that in turn have the ability to inhibit the T cell response, a phenomenon that has been consistently reported in patients with severe COVID-19 [44]. Moreover, SARS-CoV-2 can also directly induce mitochondrial ROS production by activating the hypoxia-inducible factor 1 alpha (HIF-1 alpha) that in turn is able to promote the proinflammatory cytokine storm [45, 44]. Therefore, it is feasible that hyperglycemia and ROS production from T2D overlapping with SARS-CoV-2 infection may aggravate the cellular damage, organ failure, and progression of COVID-19.

High blood sugar levels have been shown to stimulate lactate dehydrogenase (LDH) activity and increase lactate production [46]. This phenomenon is of particular interest in COVID-19, where it has been reported that increased LDH levels are accurate predictors for mortality in patients with severe SARS-CoV-2 infection [47]. Notably, increased lactate levels in T2D patients might delay clearance of SARS-CoV-2 by inhibiting the retinoic acid-inducible gene-I-like receptor (RIG-1-like receptor) via the mitochondrial antiviral-signaling protein (MAVS), which results in blockage of interferon production and reduced anti-viral response [46].

Natural killer (NK) cells are innate lymphocytes that eliminate virally infected cells. In T2D patients, hyperglycemia has been shown to increase NK cells that express low levels of the natural killer group 2 member D (NKG2D) and the natural cytotoxicity receptor NKp46, resulting in decreased degranulation capacity and inefficient anti-viral activity [48 -50]. In addition, hyperglycemia promotes viral replication in monocytes with concomitant inhibition of T cell activation [44]. Some studies have also found low numbers of dendritic cells (DCs) in T2D patients with poor glycemic control. In line with this finding, high glucose conditions in in vitro cultures were shown to prevent monocyte differentiation into DCs [51, 52]. Moreover, DCs from T2D patients poorly induce T cell proliferation in vitro [53]. Taken together, all of this evidence supports a decisive role for T2D in exacerbating COVID-19 progression, as patients with severe SARS-CoV-2 infection exhibit low cell counts of NK cells, functional DCs, and CD4+ and CD8+ T cells [54].

Hyperglycemia in T2D patients has deleterious effects on numerous neutrophil functions including migration, phagocytosis, and bacterial killing [55]. Additionally, T2D patients have low numbers of IFN-gamma-producing cells that affect viral clearance in multiple infections such as those provoked by cytomegalovirus, Epstein-Barr virus, and influenza [56]. Moreover, hyperglycemia reduces antibody titers and the ability to kill Staphylococcus pneumoniae and Staphylococcus aureus, each of which is capable of invading the lungs and producing pneumonia [57]. Considering the immune effects of hyperglycemia within the context of SARS-CoV-2 infection, the above-mentioned mechanisms might partially explain why T2D patients are at a higher risk of developing severe COVID-19 and adverse outcomes including increased viral load, sepsis, and death. For this reason, uncontrolled hyperglycemia should be considered as a crucial risk factor for COVID-19 progression in T2D patients.

Hyperinsulinemia and Pancreatic β-Cell Exhaustion

Pancreatic β-cells secrete insulin and have a primary function in the control of blood glucose levels [58]. During the evolution of T2D, pancreatic β-cells are exposed to glucotoxicity, ROS, and endoplasmic reticulum stress, which all increase β-cell apoptosis and dysfunction [58, 59]. Additionally, chronic hyperglycemia increases M1-like macrophage infiltration in pancreatic islets, where these immune cells can secrete IL-1 beta, IL-6, and TNF-alpha and promote islet inflammation and β-cell malfunction and apoptosis [60, 61]. In response, pancreatic β-cells enhance insulin secretion in order to counteract persistently high glucose levels, thus leading to a state of hyperinsulinemia [62]. Nevertheless, β-cell mass is eventually exhausted, resulting in impaired insulin production. By the time diabetes is typically diagnosed, β-cells show less than fifty percent activity and are no longer able to secrete enough insulin to effectively maintain blood glucose levels [63, 64].

β-cell dysfunction can be aggravated during COVID-19 due to the ability of SARS-CoV-2 to enter the human pancreatic islets via angiotensin-converting enzyme 2 (ACE2) [64]. After SARS-CoV-2 invasion, inflammatory cells are recruited to pancreatic tissue, where they intensify local inflammation and injury resulting in increased peri- and intraislet fibrosis, β-cell mass loss, and hyperglycemia in both non-diabetic and diabetic patients [65]. Additionally, COVID-19 is characterized by persistent acute hypoxia that can affect numerous organs, including pancreatic islets, and provoke β-cell apoptosis directly [66]. In this way, SARS-CoV-2-induced pancreatic damage contributes to impaired insulin secretion that in turn may accelerate diabetes pathogenesis and/or aggravate preexisting diabetes [67]. Similarly, β-cell dysfunction that leads to chronic hyperglycemia is accompanied by ROS release, advanced glycation end product (AGE) formation, mitochondrial oxidative stress, and low antioxidant activity, thus compounding pancreatic β-cell damage during COVID-19 [68-71].

Interestingly, a growing body of evidence suggests that SARS-CoV-2 infection is not only able to affect the endocrine pancreas but also the exocrine pancreas [72]. In fact, SARS-CoV-2 has been shown to bind and enter the exocrine pancreatic ductal cells via ACE2 [73]. Some studies estimate that the prevalence of the development of acute pancreatitis in patients with severe COVID-19 is as high as 17% [74].

All the mechanisms discussed above highlight the importance of inflammation, hyperglycemia, and pancreatic dysfunction as potential contributors to the development of severe COVID-19 in patients with type 2 diabetes [75].

Insulin Resistance

Insulin resistance refers to the inability of insulin to exert its functions in insulin-dependent tissue such as liver, adipose tissue, and skeletal muscle [76]. Insulin resistance is by far the most important etiological factor contributing to the development of T2D [77]. There are multiple mechanisms whereby insulin resistance occurs in humans, however, systemic inflammation is one of the most recently studied. Obese subjects show a constant systemic proinflammatory state characterized by abnormally high circulating levels of TNF-alpha, IL-1 beta, IL-6, IL-12, and monocyte chemoattractant protein-1 (MCP-1) [78, 79]. In adipose tissue, TNF-alpha has been shown to induce insulin resistance by activating protein-tyrosine phosphatase 1B (PTP1B) that in turn can dephosphorylate the insulin receptor substrate-2 (IRS-2) resulting in glucose transport arrest and hyperglycemia [80, 81]. IL-6 and the nuclear factor-kappa B (NF-kappa B) pathway are also involved in insulin resistance and progressive loss of normal glucose tolerance [82, 83]. In diabetic patients, NF-kappa B upregulates protein kinase C-theta (PKC-θ), activator protein-1 (AP-1), and c-Jun kinase (JNK), which all act together to inhibit the insulin receptor via serine/threonine phosphorylation of the IRS [84].

SARS-CoV-2 infection in pancreatic β-cells does not only reduce insulin secretion, but also provokes a proinflammatory cytokine storm that exacerbates insulin resistance [15]. It is well known that even mild SARS-CoV-2 infection can trigger a proinflammatory cascade that mainly increases TNF-alpha, IL-6, MCP-1, and IL-1 beta in the lungs and blood stream [85]. Similarly, SARS-CoV-2 infection produces high levels of IFN-gamma inducible protein-10 (IP-10) that is capable of lowering insulin sensitivity by itself [86]. Thus, release of proinflammatory molecules in non-diabetic and diabetic patients with untreated insulin resistance might aggravate COVID-19 and elevate its severity.

An important link between insulin resistance and severe COVID-19 is ACE2, as it acts as the main SARS-CoV-2 entry point [87]. Under normal conditions, ACE2 converts angiotensin II into angiotensin 1-7 in order to prevent angiotensin II-related physiological disturbances including vasoconstriction, inflammation, oxidative stress, and insulin resistance [88]. In mice fed a high-sucrose diet, ACE2 is upregulated to remove excess angiotensin II and mitigate its negative effects on insulin sensitivity and glucose transport via the glucose transporter protein family (GLUT) [89, 90]. T2D patients demonstrate increased ACE2 receptor levels that in turn may help SARS-CoV-2 extend cellular binding, thus boosting viral load and severity of infection.

IFN-gamma is increased in patients with severe SARS-CoV-2 infection and negatively affects insulin sensitivity via IP-10 [91]. Also, IFN-gamma produced in response to multiple viral infections can cause insulin resistance in skeletal muscle and adipose tissue via PI3K down-regulation [92, 93]. It is thus reasonable to speculate that increased IFN-gamma production during COVID-19 may aggravate pre-existing insulin resistance in both non-diabetic and diabetic patients.

Insulin resistance has been also studied for its apparent role in preventing the anti-inflammatory T-helper type 2 (Th2) differentiation of CD4+ lymphocytes via the extracellular signal-regulated kinase (ERK) [94]. In fact, CD4+ T cells have been shown to induce abnormal responses to insulin in conditions characterized by insulin resistance such as obesity and T2D [95]. Insulin resistance can also influence macrophages, an immune cell type with key roles in preventing COVID-19-related organ damage [96]. In this way, monocytes and macrophages lacking insulin signaling show an impaired ability to respond to a variety of pathogens [97]. In T2D patients, insulin resistance is also associated with high blood neutrophil count, which is of particular importance in severe COVID-19 that is characterized by neutrophilia and monocytopenia [18, 98].

Formation of Advanced Glycation End Products

The non-enzymatic covalent attachment of glucose to molecules such as proteins, lipids, and/or nucleic acids to form AGEs characterizes the pathogenesis of T2D [99]. Besides having deleterious effects on the insulin signaling pathway, AGEs have been shown to bind several surface receptors such as CD36, scavenger receptors type I and II (SR-A1 and SR-A2), and galectin-3 [100, 101]. Upon receptor recognition, AGEs stimulate the release of pro-inflammatory cytokines in lymphocytes, monocytes, and macrophages and promote vascular inflammation and endothelial dysfunction [101]. AGEs can also directly bind to the receptor for advanced glycation end products (RAGE), a multi-ligand binding protein with prominent actions in sustained inflammatory responses [102]. It is worth mentioning that the lungs can express high RAGE levels, which may increase pulmonary inflammation in T2D patients in whom a wide variety of AGEs are produced [103]. RAGE is expressed in alveolar epithelial cells, vascular smooth muscle pulmonary cells, airway smooth muscle cells, and endothelial cells [104-107]. The AGE-RAGE interaction activates the NLR family pyrin domain containing 3 (NRLP3) inflammasome pathway that in turn is able to polarize macrophages toward M1, induce neutrophil extracellular trap formation, and increase the Th17 lymphocyte population. Altogether, these inflammatory actions might perpetuate the cytokine storm, leading to pulmonary inflammation and fibrosis in patients with COVID-19 [108,103]. This hypothesis is supported by the fact that RAGE-dependent inflammatory pathways have been shown to play a detrimental role in pneumonia and ARDS [109].

Advanced glycation end products activate the classical complement pathway by recognizing C1q, which in turn inactivates CD59 and increases vascular injury in blood vessels of T2D patients [110]. In agreement with this concept, membrane attack complex deposits present in lung tissue from patients with severe COVID-19 has revealed complement-mediated damage which in turn induces vascular inflammation and results in extended lung damage [111].

It is well known that excess glucose can be non-enzymatically attached to hemoglobin to form HbA1c. This information is of special relevance in T2D and COVID-19 due to the ability of SARS-CoV-2 to alter the 1-beta chain of hemoglobin. This causes iron dissociation and porphyrin formation, thus affecting oxygen affinity and bioavailability in peripheral tissues [112]. It follows that excess glycation of hemoglobin in T2D patients may contribute to breathing difficulty that in turn progresses to ARDS, a key pathophysiological component of severe COVID-19. In parallel, a recent study reported that ACE2 can be glycated in hyperglycemic conditions [113]. Interestingly, glycation in ACE2 appears to increase SARS-CoV-2 affinity and entry in pancreatic and lung tissue [114, 115]. Moreover, good glycemic control has been shown to lower the amount of glycated ACE2 in lung tissue, ameliorating pneumonia and COVID-19 severity presumably by reducing the availability of viral entry points [115]. Conversely, uncontrolled hyperglycemia leads to aberrant formation of glycated ACE2 not only in lungs but also in nasal airways, tongue, and oropharynx, which may increase viral entry points and disease severity [115].

Finally, CD147 is a glycoprotein expressed in type II pneumocytes that binds the spike S1 protein, thus favoring SARS-CoV-2 entry in lung cells [116]. Evidence in T2D patients suggests that CD147 can be glycated in hyperglycemic conditions, which is linked to metalloproteinase upregulation and loss of tight junctions that may favor cell entry of SARS-CoV-2 and increase viral load [117]. As we have outlined, formation of AGEs appears to play a key role in the severity of COVID-19, which becomes more relevant in T2D patients with poor glycemic control, a condition that favors protein glycation.

Endothelial Dysfunction, Vascular Damage, and Prothrombotic State

Vascular homeostasis is maintained by modulating blood flow, fybrinolysis, coagulation, platelet adherence, and immune cell trafficking in response to cell injury [118]. The function of vascular endothelium is impaired in patients with Diabetes Mellitus, and this impaired function is considered a risk factor for cardiovascular disease [119]. Emerging evidence suggests that COVID-19 aggravates vascular pathology due to proliferation of SARS-CoV-2 in endothelial cells. This induces cellular damage, apoptosis, and disruption of the vascular barrier, which becomes more relevant in T2D patients that show impaired angiogenesis [120- 122]. Notably, endothelial dysfunction is a central feature in COVID-19 pathogenesis [123]. Patients with COVID-19 have nitric oxide (NO) deficiencies that lead to increased vascular contraction and reduced ROS neutralization [124, 125]. Upon SARS-CoV-2 infection, the vascular endothelium experiences vascular leakage, enhanced blood clotting, and recruitment of immune mediators resulting in inflammation that perpetuates tissue damage and vascular impairment [126]. As mentioned above, COVID-19 is accompanied by a high number of neutrophils, proinflammatory immune cells that also act as key contributors to vascular damage in T2D. Diabetic patients have neutrophils with enhanced oxidative activity that produce high free radical levels and neutrophil extracellular traps (NETs) which can cause direct injury to blood vessels [127, 128]. Neutrophils are also major producers of myeloperoxidase (MPO), a peroxidase enzyme that binds to the vascular endothelium and increases blood vessel damage in T2D patients [129]. These lines of evidence support a deleterious synergistic role between COVID-19 and T2D, where hyper-reactive neutrophils may directly injure the vascular endothelium and worsen the patient's outcome [98]. Interestingly, SARS-CoV-2 is able to infect neutrophils that in turn increase NET release, which can then directly injure the lung epithelium.

Furthermore, proinflammatory cytokines play a decisive role in endothelial dysfunction. It is well known that IL-6 is upregulated in T2D, and is associated with endothelial damage and atherosclerosis [130]. In patients with severe COVID-19, IL-6 induces chemokine expression, leukocyte trafficking, immune cell extravasation toward arterial walls, NO reduction, increased oxidative stress, and exacerbated inflammation of blood vessels [131]. TNF-alpha, another important proinflammatory cytokine involved in the COVID-19 cytokine storm, accelerates atherosclerosis via vascular cell adhesion molecule-1 (V-CAM), E-selectin, and MCP1, which impairs vasodilatation and promotes endothelial cell apoptosis [83]. Post-mortem examination of lung tissue from patients severely infected with SARS-CoV-2 revealed massive mononuclear and polymorphonuclear cell infiltration, supporting the role of immune cell recruitment in COVID-19 progression [120]. These findings indicate that chemoattraction and recruitment of immune cells act together with endothelial dysfunction to induce vascular damage in patients with type 2 diabetes, which may worsen the severity of COVID-19 by increasing vascular leaking and prothrombosis.

Damage of vascular integrity promotes basement membrane exposure, coagulopathy, D-dimer release, and fibrinogen and platelet activation, all of which are important biomarkers for poor prognosis in COVID-19 [132-135]. It is well known that the prothrombotic state common in patients with T2D may increase the occurrence of severe coagulopathy in patients critically ill with COVID-19 [136]. In fact, the severity of COVID-19 increases as pulmonary embolism, microcirculatory malfunction, and disseminated intravascular coagulation increase [137, 138]. Microvascular and macrovascular thromboembolic events have been documented in the kidneys, lungs, spleen, and brains of SARS-CoV-2 infected patients [139–142]. Thrombotic incidence of about 30% has been reported in lungs [143] and incidence of deep venous thrombosis as occurs in the lower limbs of T2D patients has been reported at 46% [144].

The mechanisms underlying thromboembolic events in T2D patients with severe COVID-19 remain unclear, but persistent inflammation has now emerged as a potential contributor [145]. As described above, levels of several proinflammatory cytokines are increased in patients with COVID-19 that require hospitalization, including TNF-alpha, IL-6, and IL-8 [146, 147]. Interestingly, some of these cytokines have prothrombotic effects by themselves [148]. For instance, there is a positive association between elevated IL-6 and increased fibrinogen levels [148]. During sepsis, monocytes and macrophages release TNF-alpha as well as tissue factors that activate clotting pathways [149]. Besides inducing proinflammatory cytokine production, SARS-CoV2 has been shown to induce expression of procoagulant genes such as fibrinogen, tissue factor, factor II, and factor X [150, 151] in vitro culture models. During numerous viral infections and sepsis, activation of the innate immune system leads to increased activation of the complement system, von Willebrand factor, tissue factor, and factor VIIa [150, 153, 154]. Likewise, complement activation during COVID-19-related sepsis intensifies the cytokine storm and perpetuates microvascular damage [152].

Due to the preexisting prothrombotic state, T2D increases mortality risk in COVID-19, where hyperglycemia by itself appears to play a contributing role [154]. Human aortal endothelial cells cultured in high glucose in vitro were shown to trigger both inflammatory and prothrombotic pathways [155]. Similarly, hyperglycemia acts in synergy with neutrophils to release calprotectin, a protein that can bind RAGE on Kupffer cells and induce IL-6 synthesis. Then, IL-6 increases thrombopoietin production that in turn enhances proliferation and expansion of thrombotic precursors, resulting in thrombocytosis [156]. Similarly, P2Y12, a receptor expressed on the surface of platelets that plays essential roles in platelet activation, may be elevated in T2D patients and facilitate platelet adhesion to vascular endothelium [157, 158]. Consistent with these findings, numerous reports have demonstrated that the prothrombotic state can be decreased by lowering blood glucose concentration [156]. Thus, several factors associated with T2D including endothelial dysfunction, vascular damage, systemic inflammation, and hyperglycemia can directly aggravate the prothrombotic state and increase mortality risk in patients with severe COVID-19 [140, 143, 159].

Miscellaneous

In addition to the pathophysiological mechanisms described above, there are other components of T2D that can impair immunity and aggravate COVID-19.

Glucagon is a hormone produced by α-pancreatic cells with central functions in gluconeogenesis in response to hypoglycemia [160]. Normal immunity response is affected by high levels of glucagon in diabetic patients. Glucagon activates the liver NLR3P inflammasome pathway, augmenting levels of C3 complement, fibrinogen, and C-reactive protein [21], all of which are molecules intimately associated with COVID-19 severity. The lack of insulin and the inflammatory response that occur in T2D promote glucagon overproduction that in turn leads to hepatic gluconeogenesis and finally increases blood glucose levels, enhancing glucotoxicity and hyperglycemia, which together perpetuate a negative feed-back in diabetic patients [161].

Adipose tissue is an endocrine with the ability to storage lipids [83]. Adipocytes produce a variety of adipocytokines including TNF-alpha, IL-6, IL-1 beta, adiponectin, and leptin [162]. As mentioned above, the role of TNF-alpha, IL-6 and IL-1 beta is a central determinant in turning protective immune response into a risk factor for mortality [163]. Similarly, adiponectin is a key regulator of the immune system that plays a pivotal role in inflammation progression [164]. Adiponectin mainly acts as an anti-inflammatory adipocytokine that inhibits neutrophil phagocytosis and apoptosis [165, 166]. Adiponectin promotes monocyte-to-fibroblast transition and overproduction of extracellular matrix proteins, leading to fibrosis [167, 168]. Moreover, high adiponectin levels are also independently associated with ARDS, an important element of severe SARS-CoV-2 infection [169].

Dyslipidemia is one of the factors more frequently found in diabetic patients that also show macrovascular complications [170]. Dyslipidemia promotes inflammation, endothelial dysfunction, and platelet hyperactivation, leading to accumulation of lipids in the arterial intimae, immune cells and the extracellular matrix, causing tissue ischemia and/or thrombocytosis [171]. Diabetes concurs with enhanced formation of oxidized Low Density Lipoproteins (ox-LDL) through oxidative stress [172]. Ox-LDL has atherogenic capacity by interacting with receptors in the vascular endothelium such as lectin-like ox-LDL receptor-1 (LOX-1) and CD36. Upon activation, these receptors orchestrate ROS production and activate the NF-kappa B-dependent signaling pathways [173] that triggers monocyte migration to the inner wall of arteries, wherein these immune cells will differentiate to macrophages and then foam cells. Macrophage-derived foam cells promote inflammation through secretion of chemokines, ROS, metalloproteinases, and proinflammatory cytokines such as IL-6 and TNF-alpha.

Glomerular damage and renal injury is a common feature in at least one in three diabetic patients [174] and is characterized by thickening of the glomerular basement membrane, mesangial expansion, and arteriole hyalinosis [175]. As mentioned, high levels of glucose are able to activate NLRP3 inflammasome and NF-kB-dependent signal pathways, resulting in production of IL-1 beta and IL-6 [176] and M1 proinflammatory activation of macrophages [177]. Interestingly, people with diabetic nephropathy display significant higher levels of IL-1 beta, IL-4, IL-5, IL-6, IL-13, TNF-alpha, MCP-1, all of which showing key roles in the cytokine storm associated with severe COVID-19 [174, 178].

Conclusion

Management of the synergistic action of the two current pandemics of COVID-19 and T2D has become a major contemporary public health challenge. Herein, we outlined the main mechanisms through which T2D acts in synergy with SARS-CoV-2 infection to increase the progression, severity, and mortality of COVID-19. Also, we discussed the mechanisms whereby hyperglycemia contributes to the cytokine storm that frequently accompanies severe SARS-CoV-2 infection by stimulating monocytes and macrophages to produce IL-1 beta, IL-6, and TNF-alpha in the airway epithelium. The key mechanisms through which hyperglycemia promotes ROS release, inhibition of T cell activation, and NETs formation in the lungs of patients with severe

SARS-CoV-2 infection have been also studied. We reviewed the molecular mechanisms by which proinflammatory cytokines induce insulin resistance and exert deleterious effects on pancreatic β-cell exhaustion in T2D patients critically ill with COVID-19. We studied the effect of excess glucose on AGEs formation and the role of AGEs in perpetuating pneumonia and ARDS. We also discussed the contribution of preexisting endothelial dysfunction secondary to diabetes in the development of neutrophil trafficking, vascular leaking, and thrombotic events in patients with severe SARS-CoV-2 infection. Finally, we addressed the possible contribution of other T2D components to the severity of COVID-19 including hyperglucagonemia, adiposity, dyslipidemia, and glomerular damage (Figure 2). A better understanding of the immunometabolic mechanisms of T2D that contribute to the severity of COVID-19 will improve our ability to identify patients at much higher mortality risk.

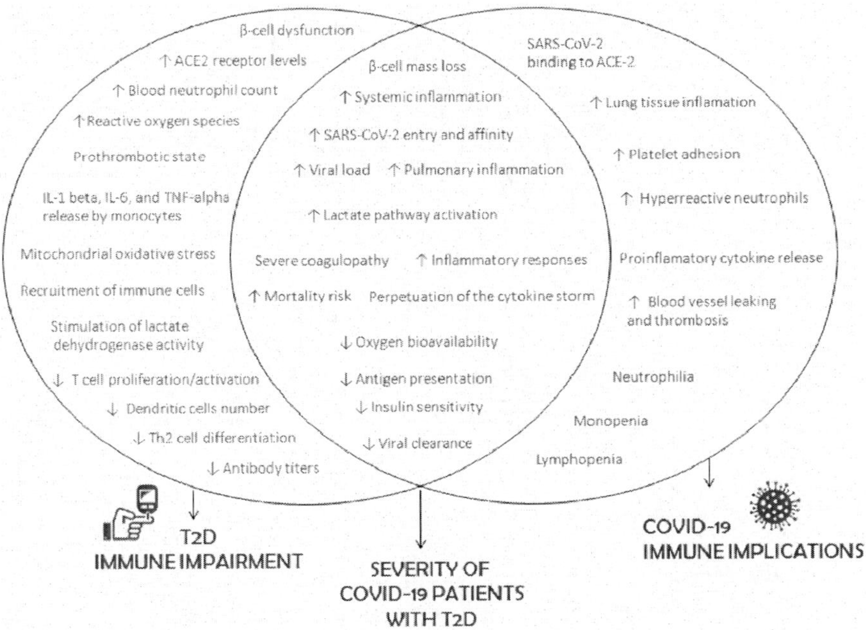

Figure 2. Synergistic interactions of type 2 diabetes with the immune system lead to impaired immune response and aggravate the severity of COVID-19.

References

[1] Zhu N, Zhang D, Wang W, Li X, Yang B, Song J, et al. A Novel Coronavirus from Patients with Pneumonia in China, 2019. *N Engl J Med* [Internet]. 2020 Feb 20 [cited 2020 Dec 22];382(8):727–33.

[2] Ceccarelli M, Berretta M, Rullo EV, Nunnari G, Cacopardo B. Differences and similarities between Severe Acute Respiratory Syndrome (SARS)-CoronaVirus (CoV) and SARS-CoV-2. Would a rose by another name smell as sweet? *European Review for Medical and Pharmacological Sciences*. 2020;24(5):2781–3.

[3] Guan W, Ni Z, Hu Y, Liang W, Ou C, He J, et al. Clinical Characteristics of Coronavirus Disease 2019 in China. *N Engl J Med* [Internet]. 2020 Apr 30 [cited 2020 Dec 21];382(18):1708–20.

[4] *COVID-19 Testing Overview* | CDC [Internet]. [cited 2020 Dec 21]. Available from: https://www.cdc.gov/coronavirus/2019-ncov/symptoms-testing/testing.html.

[5] Adhikari SP, Meng S, Wu YJ, Mao YP, Ye RX, Wang QZ, et al. *Epidemiology, causes, clinical manifestation and diagnosis, prevention and control of coronavirus disease (COVID-19) during the early outbreak period: A scoping review* [Internet]. Vol. 9, Infectious Diseases of Poverty. BioMed Central Ltd.; 2020 [cited 2020 Dec 21]. p. 29.

[6] Zou L, Ruan F, Huang M, Liang L, Huang H, Hong Z, et al. SARS-CoV-2 Viral Load in Upper Respiratory Specimens of Infected Patients. *N Engl J Med.* 2020 Mar 19;382(12):1177–9.

[7] Liu Y, Yan LM, Wan L, Xiang TX, Le A, Liu JM, et al. Viral dynamics in mild and severe cases of COVID-19. Vol. 20, *The Lancet Infectious Diseases.* Lancet Publishing Group; 2020. p. 656–7.

[8] Lai CC, Shih TP, Ko WC, Tang HJ, Hsueh PR. Severe acute respiratory syndrome coronavirus 2 (SARS-CoV-2) and coronavirus disease-2019 (COVID-19): The epidemic and the challenges [Internet]. Vol. 55, *International Journal of Antimicrobial Agents.* Elsevier B.V.; 2020

[9] Khan M, Khan H, Khan S, Nawaz M. Epidemiological and clinical characteristics of coronavirus disease (COVID-19) cases at a screening clinic during the early outbreak period: a single-centre study. *J Med Microbiol* [Internet]. 2020 Aug 12 [cited 2021 Jan 28];69(8):1114–23.

[10] Zhou F, Yu T, Du R, Fan G, Liu Y, Liu Z, et al. Clinical course and risk factors for mortality of adult inpatients with COVID-19 in Wuhan, China: a retrospective cohort study. *Lancet.* 2020 Mar 28;395(10229):1054–62.

[11] Yang JK, Feng Y, Yuan MY, Yuan SY, Fu HJ, Wu BY, et al. Plasma glucose levels and diabetes are independent predictors for mortality and morbidity in patients with SARS. *Diabet Med.* 2006 Jun;23(6):623–8.

[12] Li B, Yang J, Zhao F, Zhi L, Wang X, Liu L, et al. Prevalence and impact of cardiovascular metabolic diseases on COVID-19 in China. Vol. 109, *Clinical Research in Cardiology.* Springer; 2020. p. 531–8.

[13] Knapp S. Diabetes and infection: Is there a link? - A mini-review. Vol. 59, *Gerontology.* 2013. p. 99–104.

[14] Wu Z, McGoogan JM. Characteristics of and Important Lessons from the Coronavirus Disease 2019 (COVID-19) Outbreak in China: Summary of a Report of 72314 Cases from the Chinese Center for Disease Control and Prevention. Vol. 323, *JAMA - Journal of the American Medical Association.* American Medical Association; 2020. p. 1239–42.

[15] Onder G, Rezza G, Brusaferro S. Case-Fatality Rate and Characteristics of Patients Dying in Relation to COVID-19 in Italy. Vol. 323, *JAMA - Journal of the American Medical Association.* American Medical Association; 2020. p. 1775–6.

[16] DeFronzo RA, Ferrannini E, Groop L, Henry RR, Herman WH, Holst JJ, et al. Type 2 diabetes mellitus. *Nat Rev Dis Prim* [Internet]. 2015 Jul 23 [cited 2020 Dec 25];1. Available from: https://pubmed.ncbi.nlm.nih.gov/27189025/.

[17] Daryabor G, Kabelitz D, Kalantar K. An update on immune dysregulation in obesity-related insulin resistance. *Scand J Immunol* [Internet]. 2019 Apr 29 [cited 2020 Dec 25];89(4):e12747.

[18] Feng YM, Zhao D, Zhang N, Yu CG, Zhang Q, Thijs L, et al. Insulin resistance in relation to lipids and inflammation in type-2 diabetic patients and non-diabetic people. Kiechl S, editor. *PLoS One* [Internet]. 2016 Apr 1 [cited 2021 Jan 4];11(4):e0153171.

[19] Sattar N, Gill JMR. Type 2 diabetes as a disease of ectopic fat? *BMC Med* [Internet]. 2014 Aug 26 [cited 2020 Dec 27];12(1):123.

[20] Yagihashi S, Mizukami H, Sugimoto K. Mechanism of diabetic neuropathy: Where are we now and where to go? *J Diabetes Investig.* 2011;2(1):18–32.

[21] Fatta C Di, Spiga R, Mancuso E, Fuoco A, Averta C, Mannino GC, et al. Glucagon induces the expression of inflammatory markers in the liver through NLRP3 inflammasome. *Diabetol* 61. 2018.

[22] Leite NC, Salles GF, Araujo ALE, Villela-Nogueira CA, Cardoso CRL. Prevalence and associated factors of non-alcoholic fatty liver disease in patients with type-2 diabetes mellitus. *Liver Int* [Internet]. 2009 Jan 1 [cited 2020 Dec 25];29(1):113–9.

[23] Nielsen TB, Pantapalangkoor P, Yan J, Luna BM, Dekitani K, Bruhn K, et al. Diabetes exacerbates infection via hyperinflammation by signaling through TLR4 and RAGE. *MBio* [Internet]. 2017 Jul 1 [cited 2021 Jan 3];8(4):818–35.

[24] Blair M. Diabetes Mellitus Review. *Urol Nurs* [Internet]. 2016 Jan 1 [cited 2021 Jan 6];36(1):27–36.

[25] Muniyappa R, Gubbi S. COVID-19 pandemic, coronaviruses, and diabetes mellitus [Internet]. Vol. 318, *American Journal of Physiology - Endocrinology and Metabolism.* American Physiological Society; 2020 [cited 2021 Jan 3]. p. E736–41.

[26] Ye Q, Wang B, Mao J. The pathogenesis and treatment of the 'Cytokine Storm'' in COVID-19' [Internet]. Vol. 80, *Journal of Infection.* W. B. Saunders Ltd; 2020 [cited 2021 Jan 6]. p. 607–13.

[27] Mangalmurti N, Hunter CA. Cytokine Storms: Understanding COVID-19 [Internet]. Vol. 53, *Immunity.* Cell Press; 2020 [cited 2021 Jan 6]. p. 19–25.

[28] Hu B, Huang S, Yin L. The cytokine storm and COVID-19 [Internet]. *Journal of Medical Virology.* John Wiley and Sons Inc; 2020 [cited 2020 Dec 16]. Available from: /pmc/articles/PMC7361342/?report=abstract.

[29] Chi Y, Ge Y, Wu B, Zhang W, Wu T, Wen T, et al. Serum cytokine and chemokine profile in relation to the severity of coronavirus disease 2019 in China. *J Infect Dis* [Internet]. 2020 Sep 1 [cited 2020 Dec 16];222(5):746–54.

[30] Wang J, Jiang M, Chen X, Montaner LJ. Cytokine storm and leukocyte changes in mild versus severe SARS-CoV-2 infection: Review of 3939 COVID-19 patients in China and emerging pathogenesis and therapy concepts [Internet]. Vol. 108, *Journal of Leukocyte Biology.* John Wiley and Sons Inc.; 2020 [cited 2020 Dec 15]. p. 17–41.

[31] Ye Q, Wang B, Mao J. The pathogenesis and treatment of the 'Cytokine Storm'' in COVID-19' [Internet]. Vol. 80, *Journal of Infection.* W.B. Saunders Ltd; 2020 [cited 2021 Jan 3]. p. 607–13.

[32] Yang X, Yu Y, Xu J, Shu H, Xia J, Liu H, et al. Clinical course and outcomes of critically ill patients with SARS-CoV-2 pneumonia in Wuhan, China: a single-centered, retrospective, observational study. *Lancet Respir Med.* 2020 May 1;8(5):475–81.

[33] Zhang C, Wu Z, Li JW, Zhao H, Wang GQ. Cytokine release syndrome in severe COVID-19: interleukin-6 receptor antagonist tocilizumab may be the key to reduce mortality. *Int J Antimicrob Agents.* 2020 May 1;55(5):105954.

[34] Bode B, Garrett V, Messler J, McFarland R, Crowe J, Booth R, et al. Glycemic Characteristics and Clinical Outcomes of COVID-19 Patients Hospitalized in the United States. *J Diabetes Sci Technol* [Internet]. 2020 [cited 2021 Jan 3];14(4):813–21.

[35] Ceriello A, Zarich SW, Testa R. Lowering Glucose to Prevent Adverse Cardiovascular Outcomes in a Critical Care Setting. Vol. 53, *Journal of the American College of Cardiology*. Elsevier; 2009. p. S9–13.

[36] Ceriello A. Acute hyperglycaemia: a 'new' risk factor during myocardial infarction. *Eur Heart J* [Internet]. 2005 Feb 1 [cited 2021 Jan 2];26(4):328–31.

[37] Singh AK, Singh R. Hyperglycemia without diabetes and new-onset diabetes are both associated with poorer outcomes in COVID-19 [Internet]. Vol. 167, *Diabetes Research and Clinical Practice*. Elsevier Ireland Ltd; 2020 [cited 2021 Jan 2]. Available from: https://pubmed.ncbi.nlm.nih.gov/32853686/.

[38] Liao WI, Wang JC, Chang WC, Hsu CW, Chu CM, Tsai SH. Usefulness of Glycemic Gap to Predict ICU Mortality in Critically Ill Patients With Diabetes. *Medicine* (Baltimore) [Internet]. 2015 Sep 1 [cited 2021 Jan 2];94(36):e1525. Available from: http://journals.lww.com/00005792-201509020-00043.

[39] Iacobellis G, Penaherrera CA, Bermudez LE, Bernal Mizrachi E. Admission hyperglycemia and radiological findings of SARS-CoV2 in patients with and without diabetes. *Diabetes Res Clin Pract* [Internet]. 2020 Jun 1 [cited 2021 Jan 2];164.

[40] Yang JK, Feng Y, Yuan MY, Yuan SY, Fu HJ, Wu BY, et al. Plasma glucose levels and diabetes are independent predictors for mortality and morbidity in patients with SARS. *Diabet Med* [Internet]. 2006 Jun [cited 2021 Jan 2];23(6):623–8.

[41] Robertson RP, Harmon J, Tran PO, Tanaka Y, Takahashi H. Glucose toxicity in β-cells: Type 2 diabetes, good radicals gone bad, and the glutathione connection [Internet]. Vol. 52, *Diabetes*. American Diabetes Association; 2003 [cited 2020 Dec 25]. p. 581–7.

[42] Zhang ZY, Miao LF, Qian LL, Wang N, Qi MM, Zhang YM, et al. Molecular Mechanisms of Glucose Fluctuations on Diabetic Complications [Internet]. Vol. 10, *Frontiers in Endocrinology*. Frontiers Media S.A.; 2019 [cited 2020 Dec 25]. p. 640.

[43] Lowell BB, Shulman GI. Mitochondrial dysfunction and type 2 diabetes [Internet]. Vol. 307, *Science*. American Association for the Advancement of Science; 2005 [cited 2020 Dec 25]. p. 384–7.

[44] Codo AC, Davanzo GG, Monteiro L de B, de Souza GF, Muraro SP, Virgilio-da-Silva JV, et al. Elevated Glucose Levels Favor SARS-CoV-2 Infection and Monocyte Response through a HIF-1α/Glycolysis-Dependent Axis. *Cell Metab* [Internet]. 2020 Sep 1 [cited 2021 Jan 3];32(3):437-446.e5.

[45] Ceriello A. Hyperglycemia and COVID-19: What was known and what is really new? *Diabetes Res Clin Pract* [Internet]. 2020 [cited 2021 Jan 2];167:108383.

[46] Zhang W, Wang G, Xu ZG, Tu H, Hu F, Dai J, et al. Lactate Is a Natural Suppressor of RLR Signaling by Targeting MAVS. *Cell* [Internet]. 2019 Jun 27 [cited 2021 Jan 3];178(1):176-189.e15.

[47] Li X, Xu S, Yu M, Wang K, Tao Y, Zhou Y, et al. Risk factors for severity and mortality in adult COVID-19 inpatients in Wuhan. *J Allergy Clin Immunol* [Internet]. 2020 Jul 1 [cited 2021 Jan 3];146(1):110–8.

[48] Piątkiewicz P, Miłek T, Bernat-Karpińska M, Ohams M, Czech A, Ciostek P. The dysfunction of nk cells in patients with type 2 diabetes and colon cancer. *Arch Immunol Ther Exp* (Warsz) [Internet]. 2013 [cited 2021 Jan 6];61(3):245–53.

[49] Berrou J, Fougeray S, Venot M, Chardiny V, Gautier J-F, Dulphy N, et al. Natural Killer Cell Function, an Important Target for Infection and Tumor Protection, Is Impaired in Type 2 Diabetes. Ahlenstiel G, editor. *PLoS One* [Internet]. 2013 Apr 25 [cited 2020 Dec 25];8(4):e62418.

[50] Angelidi AM, Belanger MJ, Mantzoros CS. Commentary: COVID-19 and diabetes mellitus: What we know, how our patients should be treated now, and what should happen next. *Metabolism* [Internet]. 2020 Jun 1 [cited 2021 Jan 3];107.

[51] Summers KL, Marleau AM, Mahon JL, McManus R, Hramiak I, Singh B. Reduced IFN-α secretion by blood dendritic cells in human diabetes. *Clin Immunol* [Internet]. 2006 Oct [cited 2020 Dec 29];121(1):81–9.

[52] Gilardini Montani MS, Granato M, Cuomo L, Valia S, Di Renzo L, D'Orazi G, et al. High glucose and hyperglycemic sera from type 2 diabetic patients impair DC differentiation by inducing ROS and activating Wnt/β-catenin and p38 MAPK. *Biochim Biophys Acta - Mol Basis Dis.* 2016 Apr 1;1862(4):805–13.

[53] Macia L, Delacre M, Abboud G, Ouk T-S, Delanoye A, Verwaerde C, et al. Impairment of Dendritic Cell Functionality and Steady-State Number in Obese Mice. *J Immunol* [Internet]. 2006 Nov 1 [cited 2020 Dec 29];177(9):5997–6006.

[54] Zhong J, Tang J, Ye C, Dong L. The immunology of COVID-19: is immune modulation an option for treatment? [Internet]. Vol. 2, *The Lancet Rheumatology*. Lancet Publishing Group; 2020 [cited 2021 Jan 4]. p. e428–36.

[55] Alba-Loureiro TC, Munhoz CD, Martins JO, Cerchiaro GA, Scavone C, Curi R, et al. Neutrophil function and metabolism in individuals with diabetes mellitus. *Brazilian J Med Biol Res* [Internet]. 2007 [cited 2020 Dec 25];40(8):1037–44.

[56] Kalantar F, Dabbaghmanesh MH, Martinuzzi E, Moghadami M, Amirghofran Z. Islet amyloid polypeptide is not a target antigen for CD8+ T-cells in type 2 diabetes. *Iran J Immunol.* 2014;1–12.

[57] Mathews CE, Brown EL, Martinez PJ, Bagaria U, Nahm MH, Burton RL, et al. Impaired function of antibodies to pneumococcal surface protein A but Not to capsular polysaccharide in Mexican American adults with type 2 diabetes mellitus. *Clin Vaccine Immunol* [Internet]. 2012 Sep 1 [cited 2020 Dec 25];19(9):1360–9.

[58] Butler AE, Janson J, Bonner-Weir S, Ritzel R, Rizza RA, Butler PC. β-cell deficit and increased β-cell apoptosis in humans with type 2 diabetes. *Diabetes* [Internet]. 2003 Jan 1 [cited 2020 Dec 25];52(1):102–10.

[59] Weir GC. Glucolipotoxicity, β-cells, and diabetes: The emperor has no clothes. *Diabetes* [Internet]. 2020 Mar 1 [cited 2020 Dec 23];69(3):273–8.

[60] Inoue H, Shirakawa J, Togashi Y, Tajima K, Okuyama T, Kyohara M, et al. Signaling between pancreatic cells and macrophages via S100 calcium-binding protein A8 exacerbates β-cell apoptosis and islet inflammation. *J Biol Chem* [Internet]. 2018 Apr 20 [cited 2020 Dec 25];293(16):5934–46.

[61] Westwell-Roper CY, Ehses JA, Verchere CB. Resident macrophages mediate islet amyloid polypeptide-induced islet IL-1β production and β-cell dysfunction. *Diabetes* [Internet]. 2014 May 1 [cited 2020 Dec 25];63(5):1698–711.

[62] Guardado-Mendoza R, Davalli AM, Chavez AO, Hubbard GB, Dick EJ, Majluf-Cruz A, et al. Pancreatic islet amyloidosis, β-cell apoptosis, and α-cell proliferation are determinants of islet remodeling in type-2 diabetic baboons. *Proc Natl Acad Sci U S A* [Internet]. 2009 Aug 18 [cited 2020 Dec 28];106(33):13992–7.

[63] Jallut D, Golay A, Munger R, Frascarolo P, Schutz Y, Jéquier E, et al. Impaired glucose tolerance and diabetes in obesity: A 6-year follow-up study of glucose metabolism. *Metabolism* [Internet]. 1990 Oct 1 [cited 2020 Dec 27];39(10):1068–75.

[64] Hayden MR. *An Immediate and Long-Term Complication of COVID-19 May Be Type 2 Diabetes Mellitus: The Central Role of β-Cell Dysfunction, Apoptosis and Exploration of Possible Mechanisms*.

[65] Yang JK, Lin SS, Ji XJ, Guo LM. Binding of SARS coronavirus to its receptor damages islets and causes acute diabetes. *Acta Diabetol.* 2010 Sep;47(3):193–9.

[66] Zheng X, Zheng X, Wang X, Ma Z, Gupta Sunkari V, Botusan I, et al. Acute hypoxia induces apoptosis of pancreatic β-cell by activation of the unfolded protein response and upregulation of CHOP. *Cell Death Dis* [Internet]. 2012 Jun 14 [cited 2021 Jan 4];3(6):e322–e322.

[67] Maddaloni E, Buzzetti R. Covid-19 and diabetes mellitus: unveiling the interaction of two pandemics. Vol. 36, *Diabetes/Metabolism Research and Reviews*. John Wiley and Sons Ltd; 2020.

[68] Hayden MR, Sowers JR. Isletopathy in type 2 diabetes mellitus: Implications of islet RAS, islet fibrosis, islet amyloid, remodeling, and oxidative stress [Internet]. Vol. 9, *Antioxidants and Redox Signaling*. Mary Ann Liebert, Inc. 2 Madison Avenue Larchmont, NY 10538 USA; 2007 [cited 2021 Jan 4]. p. 891–910.

[69] Pina AF, Patarrão RS, Ribeiro RT, Penha-Gonçalves C, Raposo JF, Gardete-Correia L, et al. Metabolic Footprint, towards Understanding Type 2 Diabetes beyond Glycemia. *J Clin Med* [Internet]. 2020 Aug 10 [cited 2021 Jan 4];9(8):2588.

[70] Hayden MR, Sowers JR. Redox Imbalance in Diabetes. *Antioxid Redox Signal* [Internet]. 2007 Jul 17 [cited 2021 Jan 4];9(7):865–7.

[71] Hayden MR. Endothelial activation and dysfunction in metabolic syndrome, type 2 diabetes and coronavirus disease 2019 [Internet]. Vol. 48, *Journal of International Medical Research.* SAGE Publications Ltd; 2020 [cited 2021 Jan 4].

[72] Hayden MR, Sowers JR. Pancreatic renin-angiotensin-aldosterone system in the cardiometabolic syndrome and type 2 diabetes mellitus. [Internet]. Vol. 3, Journal of the cardiometabolic syndrome. *J Cardiometab Syndr;* 2008 [cited 2021 Jan 4]. p. 129–31.

[73] Hayden MR, Karuparthi PR, Habibi J, Wasekar C, Lastra G, Manrique C, et al. Ultrastructural Islet Study of Early Fibrosis in the Ren2 Rat Model of Hypertension. Emerging Role of the Islet Pancreatic Pericyte-Stellate Cell [Internet]. Vol. 8, JOP. *Journal of the Pancreas. Journal of the Pancreas;* 2007 [cited 2021 Jan 4].

[74] Aloysius MM, Thatti A, Gupta A, Sharma N, Bansal P, Goyal H. COVID-19 presenting as acute pancreatitis. *Pancreatology* [Internet]. 2020 Jul 1 [cited 2021 Jan 4];20(5):1026–7.

[75] Ceriello A, De Nigris V, Prattichizzo F. Why is hyperglycaemia worsening COVID-19 and its prognosis? [Internet]. Vol. 22, *Diabetes, Obesity and Metabolism.* Blackwell Publishing Ltd; 2020 [cited 2021 Jan 2]. p. 1951–2.

[76] Garbossa SG, Folli F. Vitamin D, sub-inflammation and insulin resistance. A window on a potential role for the interaction between bone and glucose metabolism [Internet]. Vol. 18, *Reviews in Endocrine and Metabolic Disorders.* Springer New York LLC; 2017 [cited 2020 Dec 27]. p. 243–58.

[77] Tabák AG, Jokela M, Akbaraly TN, Brunner EJ, Kivimäki M, Witte DR. Trajectories of glycaemia, insulin sensitivity, and insulin secretion before

diagnosis of type 2 diabetes: an analysis from the Whitehall II study. *Lancet* [Internet]. 2009 Jun 27 [cited 2020 Dec 27];373(9682):2215–21.

[78] Daniele G, Guardado Mendoza R, Winnier D, Fiorentino T V., Pengou Z, Cornell J, et al. The inflammatory status score including IL-6, TNF-α, osteopontin, fractalkine, MCP-1 and adiponectin underlies whole-body insulin resistance and hyperglycemia in type 2 diabetes mellitus. *Acta Diabetol* [Internet]. 2014 Feb [cited 2020 Dec 28];51(1):123–31.

[79] Fève B, Bastard JP. The role of interleukins in insulin resistance and type 2 diabetes mellitus [Internet]. Vol. 5, *Nature Reviews Endocrinology.* Nature Publishing Group; 2009 [cited 2020 Dec 29]. p. 305–11.

[80] Hotamisligil GS, Peraldi P, Budavari A, Ellis R, White MF, Spiegelman BM. IRS-1-mediated inhibition of insulin receptor tyrosine kinase activity in TNF-α- and obesity-induced insulin resistance. *Science* (80-) [Internet]. 1996 Feb 2 [cited 2020 Dec 26];271(5249):665–8.

[81] Méndez-García LA, Trejo-Millán F, Martínez-Reyes CP, Manjarrez-Reyna AN, Esquivel-Velázquez M, Melendez-Mier G, et al. Infliximab ameliorates tumor necrosis factor-alpha-induced insulin resistance by attenuating PTP1B activation in 3T3L1 adipocytes in vitro. *Scand J Immunol* [Internet]. 2018 Nov 1 [cited 2021 Jan 28];88(5).

[82] Akbari M, Hassan-Zadeh V. IL-6 signalling pathways and the development of type 2 diabetes [Internet]. Vol. 26, *Inflammopharmacology.* Birkhauser Verlag AG; 2018 [cited 2020 Dec 28]. p. 685–98.

[83] Antuna-Puente B, Feve B, Fellahi S, Bastard JP. Adipokines: The missing link between insulin resistance and obesity. Vol. 34, *Diabetes and Metabolism.* 2008. p. 2–11.

[84] Bastard JP, Maachi M, Lagathu C, Kim MJ, Caron M, Vidal H, et al. *Recent advances in the relationship between obesity, inflammation, and insulin resistance.*

[85] Pal R, Bhadada SK. COVID-19 and diabetes mellitus: An unholy interaction of two pandemics. *Diabetes Metab Syndr Clin Res Rev* [Internet]. 2020 Jul 1 [cited 2021 Jan 4];14(4):513–7.

[86] Kassir R. Risk of COVID-19 for patients with obesity [Internet]. Vol. 21, *Obesity Reviews.* Blackwell Publishing Ltd; 2020 [cited 2020 Dec 22].

[87] Kuba K, Imai Y, Rao S, Gao H, Guo F, Guan B, et al. A crucial role of angiotensin converting enzyme 2 (ACE2) in SARS coronavirus-induced lung injury. *Nat Med.* 2005 Aug;11(8):875–9.

[88] Underwood PC, Adler GK. The renin angiotensin aldosterone system and insulin resistance in humans [Internet]. Vol. 15, *Current Hypertension Reports.* Springer; 2013 [cited 2021 Jan 4]. p. 59–70.

[89] Takeda M, Yamamoto K, Takemura Y, Takeshita H, Hongyo K, Kawai T, et al. Loss of ACE2 exaggerates high-calorie diet-induced insulin

resistance by reduction of GLUT4 in mice. *Diabetes* [Internet]. 2013 Jan 1 [cited 2021 Jan 4];62(1):223–33.

[90] Rao S, Lau A, So H-C. Exploring diseases/traits and blood proteins causally related to expression of ACE2, the putative receptor of 2019-nCov: A Mendelian Randomization analysis. *Diabetes Care* [Internet]. 2020 May 12 [cited 2021 Jan 4];2020.03.04.20031237.

[91] Lee JS, Park S, Jeong HW, Ahn JY, Choi SJ, Lee H, et al. Immunophenotyping of covid-19 and influenza highlights the role of type i interferons in development of severe covid-19. *Sci Immunol* [Internet]. 2020 Jul 1 [cited 2021 Jan 4];5(49):1554.

[92] Šestan M, Marinović S, Kavazović I, Cekinović Đ, Wueest S, Turk Wensveen T, et al. Virus-Induced Interferon-γ Causes Insulin Resistance in Skeletal Muscle and Derails Glycemic Control in Obesity. *Immunity.* 2018 Jul 17;49(1):164-177.e6.

[93] Duffaut C, Zakaroff-Girard A, Bourlier V, Decaunes P, Maumus M, Chiotasso P, et al. Interplay between human adipocytes and T lymphocytes in obesity: CCL20 as an adipochemokine and T lymphocytes as lipogenic modulators. *Arterioscler Thromb Vasc Biol* [Internet]. 2009 Oct [cited 2021 Jan 4];29(10):1608–14.

[94] Viardot A, Grey ST, Mackay F, Chisholm D. Potential antiinflammatory role of insulin via the preferential polarization of effector T cells toward a T helper 2 phenotype. *Endocrinology* [Internet]. 2007 [cited 2021 Jan 4];148(1):346–53.

[95] Viardot A, Heilbronn LK, Samocha-Bonet D, Mackay F, Campbell LV., Samaras K. Obesity is associated with activated and insulin resistant immune cells. *Diabetes Metab Res Rev* [Internet]. 2012 Jul [cited 2021 Jan 4];28(5):447–54.

[96] Gómez-Rial J, Rivero-Calle I, Salas A, Martinón-Torres F. Role of monocytes/macrophages in COVID-19 pathogenesis: Implications for therapy [Internet]. Vol. 13, *Infection and Drug Resistance.* Dove Medical Press Ltd.; 2020 [cited 2021 Jan 4]. p. 2485–93.

[97] Ieronymaki E, Daskalaki MG, Lyroni K, Tsatsanis C. Insulin signaling and insulin resistance facilitate trained immunity in macrophages through metabolic and epigenetic changes [Internet]. Vol. 10, *Frontiers in Immunology.* Frontiers Media S.A.; 2019 [cited 2021 Jan 4]. p. 1330.

[98] Rizo-Téllez SA, Méndez-García LA, Flores-Rebollo C, Alba-Flores F, Alcántara-Suárez R, Manjarrez-Reyna AN, et al. The neutrophil-to-monocyte ratio and lymphocyte-to-neutrophil ratio at admission predict in-hospital mortality in mexican patients with severe sars-cov-2 infection (Covid-19). *Microorganisms* [Internet]. 2020 Oct 1 [cited 2021 Jan 4];8(10):1–17.

[99] Maessen DEM, Stehouwer CDA, Schalkwijk CG. The role of methylglyoxal and the glyoxalase system in diabetes and other age-related diseases [Internet]. Vol. 128, *Clinical Science*. Portland Press Ltd; 2015 [cited 2020 Dec 25]. p. 839–61.
[100] Hegab Z. Role of advanced glycation end products in cardiovascular disease. *World J Cardiol* [Internet]. 2012 [cited 2020 Dec 25];4(4):90.
[101] Ramasamy R, Yan SF, Herold K, Clynes R, Schmidt AM. *Receptor for Advanced Glycation End Products. Ann N Y Acad Sci* [Internet]. 2008 Apr 1 [cited 2020 Dec 25];1126(1):7–13.
[102] Fehrenbach, H, M Kasper, T Tschernig, M S Shearman, D Schuh MM. Receptor for advanced glycation endproducts (RAGE) exhibits highly differential cellular and subcellular localisation in rat and human lung. *Cell Moelcular Biol.* 1998;44 (7):1147–57.
[103] Oczypok EA, Perkins TN, Oury TD. All the "RAGE" in lung disease: The receptor for advanced glycation endproducts (RAGE) is a major mediator of pulmonary inflammatory responses [Internet]. Vol. 23, *Paediatric Respiratory Reviews*. W.B. Saunders Ltd; 2017 [cited 2021 Jan 3]. p. 40–9.
[104] Katsuoka F, Kawakami Y, Arai T, Imuta H, Fujiwara M, Kanma H, et al. Type II alveolar epithelial cells in lung express receptor for advanced glycation end products (RAGE) gene. *Biochem Biophys Res Commun* [Internet]. 1997 Sep 18 [cited 2021 Jan 3];238(2):512–6.
[105] Prasad K. AGE-RAGE Stress in the Pathophysiology of Pulmonary Hypertension and its Treatment. *Int J Angiol* [Internet]. 2019 Jun 2 [cited 2021 Jan 3];28(2):71–9.
[106] Nakamura K, Sakaguchi M, Matsubara H, Akagi S, Sarashina T, Ejiri K, et al. Crucial role of RAGE in inappropriate increase of smooth muscle cells from patients with pulmonary arterial hypertension. *PLoS One* [Internet]. 2018 Sep 1 [cited 2021 Jan 3];13(9).
[107] Polverino F, Celli BR, Owen CA. COPD as an endothelial disorder: endothelial injury linking lesions in the lungs and other organs? (2017 Grover Conference Series). In: *Pulmonary Circulation* [Internet]. SAGE Publications Ltd; 2018 [cited 2021 Jan 3].
[108] Ahn M, Anderson DE, Zhang Q, Tan CW, Lim BL, Luko K, et al. Dampened NLRP3-mediated inflammation in bats and implications for a special viral reservoir host. *Nat Microbiol.* 2019 May 1;4(5):789–99.
[109] Boteanu RM, Uyy E, Suica VI, Antohe F. High-mobility group box 1 enhances the inflammatory process in diabetic lung. *Arch Biochem Biophys.* 2015 Aug 15;583:55–64.
[110] Chikazawa M, Shibata T, Hatasa Y, Hirose S, Otaki N, Nakashima F, et al. Identification of C1q as a Binding Protein for Advanced Glycation End

Products. *Biochemistry* [Internet]. 2016 Jan 26 [cited 2020 Dec 25];55(3):435–46.

[111] Manzo G. COVID-19 as an Immune Complex Hypersensitivity in Antigen Excess Conditions: Theoretical Pathogenetic Process and Suggestions for Potential Therapeutic Interventions. *Front Immunol* [Internet]. 2020 Oct 21 [cited 2021 Jan 3];11.

[112] Means C. Mechanisms of increased morbidity and mortality of SARS-CoV-2 infection in individuals with diabetes: what this means for an effective management strategy. Vol. 108, *Metabolism: Clinical and Experimental.* W.B. Saunders; 2020.

[113] Guo YR, Cao QD, Hong ZS, Tan YY, Chen SD, Jin HJ, et al. The origin, transmission and clinical therapies on coronavirus disease 2019 (COVID-19) outbreak- A n update on the status. Vol. 7, *Military Medical Research.* BioMed Central Ltd.; 2020.

[114] Zhao P, Praissman J, Grant O, Cai Y, Xiao T, Rosenbalm K, et al. Virus-Receptor Interactions of Glycosylated SARS-CoV-2 Spike and Human ACE2 Receptor. *SSRN Electron J* [Internet]. 2020 [cited 2021 Jan 6];

[115] Brufsky A. Hyperglycemia, hydroxychloroquine, and the COVID-19 pandemic. *J Med Virol* [Internet]. 2020 Jul 27 [cited 2021 Jan 2];92(7):770–5.

[116] Bao W, Min D, Twigg SM, Shackel NA, Warner FJ, Yue DK, et al. Monocyte CD147 is induced by advanced glycation end products and high glucose concentration: possible role in diabetic complications. *Am J Physiol Physiol* [Internet]. 2010 Nov [cited 2021 Jan 3];299(5):C1212–9.

[117] De Francesco EM, Vella V, Belfiore A. COVID-19 and Diabetes: The Importance of Controlling RAGE [Internet]. Vol. 11, *Frontiers in Endocrinology.* Frontiers Media S.A.; 2020 [cited 2021 Jan 3].

[118] Pober JS, Sessa WC. Evolving functions of endothelial cells in inflammation [Internet]. Vol. 7, *Nature Reviews Immunology. Nat Rev Immunol;* 2007 [cited 2020 Dec 25]. p. 803–15.

[119] Almdal T, Scharling H, Jensen JS, Vestergaard H. The independent effect of type 2 diabetes mellitus on ischemic heart disease, stroke, and death: A population-based study of 13 000 men and women with 20 years of follow-up. *Arch Intern Med* [Internet]. 2004 Jul 12 [cited 2020 Dec 25];164(13):1422–6.

[120] Varga Z, Flammer AJ, Steiger P, Haberecker M, Andermatt R, Zinkernagel AS, et al. Endothelial cell infection and endotheliitis in COVID-19 [Internet]. Vol. 395, *The Lancet.* Lancet Publishing Group; 2020 [cited 2021 Jan 4]. p. 1417–8.

[121] Kumar S. COVID-19 and vascular disease. *EBio Medicine* [Internet]. 2020 [cited 2021 Jan 3];58:102966. Available from: https://doi.org/10.1016/j.ebiom.2020.102966.

[122] Kahn MB, Yuldasheva NY, Cubbon RM, Smith J, Rashid ST, Viswambharan H, et al. Insulin resistance impairs circulating angiogenic progenitor cell function and delays endothelial regeneration. *Diabetes* [Internet]. 2011 Apr 1 [cited 2020 Dec 25];60(4):1295–303.

[123] Iba T, Levy JH, Levi M, Connors JM, Thachil J. Coagulopathy of Coronavirus Disease 2019 [Internet]. Vol. 48, *Critical Care Medicine.* Lippincott Williams and Wilkins; 2020 [cited 2021 Jan 4]. p. 1358–64.

[124] Green SJ. Covid-19 accelerates endothelial dysfunction and nitric oxide deficiency [Internet]. Vol. 22, *Microbes and Infection.* Elsevier Masson SAS; 2020 [cited 2021 Jan 4]. p. 149–50.

[125] Åkerström S, Gunalan V, Keng CT, Tan YJ, Mirazimi A. Dual effect of nitric oxide on SARS-CoV replication: Viral RNA production and palmitoylation of the S protein are affected. *Virology* [Internet]. 2009 Dec 5 [cited 2021 Jan 4];395(1):1–9.

[126] Teuwen LA, Geldhof V, Pasut A, Carmeliet P. COVID-19: the vasculature unleashed [Internet]. Vol. 20, Nature Reviews Immunology. *Nature Research*; 2020 [cited 2021 Jan 4]. p. 389–91.

[127] Llorente L, De La Fuente H, Richaud-Patin Y, Alvarado-De La Barrera C, Diaz-Borjón A, López-Ponce A, et al. Innate immune response mechanisms in non-insulin dependent diabetes mellitus patients assessed by flow cytoenzymology. *Immunol Lett* [Internet]. 2000 Nov 1 [cited 2020 Dec 26];74(3):239–44.

[128] Joshi MB, Baipadithaya G, Balakrishnan A, Hegde M, Vohra M, Ahamed R, et al. Elevated homocysteine levels in type 2 diabetes induce constitutive neutrophil extracellular traps. *Sci Rep* [Internet]. 2016 Nov 4 [cited 2020 Dec 25];6(1):1–15.

[129] Zhang C, Yang J, Jennings LK. Leukocyte-derived myeloperoxidase amplifies high-glucose-induced endothelial dysfunction through interaction with high-glucose-stimulated, vascular non-leukocyte-derived reactive oxygen species. *Diabetes* [Internet]. 2004 Nov 1 [cited 2020 Dec 25];53(11):2950–9.

[130] Andreozzi F, Laratta E, Procopio C, Hribal ML, Sciacqua A, Perticone M, et al. Interleukin-6 Impairs the Insulin Signaling Pathway, Promoting Production of Nitric Oxide in Human Umbilical Vein Endothelial Cells. *Mol Cell Biol* [Internet]. 2007 Mar 15 [cited 2020 Dec 29];27(6):2372–83.

[131] Didion SP. Cellular and oxidative mechanisms associated with interleukin-6 signaling in the vasculature [Internet]. Vol. 18, *International Journal of*

Molecular Sciences. MDPI AG; 2017 [cited 2021 Jan 4]. Available from: /pmc/articles/PMC5751166/?report=abstract.

[132] Siddiqi HK, Libby P, Ridker PM. COVID-19 – A vascular disease [Internet]. *Trends in Cardiovascular Medicine.* Elsevier Inc.; 2020 [cited 2021 Jan 4]. Available from: /pmc/articles/PMC7556303/?report=abstract.

[133] Sturtzel C. Endothelial cells. In: *Advances in Experimental Medicine and Biology* [Internet]. Springer New York LLC; 2017 [cited 2021 Jan 4]. p. 71–91.

[134] Panigada M, Bottino N, Tagliabue P, Grasselli G, Novembrino C, Chantarangkul V, et al. Hypercoagulability of COVID-19 patients in intensive care unit: A report of thromboelastography findings and other parameters of hemostasis. *J Thromb Haemost* [Internet]. 2020 Jul 1 [cited 2021 Jan 4];18(7):1738–42.

[135] Figliozzi S, Masci PG, Ahmadi N, Tondi L, Koutli E, Aimo A, et al. Predictors of adverse prognosis in COVID-19: A systematic review and meta-analysis. *Eur J Clin Invest* [Internet]. 2020 Oct 1 [cited 2021 Jan 4];50(10).

[136] McFadyen JD, Stevens H, Peter K. The Emerging Threat of (Micro)Thrombosis in COVID-19 and Its Therapeutic Implications [Internet]. Vol. 127, *Circulation Research.* Lippincott Williams and Wilkins; 2020 [cited 2021 Jan 3]. p. 571–87.

[137] Jabaudon M, Blondonnet R, Roszyk L, Bouvier D, Audard J, Clairefond G, et al. Soluble receptor for advanced glycation end-products predicts impaired alveolar fluid clearance in acute respiratory distress syndrome. *Am J Respir Crit Care Med* [Internet]. 2015 Jul 15 [cited 2021 Jan 3];192(2):191–9.

[138] Tang N, Bai H, Chen X, Gong J, Li D, Sun Z. Anticoagulant treatment is associated with decreased mortality in severe coronavirus disease 2019 patients with coagulopathy. *J Thromb Haemost* [Internet]. 2020 May 1 [cited 2021 Jan 3];18(5):1094–9.

[139] Klok FA, Kruip MJHA, van der Meer NJM, Arbous MS, Gommers D, Kant KM, et al. Confirmation of the high cumulative incidence of thrombotic complications in critically ill ICU patients with COVID-19: An updated analysis. *Thromb Res* [Internet]. 2020 Jul 1 [cited 2021 Jan 4];191:148–50.

[140] Xu X, Chang XN, Pan HX, Su H, Huang B, Yang M, et al. [Pathological changes of the spleen in ten patients with coronavirus disease 2019(COVID-19) by postmortem needle autopsy]. *Zhonghua bing li xue za zhi = Chinese J Pathol* [Internet]. 2020 [cited 2021 Jan 4];49(6):576–82.

[141] Vulliamy P, Jacob S, Davenport RA. Acute aorto-iliac and mesenteric arterial thromboses as presenting features of COVID-19 [Internet]. Vol. 189, *British Journal of Haematology.* Blackwell Publishing Ltd; 2020 [cited 2021 Jan 4]. p. 1053–4.
[142] Wichmann D, Sperhake JP, Lütgehetmann M, Steurer S, Edler C, Heinemann A, et al. Autopsy Findings and Venous Thromboembolism in Patients With COVID-19: A Prospective Cohort Study. *Ann Intern Med* [Internet]. 2020 Aug 18 [cited 2021 Jan 4];173(4):268–77.
[143] Cattaneo M, Bertinato EM, Birocchi S, Brizio C, Malavolta D, Manzoni M, et al. Pulmonary embolism or pulmonary thrombosis in COVID-19? Is the recommendation to use high-dose heparin for thromboprophylaxis justified? [Internet]. Vol. 120, *Thrombosis and Haemostasis.* Georg Thieme Verlag; 2020 [cited 2021 Jan 4]. p. 1230–2.
[144] Zhang L, Feng X, Zhang D, Jiang C, Mei H, Wang J, et al. Deep Vein Thrombosis in Hospitalized Patients with COVID-19 in Wuhan, China: Prevalence, Risk Factors, and Outcome. *Circulation* [Internet]. 2020 Jul 14 [cited 2021 Jan 4];142(2):114–28.
[145] Karakike E, Giamarellos-Bourboulis EJ. Macrophage activation-like syndrome: A distinct entity leading to early death in sepsis [Internet]. Vol. 10, *Frontiers in Immunology.* Frontiers Media S.A.; 2019 [cited 2021 Jan 4]. Available from: https://pubmed.ncbi.nlm.nih.gov/30766533/.
[146] Chen G, Wu D, Guo W, Cao Y, Huang D, Wang H, et al. Clinical and immunological features of severe and moderate coronavirus disease 2019. *J Clin Invest* [Internet]. 2020 May 1 [cited 2021 Jan 4];130(5):2620–9.
[147] Henry BM, De Oliveira MHS, Benoit S, Plebani M, Lippi G. Hematologic, biochemical and immune biomarker abnormalities associated with severe illness and mortality in coronavirus disease 2019 (COVID-19): A meta-analysis [Internet]. Vol. 58, *Clinical Chemistry and Laboratory Medicine.* De Gruyter; 2020 [cited 2021 Jan 4]. p. 1021–8.
[148] Ranucci M, Ballotta A, Di Dedda U, Bayshnikova E, Dei Poli M, Resta M, et al. The procoagulant pattern of patients with COVID-19 acute respiratory distress syndrome. *J Thromb Haemost* [Internet]. 2020 Jul 1 [cited 2021 Jan 4];18(7):1747–51.
[149] Subramaniam S, Scharrer I. Procoagulant activity during viral infections. *Front Biosci - Landmark* [Internet]. 2018 Jan 1 [cited 2021 Jan 4];23(6):1060–81.
[150] Giannis D, Ziogas IA, Gianni P. Coagulation disorders in coronavirus infected patients: COVID-19, SARS-CoV-1, MERS-CoV and lessons from the past. *J Clin Virol.* 2020 Jun 1;127:104362.
[151] Ng LFP, Hibberd ML, Ooi EE, Tang KF, Neo SY, Tan J, et al. A human in vitro model system for investigating genome-wide host responses to

SARS coronavirus infection. *BMC Infect Dis* [Internet]. 2004 Sep 9 [cited 2021 Jan 4];4.

[152] Lupu F, Keshari RS, Lambris JD, Mark Coggeshall K. Crosstalk between the coagulation and complement systems in sepsis. *Thromb Res* [Internet]. 2014 [cited 2021 Jan 4];133(SUPPL. 1). Available from: https://pubmed.ncbi.nlm.nih.gov/24759136/.

[153] Gragnano F, Sperlongano S, Golia E, Natale F, Bianchi R, Crisci M, et al. The Role of von Willebrand Factor in Vascular Inflammation: From Pathogenesis to Targeted Therapy [Internet]. Vol. 2017, *Mediators of Inflammation.* Hindawi Limited; 2017 [cited 2021 Jan 4].

[154] Guo W, Li M, Dong Y, Zhou H, Zhang Z, Tian C, et al. Diabetes is a risk factor for the progression and prognosis of COVID-19. *Diabetes Metab Res Re*v [Internet]. 2020 Oct 1 [cited 2020 Dec 22];36(7).

[155] Irshad Z, Xue M, Ashour A, Larkin JR, Thornalley PJ, Rabbani N. Activation of the unfolded protein response in high glucose treated endothelial cells is mediated by methylglyoxal. *Sci Rep* [Internet]. 2019 Dec 27 [cited 2020 Dec 25];9(1):7889.

[156] Kraakman MJ, Lee MKS, Al-Sharea A, Dragoljevic D, Barrett TJ, Montenont E, et al. Neutrophil-derived S100 calcium-binding proteins A8/A9 promote reticulated thrombocytosis and atherogenesis in diabetes. *J Clin Invest* [Internet]. 2017 Jun 1 [cited 2020 Dec 25];127(6):2133–47.

[157] Dorsam RT, Kunapuli SP. Central role of the P2Y12 receptor in platelet activation. Vol. 113, *Journal of Clinical Investigation.* The American Society for Clinical Investigation; 2004. p. 340–5.

[158] Przygodzki T, Luzak B, Kassassir H, Mnich E, Boncler M, Siewiera K, et al. Diabetes and Hyperglycemia Affect Platelet GPIIIa Expression: Effects on Adhesion Potential of Blood Platelets from Diabetic Patients under In Vitro Flow Conditions. *Int J Mol Sci* [Internet]. 2020 [cited 2021 Jan 2];21(9).

[159] Gąsecka A, Borovac JA, Rui &, Guerreiro A, Giustozzi M, Parker W. Thrombotic Complications in Patients with COVID-19: *Pathophysiological Mechanisms, Diagnosis, and Treatment*. [cited 2021 Jan 4];8:9.

[160] Rui L. Energy metabolism in the liver. *Compr Physiol* [Internet]. 2014 [cited 2021 Jan 2];4(1):177–97.

[161] Xu E, Kumar M, Zhang Y, Ju W, Obata T, Zhang N, et al. Intra-islet insulin suppresses glucagon release via GABA-GABAA receptor system. *Cell Metab* [Internet]. 2006 Jan [cited 2020 Dec 27];3(1):47–58.

[162] Scherer PE. The multifaceted roles of adipose tissue - Therapeutic targets for diabetes and beyond: The 2015 banting lecture [Internet]. Vol. 65,

Diabetes. American Diabetes Association Inc.; 2016 [cited 2020 Dec 28]. p. 1452–61.

[163] Huang S, Hu B, Yin L. The cytokine storm and COVID-19. *J Med Virol.*

[164] Luo Y, Liu M. Adiponectin: A versatile player of innate immunity [Internet]. Vol. 8, *Journal of Molecular Cell Biology*. Oxford University Press; 2016 [cited 2021 Jan 5]. p. 120–8.

[165] Rossi A, Lord J. Adiponectin Inhibits Neutrophil Phagocytosis of Escherichia coli by Inhibition of PKB and ERK 1/2 MAPK Signalling and Mac-1 Activation. *PLoS One* [Internet]. 2013 Jul 25 [cited 2021 Jan 5];8(7).

[166] Rossi A, Lord JM. Adiponectin inhibits neutrophil apoptosis via activation of AMP kinase, PKB and ERK 1/2 MAP kinase. *Apoptosis* [Internet]. 2013 Dec [cited 2021 Jan 5];18(12):1469–80.

[167] Yang J, Lin SC, Chen G, He L, Hu Z, Chan L, et al. Adiponectin promotes monocyte-to-fibroblast transition in renal fibrosis. *J Am Soc Nephrol* [Internet]. 2013 Oct [cited 2021 Jan 5];24(10):1644–59.

[168] Luo L, Li J, Liu H, Jian X, Zou Q, Zhao Q, et al. Adiponectin is involved in connective tissue growth factor-induced proliferation, migration and overproduction of the extracellular matrix in keloid fibroblasts. *Int J Mol Sci* [Internet]. 2017 May 12 [cited 2021 Jan 5];18(5):1044.

[169] Jessica A. Palakshappa BJA, JPR, MGSS, AT, TD, DC, AK, JDC, NJM. Plasma Adiponectin Levels are Associated with Acute Respiratory Distress Syndrome in Severe Sepsis. *Am J Respir Crit Care Med* [Internet]. 2020 [cited 2021 Jan 5];201(A1609).

[170] Vijayaraghavan K. Treatment of dyslipidemia in patients with type 2 diabetes [Internet]. Vol. 9, Lipids in Health and Disease. *Lipids Health Dis;* 2010 [cited 2021 Jan 2]. Available from: https://pubmed.ncbi.nlm.nih.gov/21172030/.

[171] Libby P, Buring JE, Badimon L, Hansson GK, Deanfield J, Bittencourt MS, et al. Atherosclerosis. *Nat Rev Dis Prim* [Internet]. 2019 Dec 1 [cited 2020 Dec 25];5(1). Available from: https://pubmed.ncbi.nlm.nih.gov/31420554/.

[172] Shimada K, Mokuno H, Matsunaga E, Miyazaki T, Sumiyoshi K, Kume A, et al. Predictive Value of Circulating Oxidized LDL for Cardiac Events in Type 2 Diabetic Patients with Coronary Artery Disease [5] [Internet]. Vol. 27, *Diabetes Care*. American Diabetes Association; 2004 [cited 2020 Dec 25]. p. 843–4. Available from: https://care.diabetesjournals.org/content/27/3/843.

[173] Cominacini L, Fratta Pasini A, Garbin U, Davoli A, Tosetti ML, Campagnola M, et al. Oxidized low density lipoprotein (ox-LDL) binding to ox-LDL receptor-1 in endothelial cells induces the activation of NF-κB

through an increased production of intracellular reactive oxygen species. *J Biol Chem* [Internet]. 2000 Apr 28 [cited 2020 Dec 25];275(17):12633–8.

[174] Tesch GH. Diabetic nephropathy - is this an immune disorder? [Internet]. Vol. 131, *Clinical Science*. Portland Press Ltd; 2017 [cited 2021 Jan 1]. p. 2183–99.

[175] Najafian B, Alpers CE, Fogo AB. Pathology of human diabetic nephropathy. *Contrib Nephrol* [Internet]. 2011 Jun [cited 2021 Jan 1];170:36–47.

[176] Wada J, Makino H. Innate immunity in diabetes and diabetic nephropathy [Internet]. Vol. 12, *Nature Reviews Nephrology*. Nature Publishing Group; 2016 [cited 2021 Jan 1]. p. 13–26.

[177] You H, Gao T, Cooper TK, Reeves WB, Awad AS. Macrophages directly mediate diabetic renal injury. *Am J Physiol - Ren Physiol* [Internet]. 2013 Dec 15 [cited 2021 Jan 1];305(12):F1719.

[178] Liu C, Qin L, Ding J, Zhou L, Gao C, Zhang T, et al. Group 2 innate lymphoid cells participate in renal fibrosis in diabetic kidney disease partly via TGF-β1 signal pathway. *J Diabetes Res.* 2019;2019.

Section 2. Innate Immunity and Cancer

Chapter 4

A Dual Role for Neutrophils and Myeloid-Derived Suppressor Cells in the Development of Colorectal Cancer

C. Ángel Sánchez-Barrera[1], J. Efraín Olguín[2] and Luis I. Terrazas[1,2,*]
[1]Unidad de Investigación en Biomedicina
[2]Laboratorio Nacional en Salud,
Facultad de Estudios Superiores Iztacala,
Universidad Nacional Autónoma de México,
Tlalnepantla, Estado de México, Mexico

Abstract

Worldwide, colorectal cancer (CRC) is the third most common cancer in men and the second in women, while it is the third leading cause of cancer death in men and the fourth in women. Only 15% of CRCs arise from inherited diseases such as Lynch syndrome and polyposis associated with APC and MUTYH genes. The remaining 85% arises sporadically and is associated with different factors such as obesity, sedentary lifestyle, alcohol and tobacco consumption, low fiber consumption, high consumption of red/processed meats, alterations in the microbiota, and inflammatory bowel diseases. Inflammation is now recognized as a hallmark for tumor progression, especially in CRC. Inflammation implies many cell populations of the immune system and soluble factors,

* Corresponding Author's E-mail: literrazas@unam.mx.

In: The Innate Immune System in Health and Disease
Editor: Jorge Morales-Montor
ISBN: 978-1-68507-507-1

molecules, and cells that are not part of the immune system. Thus, an outstanding balance of the inflammatory response is critical to restrain or not the development of CRC. Even more, a disbalance of the inflammatory response may participate in restricting or favoring CRC development. Cells from the innate immune system and adaptive immune response play a significant role in orchestrating immunomodulation during CRC. In this chapter, we analyzed concrete evidence supporting that neutrophils and myeloid-derived suppressor cells (MDSCs) have an essential role in promoting tumor development during CRC and regulating its growth and metastasis. Most of the studies on MDSCs focus on how these cells mediate an increasing suppressive capacity, which controls the effector immune response generating protection for tumors. Controversial and more recent reports indicate that MDSCs may play a role in controlling excessive immune responses and consequently inflammation. Similarly, neutrophils have been considered a bad prognostic marker in patients suffering from CRC, but as recently described, the N1 type neutrophils may play an opposite role. These cell populations are now recognized and may go through a process called "phenotypic plasticity," where they co-express transcription factors and membrane molecules that promote an inflammatory profile. We focus on evidence that describes the interaction between the different effector populations of the immune response and its modulation by neutrophils and MDSCs adapted to the tumor microenvironment. Including the mechanisms used by neutrophils and MDSCs to modulate putative protective or deleterious immune responses, as well as the different subpopulations of both cell types participating in tumor progression, generating susceptibility or resistance during CRC development. Finally, we discuss whether innate immune cells and neutrophils or MDSCs might or might not be a therapeutic target for effectively reducing the morbidity and mortality caused by CRC.

Keywords: colorectal cancer, neutrophils, myeloid suppressor cells, inflammation, metastases

Abbreviations

AAMs	Alternatively activated macrophages
AKT	Protein kinase B
AOM	Azoxymethane
APC	Adenomatous polyposis coli
APCs	Antigen-presenting cells
CAC	Colitis associated colon cancer

CAMs	Classically activated macrophages
CD	Crohn's disease
cDCs	Conventional DCs
CIMP	CpG island methylator phenotype
CIN	Chromosomal instability
CMS	Consensus molecular subtypes
C-Myc	Cellular Myc signaling pathway
CRC	Colorectal Cancer
CRCSC	CRC subtyping consortium
CTLs	Cytotoxic T lymphocytes
CXCL	Chemokine (C-X-C motif)
CXCR	Chemokine (C-X-C motif)
DCs	Dendritic cells
DFS	Disease-free survival
DSS	Dextran sulfate sodium
EMT	Epithelial-mesenchymal transition
FAP	Familial adenomatous polyposis
FATP2	Fatty acid transport protein 2
G-CSF	Granulocyte colony-stimulating factor
G-MDCs	Granulocytic-myeloid derived suppressor cells
HDI	Human development index
HDN	High-density neutrophils
HNCC	Hereditary nonpolyposis colorectal cancer
IBD	Inflammatory bowel diseases
IFN-β	Interferon-beta
IFN-γ	Interferon gamma
IHC	Immunohistochemistry
IL	Interleukin
JAK	Janus kinase
Kras	Kirsten rat sarcoma viral oncogene
LDN	Low density neutrophils
LOX1	Lectin-type oxidized LDL receptor 1
LOXL4	Lysyl oxidase-like 4
LPS	Lipopolysaccharide
MAP	Multiple Colorectal Adenomas or MUTYH-associated polyposis
MAPK	Mitogen-activated protein kinases
MDSCs	Myeloid-derived suppressor cells
MHC-I	Major histocompatibility complex class I

M-MDSCs	Monocytic MDSCs
MMP9	Metalloproteinase 9
moDCs	Monocyte DCs
MPO	Myeloperoxidase
MSI	Microsatellite instability
NETs	Neutrophil extracellular traps
NF-κB	Nuclear factor kappa B
NLR	Neutrophil/lymphocyte ratio
NO	Nitric oxide
OS	Overall survival
OSM	Oncostatin 6
PD1	Programmed death molecule
pDCs	Plasmacytoid DCs
PDL	Programmed death-ligand
PFS	Progression-free survival
PI3K	Phosphatidylinositol 3-Kinase
RNI	Reactive nitrogen intermediates
ROS	Reactive oxygen species
SMAD4	SMAD family member 4
SPS	Serrated Polyposis Syndrome
STAT	Signal transducer and activator of transcription
TAMs	Tumor-associated macrophages
TANs	Tumor-associated neutrophils
TGF-β	Tumor Growth Factor beta
TILs	Tumor-infiltrating lymphocytes
TME	Tumor microenvironment
TNF-α	Tumor necrosis factor-alpha
TRAIL	TNF-related apoptosis-inducing ligand
Treg	Regulatory T cells
TRMP2	Transient receptor potential cation channel, subfamily M, member 2
UC	Ulcerative colitis
VEGF	Vascular endothelial growth factor
WNT	Wingless and Int-1 signaling pathway

Introduction

Incidence and mortality of colorectal cancer (CRC) have increased worldwide over time; specifically, in North America, a steady increase in CRC has been maintained for 20 years (Figure 1). According to 2020 GLOBOCAN statistics, CRC is the third most common cancer globally with 1.93 million cases (10% of total cancer cases) and is the second leading cause of death due to oncological pathology, accounting for 9.4% of total cancer deaths (935,173). Most CRC cases occur in countries with a high human development index (HDI). The incidence of CRC in countries with very high HDI is almost 1.5 times greater than countries with high HDI and almost 2.5 times greater than countries with low and medium HDI. Similar patterns are observed in mortality (World Health Organization (WHO) 2020; Sung et al. 2021). GLOBOCAN 2020 prediction estimates that, by the year 2030, there will be 2.2 million new cases of CRC and 1.09 million new deaths caused by CRC. Also, the incidence rate will increase by 42% in countries with low HDI, followed by 33.2% in countries with medium HDI and 33.9% in countries with high HDI, and, finally, 18.4% in countries with very high HDI. Mortality rates are expected to increase 42.1% in low HDI countries, 34.7% in countries with medium HDI, 39.8% in high HDI countries, and 22.4% in countries with very high HDI (WHO 2020).

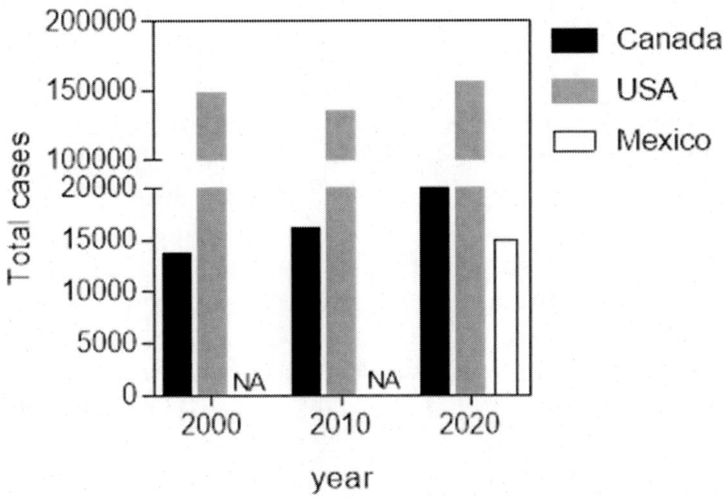

Figure 1. Total (both sexes) new cases reported for CRC in North America over the decades 2000-2020. NA: Not available.

The increased incidence of CRC is closely related to hereditary risk factors, personal medical history, and modifiable risk factors (American Cancer Society 2021). Approximately 20 to 30% of CRC patients have a family history of the disease; however, only 5% to 10% are related to a well-established inherited genetic disorder (American Cancer Society 2021; Jasperson et al. 2010; Wells and Wise 2017). Another 70 to 80% of cases of CRC arise sporadically by either environmental factors or modifiable risk factors such as obesity, sedentary lifestyle, alcohol and tobacco consumption, low fiber consumption, a diet rich in processed food and red meats, and microbiome alteration (C. M. Johnson et al. 2013). The remaining 15 to 20% is associated with chronic inflammatory bowel diseases (IBD) such as Crohn's disease (CD) and ulcerative colitis (UC) (Stidham and Higgins 2018).

Classification of CRC Subtypes

In most cases of CRC, tumors develop from non-cancerous polyps that arise in the mucosal layer (inner layer) of the colon or rectum. Although all polyps represent an increased risk of CRC, adenomatous and serrated polyps are considered the main precursors of colon tumors. Interestingly, polyps are frequently found in people over 50 years old and older. However, only less than 10% progresses to invasive cancer (American Cancer Society 2021; Ismail and Lance 2004; Rex et al. 2012; Øines et al. 2017).

Three primary forms of CRC origin have been described. First, the inherited pathway is divided into familial syndromes without the presence of colonic polyps, hereditary nonpolyposis colorectal cancer (HNCC or Lynch Syndrome), and diseases with the presence of colonic polyps such as familial adenomatous polyposis (FAP), multiple colorectal adenomas or MUTYH-associated polyposis (MAP), hamartomatous polyposis syndromes (Peutz-Jeghers, juvenile polyposis), Cowden syndrome and serrated polyposis syndrome (SPS) (American Cancer Society 2021; Jasperson et al. 2010; Wells and Wise 2017; Malki et al. 2020).

The second pathway involves sporadic colon tumor formation supported by either environmental factors or modifiable risk factors. According to Faeron's and Vogelstein's adenoma-to-carcinoma sequence model, sporadic colorectal tumors have a monoclonal origin, and they arise from the accumulation of mutations in oncogenes and tumor suppressor genes as well as chromosomal deletions and epigenetic changes (Fearon and Vogelstein 1990). This pathway is subdivided into four named 1) chromosomal instability

(CIN) or traditional pathway, 2) CpG island methylator phenotype (CIMP) or alternative pathway, 3) microsatellite instability (MSI) or serrated pathway, and 4) *de novo* (Worthley and Leggett 2010; Yamagishi et al. 2016).

Finally, IBD, such as UC and CD, promotes the development of colitis-associated colon cancer (CAC), a subtype of CRC where inflammation plays a critical role by favoring the initiation and progression of the tumor (Terzić et al. 2010). UC patients showed CRC increased risk about 2% at ten years after diagnosis, about 8% at 20 years, and about 20% at 30 years after IBD diagnosis (Eaden 2001).

The heterogeneity of CRC tumors plays an essential role in the prognosis and treatment of the disease. Correct classification of the tumor subtype can facilitate treatment and improve the accuracy of predictions (Fessler and Medema 2016). Over time, several classifications have been made based on colon tumors' clinical and molecular features such as localization, origin, gene expression, molecular markers, and microenvironmental regulation. Recently, the CRC subtyping consortium (CRCSC) unified 18 data sets and classified human colorectal carcinomas into four consensus molecular subtypes (CMS) (Fessler and Medema 2016; Guinney et al. 2015; Malki et al. 2020).

- MSI immune tumors (CMS1) account for 14% of CRC cases. They are hypermutated with MSI and strong immune activation.
- Canonical tumors (CMS2) represent about 37% of CRC cases. They are epithelial with strong wingless and Int-1 (WNT) and cellular-Myc (C-myc) signaling activation.
- Metabolic tumors (CMS3): account for 13% of CRC cases. They are epithelial with metabolic dysregulation.
- Mesenchymal tumors (CMS4) represent about 23% of CRC cases. They are tumor growth factor-beta (TGF-β) activated with stromal invasion and angiogenesis.

According to the CRCSC, the genomic instability caused by MSI favors the antigenicity in CMS1 tumors. These tumors show increased immune infiltrate and higher expression of immune checkpoints. The use of immune checkpoints inhibitors improves the prognosis for the patients. However, after recurrence, patients may have a worse prognosis. CMS2 tumors present an immune desert phenotype, while CMS3 tumors are immune excluded tumors. Finally, CMS4 tumors show a TGF-β rich environment associated with immunosuppressive effects (Guinney et al. 2015; Fidelle et al. 2020; Malki et al. 2020).

Immune Response in Colorectal Cancer

The immune system is critical and plays a complex role in CRC. Throughout CRC, the immune response varies depending on the tumor's microenvironment. These changes can delay and/or promote tumor progression (Guo et al. 2020). The tumor microenvironment (TME) is composed of a variety of cells of different origins and soluble factors. In the TME are grouped innate immune cells, including macrophages, neutrophils, natural killer cells (NK), dendritic cells (DCs), as well as adaptive immune cells such as T cells, B cells, and nests of fibroblasts, blood vessels that surround tumor cells (Labani-Motlagh, Ashja-Mahdavi, and Loskog 2020). The interaction among these cells is complicated by the secretion of a myriad of soluble factors such as cytokines, chemokines, growth factors, remodeling enzymes of the extracellular matrix, which can be originated even from the tumor itself. Thus, this complex microenvironment may take advantage of the tumor cells to keep themselves growing. Moreover, this TME is not static; in contrast, it is a dynamic process that continuously changes according to the tumor's stage, making it critical to know and understand the different cell populations participating in this microenvironment, mainly those associated cells with the immune response.

The innate immune response cells have a significant role in protection during the early stages of CRC, favoring the immunosurveillance and inhibiting tumor growth (Reid et al. 2021). Loss of major histocompatibility complex class I (MHC-I) favors the destruction of CRC-initiating cells by cytotoxic NK cells (Watson et al. 2006; Tallerico et al. 2013). Increased infiltration of NK cells is related to a higher overall survival (OS) and higher disease-free survival (DFS) in CRC patients (Coca et al. 1997). In addition, high blood levels of NK cells are associated with higher OS (Tang et al. 2020). On the other hand, increased expression of MHC-I promotes tumor cell destruction by cytotoxic T cells (Watson et al. 2006). An immunoscore system was established in stage I-III of colorectal tumors, based on the association between cytotoxic cell infiltration and disease prognosis (Guo et al. 2020). High immunoscore values are associated with the lowest risk of recurrence, increased OS, and increased DFS in CRC patients (Lanzi et al. 2020). Moreover, increased tumor-infiltrating lymphocytes (TILs) improve prognosis in CRC patients, however, activated TILs decrease in late stages of CRC (Koch et al. 2006). These observations highlight the importance of innate and adaptative cytotoxicity to induce protection during CRC.

Over time, strong evidence suggests that some colorectal tumors become resistant to the immune response by selecting immunoediting processes. Several mutations were found in immune-modulating pathways and antigen presentation machinery in tumors with high MSI (Grasso et al. 2018). This finding is consistent with the evidence suggesting that immune tumors (CSM1) have a worse prognosis after recurrence (Guinney et al. 2015; Fidelle et al. 2020; Malki et al. 2020). Furthermore, tumor cells also promote immunosuppression through the recruitment of immunosuppressive cells such as regulatory T (Treg) cells (Olguín et al. 2020) and MDSCs (Pernot 2014), which play a central role in CRC progression and metastasis.

Inflammation is the major event that natural selection and immune response developed to clear and resolve injuries. In fact, inflammation inhibits tumor progression. However, defects in this process without resolution of the initial steps lead to a state of chronic inflammation, where activation of tissue repair mechanisms can promote the initiation and progression of cancer (Medina-Andrade et al. 2021). For example, in CAC, inflammatory mediators such as reactive oxygen species (ROS) and reactive nitrogen intermediates (RNI) generate mutations specifically by DNA oxidation (Frick et al. 2018). Also, inflammatory cytokines such as interleukin (IL)-1β, IL-6, and tumor necrosis factor-alpha (TNF-α) activate transcription factors that favor tumor progression (e.g., *β-catenin, NF-κB*, and *STAT3*) (West et al. 2015; Lasry, Zinger, and Ben-Neriah 2016; Terzić et al. 2010). Thus, it is necessary to understand "the thin line" between protection and aggression caused by inflammation and all the immune mechanisms and cells involved in these initial steps.

The Innate Immune Response in Colorectal Cancer

Immune cells infiltrating tumors during CRC initiation and progression vary in composition, phenotype, and function. This chapter focuses on innate immune populations, including macrophages, DCs, neutrophils, and MDSCs, which constitute an important part of this infiltrate.

Macrophages

Macrophages develop from either circulating monocytes or tissue-resident macrophages, which differentiate during embryogenesis (Zhong, Chen, and

Yang 2018). According to the necessity for activation of immune responses, macrophages are classified into two main subtypes:

- The M1 or classically activated macrophages (CAMs) elicit in response to inflammatory stimuli such as interferon-gamma (IFN-γ) or bacterial products such as lipopolysaccharide (LPS) (Martinez and Gordon 2014). These cells exert a robust inflammatory response related to pathogen and tumoral cells destruction through the production of inflammatory mediators such as IL-1β, IL-6, TNF-α, nitric oxide (NO), and ROS (Zhong, Chen, and Yang 2018; Martinez and Gordon 2014).
- The M2 or alternatively activated macrophages (AAMs) are elicited in response to IL-4, IL-10, and IL-13. These cells produce anti-inflammatory cytokines (IL-10 and TGF-β), and they are mainly related to tissue repair processes (Zhong, Chen, and Yang 2018; Martinez and Gordon 2014). Although both M1 and M2 macrophages can be found in tumors, tumor-associated macrophages (TAMs) are closely related to the M2 subtype because their functions and secreted products are very similar (Zhong, Chen, and Yang 2018).

In CRC, macrophages are associated with both good and poor prognoses. Meta-analysis research carried out in 6115 patients from 27 studies showed a favorable 5-year OS in patients with a high density of TAMs, especially in those with mismatch repair system-proficient CRC (J. Li et al. 2020). Another meta-analysis research performed in 3749 patients from 17 studies showed that high density of TAMs is related to improved OS and high $CD8^+$ T cell tumor infiltration (Zhao et al. 2019).

Macrophage profiling allows a better understanding of predictive value during treatments in patients. For example, inflammatory macrophages (classified by expressing $CD68^+$ and $CD80^+$ molecules) found in less invasive tumors have a role associated with protection, whereas anti-inflammatory macrophages (classified by expressing $CD68^+$ and $CD163^+$ molecules) are frequently found in the invasive front of stage II tumors (Pinto et al. 2019). Moreover, the presence of $CD163^+$ macrophages in the invasive space of tumors was associated with epithelial-mesenchymal transition (EMT) (C. Wei et al. 2019), lymphovascular invasion, low histological differentiation, and lymph node metastasis (Inagaki et al. 2021). In addition, a lower CD80/CD163 ratio is associated with low OS (Pinto et al. 2019) and metastasis (Inagaki et al. 2021). The description and understanding of some surface markers in macrophages are necessary to better understand populations associated with either protection or aggression. Also, a unified classification could improve the conclusion about the role of macrophages during CRC (Bain and Mowat 2014).

Dendritic Cells

Dendritic cells (DCs) are considered the major population of antigen-presenting cells (APCs) and the bridge between innate and adaptative immunity. They can be classified broadly in three different subsets including, monocyte DCs (moDCs), plasmacytoid DCs (pDCs), and conventional DCs (cDCs) (Gardner, de Mingo Pulido, and Ruffell 2020; Pryczynicz et al. 2016). DCs can be identified by several markers such as S-100, HLA class II, CD208, CD86, CD1a, and others (Pryczynicz et al. 2016; Sandel et al. 2005).

In CRC, DCs infiltration in tumors is related to prolonged survival (Ambe, Mori, and Enjoji 1989). Furthermore, intraepithelial infiltration of S-100$^+$ DCs is associated with elevated DFS and higher infiltration of CD4$^+$ and CD8$^+$ T cells (Dadabayev et al. 2004). However, the maturation status of DCs can negatively influence CRC prognosis. Infiltration of mature DCs (CD208$^+$) to the epithelium decreases OS, whereas the presence of CD1a$^+$ DCs in the advancing margin of the tumor promotes short DFS (Sandel et al. 2005). In addition, infiltration of CD86$^+$ DCs in the tumoral microenvironment is related to lymph node metastasis (Pryczynicz et al. 2016). A better understanding of the maturation state of DCs, and their functionality as APCs is necessary. In fact, due to their enhanced antitumor capacity by inducing activation in CD4$^+$ and CD8$^+$ T cells, recent efforts have focused on developing treatments based on the activation of DCs (Gardner, de Mingo Pulido, and Ruffell 2020). According to the U.S National Library of Medicine, approximately 45 clinical trials were carried out in the last years using DCs "vaccine" in CRC patients, where 1 study is on phase III, 25 are on phase II, 24 are on phase I, and 1 is on early phase I (National Institutes of Health (NIH) 2021). A phase II clinical trial performed in patients with CRC and liver metastases showed that DCs vaccine increased the median of DFS from 9.53 months in the observational group to 25.26 months in the vaccinated group (Rodriguez et al. 2018). In addition, DCs vaccination has been combined with other therapies such as chemotherapy and cytokine-induced killer cells therapy. A meta-analysis of 871 patients from 4 clinical trials showed that combining chemotherapy with DCs vaccine and cytokine-induced killer cells therapy improves OS and DFS (Zhou et al. 2018). This evidence demonstrates the efficacy of DCs therapy in the treatment of CRC.

Neutrophils

Neutrophils are the most abundant leucocytes and constitute 50 to 70% of circulating white blood cells. They are small cells around 9 to 15 μm in diameter with segmented nucleus (2 to 5 lobes) and granular cytoplasm (Wickramasinghe and Erber 2011).

Neutrophils are considered markers of acute inflammation because they are among the first cells recruited to the inflammatory focus (Kolaczkowska and Kubes 2013). Initially, they were supposed as cells with a short half-life (3-24h) (Wu et al. 2019) with the unique ability to kill microbes. However, recent studies suggest that neutrophils play a critical role in chronic inflammation processes, demonstrating phenotypic plasticity (Sagiv et al. 2015) and a half-life of 5 or more days (Pillay et al. 2010). This discovery re-focusses the study of this cell population during chronic inflammatory processes, including cancer.

Neutrophil Subsets

In humans, neutrophils are identified as $CD11b^+$, $CD14^-$, $CD15^+$, $CD66b^+$ cells, whereas, in mice, they are identified as $CD11b^+$, $Ly6C^+$, $Ly6G^+$ cells. However, these markers are shared with both human and mice granulocytic-myeloid derived suppressor cells (G-MDSCs), respectively (Bronte et al. 2016; Gabrilovich 2017; Veglia, Sanseviero, and Gabrilovich 2021). In general, unlike neutrophils, G-MDSCs have a suppressive capacity. However, a suppressive subset of neutrophils has also been described (Sagiv et al. 2015). Then, neutrophil discrimination from G-MDSCs is done by density gradient, where G-MDSCs were enriched in low-density fraction, while neutrophils were identified in high-density fraction (Bronte et al. 2016; Gabrilovich 2017; Veglia, Sanseviero, and Gabrilovich 2021; Bergenfelz and Leandersson 2020). The main limitation of this method is that low-density faction also contains neutrophils (Bronte et al. 2016; Sagiv et al. 2015). Finally, expression of the lectin-type oxidized LDL receptor 1 (LOX1) was used to identify human but not mouse G-MDSCs (Bronte et al. 2016; Gabrilovich 2017; Veglia, Sanseviero, and Gabrilovich 2021). Serum levels of LOX1 are proposed as a biomarker of poor prognosis in CRC, associated with the neutrophil/lymphocyte ratio (Nakashima-Nakasuga et al. 2020). It was recently described that lysyl oxidase-like 4 (LOXL4) proteins are exclusively expressed in CRC patients' circulating neutrophils (Palmieri et al. 2020).

Currently, two classifications of neutrophils have been described based on different characteristics:

- Classification is based on cell density which involves the high-density neutrophils (HDN) and the low-density neutrophils (LDN) (Bronte et al. 2016; Veglia, Sanseviero, and Gabrilovich 2021; Sagiv et al. 2015).
- A relatively recent classification of neutrophils based on their cell functions identifies N1 neutrophils with antitumoral functions and N2 neutrophils with protumoral functions (Fridlender and Albelda 2012; Fridlender et al. 2009; Andzinski et al. 2016; Pylaeva, Lang, and Jablonska 2016).

Interestingly both classification systems have a cross-linking (Sagiv et al. 2015). For this reason, this chapter considers HDN and N1 neutrophils as one subpopulation and LDN and N2 neutrophils as another subpopulation.

Antitumor Functions of Neutrophils (HDN or N1 Neutrophils)

Neutrophils with antitumor functions have been described as small and mature cells with a segmented nucleus, a high migration capacity, highly active phagocytosis, and aggressive respiratory burst (Sagiv et al. 2015). Cytotoxic mechanisms of neutrophils involve 1) granzyme B induced apoptosis (Martin et al. 2018), 2) ROS mediated apoptosis induced by DNA damage (Yan et al. 2014) or by Ca^{2+} influx promoted by the activation of the transient receptor potential cation channel, subfamily M, member 2 (TRMP2) (Gershkovitz et al. 2018), 3) contact-dependent apoptosis mediated by TNF-related apoptosis-inducing ligand (TRAIL) (Koga et al. 2004) and Fas Ligand (Kim et al. 2000), 4) cell cycle arrest promoted by Fas Ligand (Sun et al. 2018), and 5) trogoptosis, a process where neutrophils phagocyte part of the plasma membrane of tumor cells opsonized by antibodies, causing cellular death by necrosis (Matlung et al. 2018) (Figure 2).

The antitumor functions of neutrophils are controlled mainly by the response to interferon-beta (IFN-β) (Andzinski et al. 2016) (Figure 2). IFN-β inhibited angiogenesis through the repression of genes such as vascular endothelial growth factor (VEGF) and metalloproteinase 9 (MMP9) in tumor-associated neutrophils (TANs) (Andzinski et al. 2016; Fridlender and Albelda 2012; Pylaeva, Lang, and Jablonska 2016). Likewise, this cytokine controls the recruitment of neutrophils to the tumor, inhibiting the expression of CXCR2 and CXC4 (Jablonska et al. 2014; Pylaeva, Lang, and Jablonska 2016). One clinical trial showed an increased number of neutrophils in patients on stage I to III of CRC but not in stage IV; thus, TANs were associated with protection (Galdiero et al. 2016).

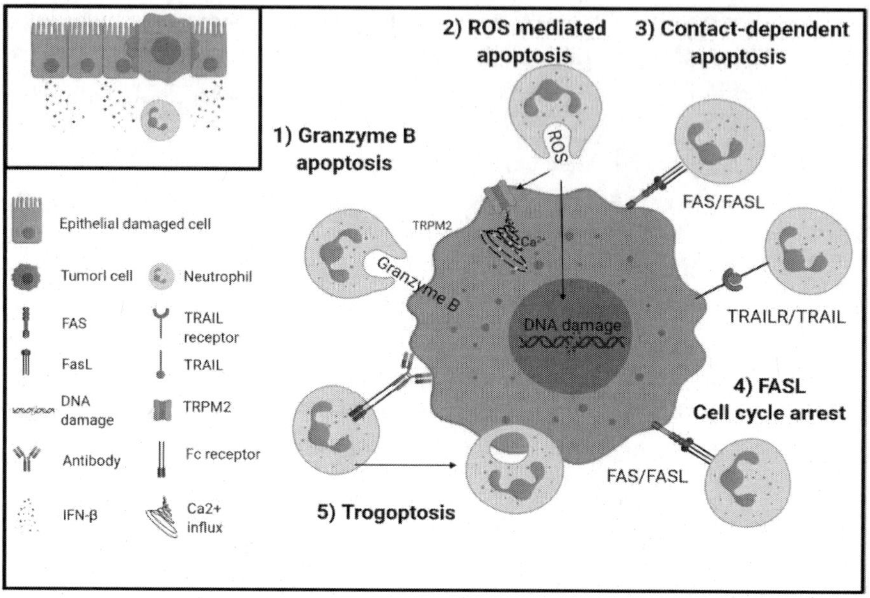

Figure 2. Antitumor functions of neutrophils. Neutrophil antitumor mechanisms include trogocytosis, induction of apoptosis, and cell cycle arrest. These functions are supported by an environment rich in IFN-β generated by damaged epithelial cells that secrete this cytokine as an alarm signal (Created with BioRender.com).

Protumoral Functions of Neutrophils (LDN or N2 Neutrophils)

Protumoral neutrophils have been identified as either mature cells with a segmented nucleus or immature cells with a band nucleus

(G-MDSCs); both populations are large, have a poor migratory capacity, reduced phagocytosis, and a weak respiratory burst (Sagiv et al. 2015). Furthermore, these neutrophils have been associated with chronic inflammatory processes due to their long half-life (Wu et al. 2019) compared to N1 neutrophils. Some protumoral functions of neutrophils are sustained by an environment rich in TGF-β (Fridlender et al. 2009; Fridlender and Albelda 2012) (Figure 3). These protumoral functions include the generation of mutations through ROS-mediated DNA damage (Knaapen 2006), the activation of transcription factors, as well as intracellular signaling pathways, that together promote proliferation and survival of tumor cells such as NF-κB (Gloire, Legrand-Poels, and Piette 2006), PI3K/AKT (Wen et al. 2019) and MAPK (Son et al. 2011), and the suppression of cytotoxic T lymphocytes (CTLs) (Kusmartsev et al. 2004).

Also, the epithelial-transformed cells in the colon induce the suppressive capacity of neutrophils over T cells via CXCR2 and TGFβR1 (Ruland 2019). Conversely, in the absence of SMAD4 protein, a signaling factor for TGF-β, there is an increased level of CXCL8 produced by neutrophils, which is associated with a shorter OS and relapse-free survival in CRC patients (Ogawa et al. 2019). Furthermore, the release of neutrophil extracellular traps (NETs) can promote metastasis by EMT, promoting the degradation of the extracellular matrix by MMP9 and neutrophil elastase (NE) (Felix and Gaida 2016) (Figure 3).

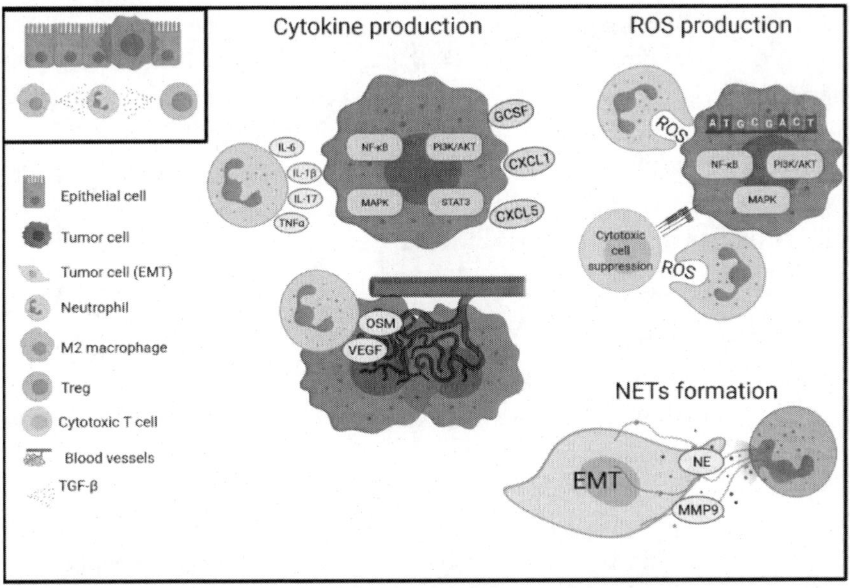

Figure 3. Protumor functions of neutrophils. The protumoral mechanisms of neutrophils include the generation of mutations through ROS-mediated DNA damage, the degradation of the extracellular matrix through the release of proteases by NETosis, the activation of protumoral factors through the production of cytokines, and suppression of the antitumor immune response. An environment rich in TGF-β generated by Treg and M2 macrophages supports these neutrophil functions are supported by (Created with BioRender.com).

Cytokine release generates an interaction between neutrophils and tumor cells. For example, IL-6, IL-17, IL-1β, and TNF-α produced by neutrophils (Tecchio, Micheletti, and Cassatella 2014), activate signaling pathways such as NF-κB, MAPK, PI3K / AKT, and STAT3 in tumor cells (Terzić et al. 2010; West et al. 2015; Lasry, Zinger, and Ben-Neriah 2016). Likewise, other neutrophil products such as oncostatin 6 (OSM) and VEGF favor the formation of new blood vessels (Queen et al. 2005). Moreover, tumor cells promote the recruitment of neutrophils to the tumor through granulocyte colony-stimulating factor (G-CSF), CXCL1, and CXCL2 (Sharma et al. 2013) (Figure 3). Kras protein mutation has been associated with CRC development; one study in mice suggests that *Kras* mutation favors the increase of NETs production by neutrophils in an IL-8 mediated way (Shang et al. 2020). Therefore, it is clear the relationship between tumoral and immune cells, which in some scenarios favored tumoral development.

Neutrophils in Colorectal Cancer

Tumor Associated-Neutrophils (TANs) have been reported in different types of cancer such as bronchioloalveolar carcinoma, metastatic melanoma, renal carcinoma, small cell lung cancer, and colorectal cancer. So far, the evidence suggests that the role of these cells is varied in promoting or inhibiting tumor development depending on their activation (Wu et al. 2019; Fridlender and Albelda 2012; Uribe-Querol and Rosales 2015). The tumoral microenvironment can switch off some immune cells' natural and protective functions, favoring their development and growth, even concluding in metastasis. TANs vary in promoting or inhibiting tumor functions depending on their activation. For this reason, neutrophils play a complex role in CRC initiation and progression summarizes the role of neutrophils and G-MDSCs in CRC development.

Table 1. Comparative protumoral and antitumoral activities reported for Neutrophils and G-MDSCs in colorectal cancer

Colorectal cancer	
Neutrophils	
Protumoral role	Antitumoral role
• High NLR values are associated with low DFS, decreasing OS and shorter PFS. • High pretreatment NLR values are related to shorter OS after treatment. • High preoperative NLR values are related to decreasing cancer-specific survival. • High preoperative NLR values are related to the lymphatic invasion. • High number of intratumoral neutrophils correlates with a more advanced tumor stage and shorter OS. • CXCL8 produced by neutrophils is associated with a shorter OS and relapse-free survival in CRC patients. • Neutrophils in colorectal cancer resected patients have reduced NETs formation and reduced apoptosis. The accumulation of these neutrophils may be harmful to patients. • $CD11b^+$ $Ly6C^+$ $Ly6G^+$ cells and IL-17 increase are related to accelerated tumor formation in $STAT1^{-/-}$ mice. • Blockade of IL-17 in $STAT1^{-/-}$ mice leads to decreasing of $CD11b^+$ $Ly6C^+$ $Ly6G^+$ cells and inhibition of tumor formation. • Neutrophils favor tumor formation by IL-1β production and mononuclear cell activation.	• High number of intratumoral neutrophils is associated with increased overall survival in stage II patients and $CD8^+$ cytotoxic T cell activation. • High number of intratumoral neutrophils predicts a better response to 5-fluorouracil-based chemotherapy. • High number of $CD177^+$ neutrophils is associated with increased OS and DFS. • $CD177^+$ neutrophil deficiency is related to an increase in tumor establishment. • The absence of neutrophils promotes the increased capacity of transformed cells to grow, proliferate and invade in an IL-17 dependent manner.

Table 1. (Continued)

Colorectal cancer	
Neutrophils	
Protumoral role	Antitumoral role
• Elimination of neutrophils leads to a reduction in the size and number of colon tumors. • Histone deacetylase inhibitor prevents the protumoral effects of AOM/DSS in a CAC model through reduction of infiltrating macrophages and neutrophils. • Tumor-infiltrating T-cell suppression by neutrophils via metalloproteinase-mediated activation of TGF-β. • Angiogenesis mediated by neutrophils confers resistance to anti-VEGF therapy. • NETs formation favors metastasis in the peritoneum and liver. • Blockade of NETs formation enhances anti-PD-1 therapy and reduces metastasis.	
G-MDSCs	
• $LOX1^+$ G-MDSCs are proposed as a biomarker of poor prognosis in CRC patients. • Suppression of the activation of T cells. • G-CSF blockade reduces G-MDSCs number and reduces tumor formation. • G-MDSCs exosomes promote CRC cell stemness and CRC progression. • CD244+ G-MDSCs are associated with the peritoneal dissemination of CRC cells.	None

Neutrophil/lymphocyte ratio (NLR) in the blood of patients with CRC has been used as a prognostic factor for the degree of tumor development and patient survival. Differences in NLR between healthy, adenomatous polyp and colorectal cancer patients have been reported. Cancer patients exhibited higher NLR values than healthy people, while patients with advanced (high-risk) polyps showed higher NLR values than low-risk polyps (W. Zhou, Chu, and An 2017). In addition, high NLR values were associated with poor clinical outcomes in CRC patients, including worse DFS (Rashtak et al. 2017), decreasing OS (Dell'Aquila et al. 2018; Naszai, Kurjan, and Maughan 2021; J. Zhang et al. 2017), and shorter progression-free survival (PFS) (Dell'Aquila et al. 2018). Moreover, high pretreatment NLR values were related to shorter OS in CRC patients after treatment (Ozdemir et al. 2014; M.-X. Li et al. 2014). High preoperative NLR values correlated with decreasing cancer-specific survival and lymphatic invasion (Shibutani et al. 2013; Jakubowska et al. 2020).

Tumor-infiltrating neutrophils in CRC patients have been the subject of discussion. Some studies using CD66b (Galdiero et al. 2016; Droeser et al. 2013; Wikberg et al. 2017; Berry et al. 2017; Rao et al. 2012; Governa et al. 2017), myeloperoxidase (MPO) (Droeser et al. 2013; Galdiero et al. 2016) and CD159 detection by immunohistochemistry (IHC) as the leading way to describe a role for neutrophils in CRC have been reported. Adaptations, like blinded to stage/outcome with different pathologists for a better approach, were described. With all these characteristics, MPO^+ cell infiltration was related to higher OS (Droeser et al. 2013), whereas neutrophils' role in protecting and increasing OS in patients in stage II of CRC was described (Berry et al. 2017). In addition, neutrophil tumor infiltration correlated with $CD8^+$ cytotoxic T cell activation (Governa et al. 2017) and better response to 5-fluorouracil-based chemotherapy (Galdiero et al. 2016). All this evidence suggests a protective role of intratumoral neutrophils in CRC patients; however, as we mentioned before, the tumor microenvironment can alter neutrophil functions. In fact, a high number of intratumoral neutrophils have also been correlated with more advanced tumor stages and shorter OS (Rao et al. 2012).

It is necessary to have a better approach with quantitative techniques as both flow and mass cytometry with specific markers to better define the role of TANs in CRC. A study using percoll gradient to enrich neutrophils from patients undergoing colorectal cancer resection showed that, after surgery, neutrophils of patients displayed reduced NETs formation and apoptosis, indicating that these neutrophils may accumulate and be harmful to patients

(Richardson et al. 2017). A study including 378 patients with CRC from China, using flow cytometry, IHC, and quantitative PCR to identify neutrophils, found an increased population of neutrophils characterized by the expression of the surface marker CD177, which correlated with an increased OS and DFS. In support of these findings, experimental and specific depletion of CD177 in mice leads to an increased susceptibility to CAC (Zhou et al. 2018). Together these studies suggest a beneficial role for neutrophils in CRC.

Animal models represent a useful approach for the study of neutrophil biology. In the CAC AOM/DSS model, the role of neutrophils is also controversial. It has been reported that these cells inhibit tumor development by controlling the intestinal microbiota (Triner et al. 2019). Likewise, $CD177^{+}$ neutrophil deficiency is related to an increase in the establishment of colon tumors (Zhou et al. 2018). On the other hand, it was shown that neutrophils favor tumor formation through IL-1β/IL-6 activation (Wang et al. 2014).

Furthermore, eliminating neutrophils leads to a reduction in the size and number of colon tumors (Shang et al. 2012). Finally, through antibiotics, the regulation of the intestinal microbiota prevents tissue damage mediated by infiltrating neutrophils (Tanaka, Ito, and Isobe 2016). Also was described that a histone deacetylase inhibitor prevents the protumoral effects of AOM/DSS in a CAC model. This protection is associated with reduced infiltration of macrophages and neutrophils near the tumor development zone. Together these results suggest a protumoral role for both neutrophils and macrophages in the murine model of CAC; however, the technique used for these conclusions was IHC (Wei et al. 2016). Using a specific Lys-cre mouse model for the depletion of neutrophils during CAC, it was found an increased capacity of the transformed cells to grow, proliferate and invade the host, that was IL-17-mediated, which suggests that neutrophils are related to protection (Triner et al. 2019).

Interestingly, in $STAT1^{-/-}$ mice, high amounts of IL-17 and $CD11b^{+}$ $Ly6C^{+}Ly6G^{+}$ cells were associated with accelerated tumor formation (Leon-Cabrera et al. 2018). As we mentioned before, IFN-β signaling plays an essential role in protumoral neutrophil functions (Andzinski et al. 2016), and STAT1 is required for IFN-β signaling. Together, these findings suggest that absent or defective STAT1 signaling in neutrophils can promote IL-17 mediated tumor growth. In fact, depletion of IL-17 in $STAT1^{-/-}$ mice leads to decreasing $CD11b^{+}$ $Ly6C^{+}$ $Ly6G^{+}$ cells and inhibiting tumor formation (Delgado-Ramirez et al. 2021).

Genetic mouse models represent another way to study neutrophils in CRC. Studies made in $APC^{fl/fl}$ mice showed tumor-infiltrating T-cell suppression by neutrophils via metalloproteinase-mediated activation of TGF-β (Germann et al. 2020). In addition, it has been reported that angiogenesis mediated by neutrophils confers resistance to anti-VEGF therapy (Itatani et al. 2020). On the other hand, mouse models of CRC induced by cell lines have allowed the study of neutrophils in metastatic processes where it was shown that NETs formation favors metastasis to the peritoneum (Al-Haidari et al. 2019) and liver (Yang et al. 2020). In fact, blockade of NETs enhances anti-PD-1 therapy and reduces metastasis of CRC cells (Onuma et al. 2020).

Myeloid-Derived Suppressor Cells in Colorectal Cancer

Myeloid-derived suppressor cells are considered a heterogeneous population of myeloid origin that has an immature state and have been associated with the ability to suppress T cell proliferation. Morphologically, MDSCs are a mixed population where monocytic and granulocytic cells can be identified yet immature. These cells were originally identified in mice expressing either CD11b and Gr-1 surface markers, but no lymphocytes or mature macrophages markers. However, antibodies recognizing the marker Gr-1 were also binding to another epitope known as Ly6C. This observation led to the development of more specific monoclonal antibodies, anti-Ly6C and anti-Ly6G, which improved the identification of MDSCs in monocytic MDSCs (M-MDSCs) and granulocytic (G-MDSCs), respectively. Currently, it is widely accepted that M-MDSCs express in their membranes $CD11b^{+}$, $Ly6C^{high}$, and $Ly6G^{-}$, whereas the granulocytic MDSCs express $CD11b^{+}$, $Ly6C^{low}$, $Ly6G^{high}$ (Youn and Gabrilovich 2010). As described in mice and humans, a similar distinction has been accepted regarding M-MDSCs and G-MDSCs. However, the surface markers are not yet well defined. Human MDSCs are usually identified as $CD14^{-}$, $CD11b^{+}$, $CD33^{+}$; plus, CD15 for defining G-MDSCs whereas the expression of CD14 and CD66 is associated with M-MDSCs (Veglia, Sanseviero, and Gabrilovich 2021). Nevertheless, in both humans and mice, the largest subpopulation of MDSCs is G-MDSCs, but this subpopulation is not necessarily the most suppressive.

Myeloid-derived suppressor cells are present in the circulation of healthy individuals just in low numbers. This cell population rapidly increases after exposure to an infection or inflammatory insult. However, the most dramatic expansion of MDSCs occurs during the development or presence of different tumors in the host. The role of these MDSCs either in circulation or in TME is mainly negative since they have been recognized for their immunosuppressive role (Kumar et al. 2016).

The suppressive ability of MDSCs includes both soluble factors and cell-contact dependent mechanisms. The main soluble factors participating in a suppressive TME are NO, which comes from M-MDSCs, and ROS from G-MDSCs. Besides these classical suppressor mechanisms, MDSCs can also inhibit immune responses by depleting essential amino acids, such as L-arginine, through the overexpression and activity of arginase-1. Also, IDO-mediated depletion of tryptophan in TME is another mechanism to inhibit T cell responses to tumors used by MDSCs. Other soluble factors released by MDSCs with adverse effects on the immune response are cytokines such as IL-10, TGF-b and IL-17, and chemokines such as CCL12, CXCL8, and CXCL12 (Umansky et al. 2016)· The secretion of exosomes from MDSCs can be considered another soluble factor with immunosuppressive activity (Rashid et al. 2021). In the cell-cell interaction to induce immune response suppression, MDSCs may increase the expression of galectin-9 and FAS-L, which bind to TIM-3 and FAS expressed on activated T cells (either CD4 or CD8), both interactions lead to inhibit T cell proliferation and survival (Dardalhon et al. 2010; Sinha et al. 2011). More recently, PD-L1 was also reported to be expressed on MDSCs, and it is involved in blocking T cell activity by binding to CD4 and CD8 activated T cells expressing PD1 (Ballbach et al. 2017).

Given its immunosuppressive capacity, the increased presence of MDSCs in circulation and colonic tissue has been associated with CRC as a poor prognostic marker in both human CRC patients and in different mouse models of colon cancer (Sieminska and Baran 2020; Ren et al. 2021; Itatani et al. 2014; Wang et al. 2020). Specifically, an increase in circulating $CD33^+$, $CD11b^+$ MDSCs as well as its infiltration in the colonic tissue, has been detected in CRC patients, which positively correlated with poor prognosis (Zhang et al. 2013). Unfortunately, both MDSCs subpopulations, M-MDSCs and G-MDSCs, have been clinically associated with poor prognosis in CRC. Recently, a clinical study demonstrated an increased expression of CD38 on M-MDSCs in CRC patients. Such M-MDSCs $CD38^+$ subpopulation expanded twice and was associated with patients already receiving treatment against CRC but not in CRC patients without treatment. This study indicates that a delicate balance of these MDSCs can be negatively altered after medication in CRC patients. Moreover, M-MDSCs $CD38^+$ displayed a more immunosuppressive activity than monocytes from the same CRC patients (Karakasheva et al. 2018).

The Jak/STAT signaling pathway appears to be a critical factor in the MDSCs role in CRC, where STAT3 activation (phosphorylation) enhances MDSCs suppressive activity by increasing ROS production and promoting the survival of these cells in a mouse model of CAC (Callejas et al. 2019). STAT3 activation also is increased in colon tissue from patients suffering CRC (Corvinus et al. 2005). Other STATs seem to have similar roles in expanding MDSCs in CRC. For example, different research groups have reported that STAT6 activation can favor the expansion and recruitment of MDSCs in a mouse model of colon cancer and subsequently favor the growth and proliferation of colon tumors (Jayakumar and Bothwell 2017), interestingly mice lacking STAT6 display lower numbers of MDSCs and become resistant to colon cancer development (Leon-Cabrera et al. 2017). Even more, the in vivo inhibition of STAT6 activation is enough to down-modulate the enhancement and recruitment of MDSCs in the same mouse model (Mendoza-Rodríguez et al. 2020). Another critical STAT signaling factor involved in the role of MDSCs on CRC appears to be STAT1. STAT1 is critical for IFN-γ signaling and activating several immune cells such as macrophages, NK, CD4, and CD8 cells (Meiss et al. 2017). MDSCs isolated from mice injected with a colorectal human cell line displayed a deficient STAT1 activation. Whereas mice lacking STAT1 displayed enhanced susceptibility to colon cancer and expanded G-MDSCs subpopulation and a significant reduction of M-MDSCs (Leon-Cabrera et al. 2018), demonstrating that STAT1 is a crucial regulator of MDSCs in colon cancer because it at least alters the balance between M-MDSCs and G-MDSCs. However, their suppressive function in the absence of STAT1 still needs to be analyzed.

Granulocytic-Myeloid-Derived Suppressor Cells in Colorectal Cancer

The similarities between neutrophil and G-MDSCs complicate the discrimination of both populations. The arising question is whether both neutrophil and G-MDSCs are different cells or a single cell population with phenotypic plasticity. Recently, new markers in both populations have been described to identify these cells better (Figure 4). For example, G-MDSCs isolated from tumor-bearing mice and neutrophils isolated from healthy mice were characterized, allowing CD115 and CD244 expression recognition in G-MDSCs but not in neutrophils (Youn et al. 2012). Lipid metabolism of G-MDSCs may be another difference concerning neutrophils, in fact, lipid uptake by the fatty acid transport protein 2 (FATP2) plays an important role in the suppressive activity of both human and mouse G-MDSCs (Veglia et al. 2019; Veglia, Sanseviero, and Gabrilovich 2021). In addition, LOX-1 was identified as a marker of immunosuppressive neutrophils with G-MDCSs gene signature in cancer patients (Condamine et al. 2016).

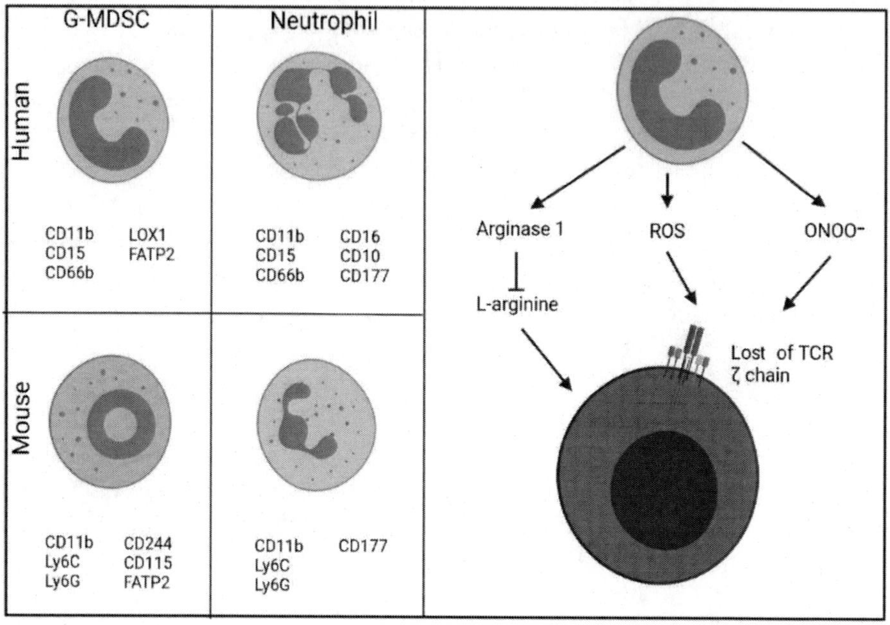

Figure 4. Currently proposed discrimination markers of Neutrophils and G-MDSCs. Right panel: main suppression mechanism used by G-MDSCs (Created with BioRender.com).

Neutrophil age differentiation is another way to solve this arising question. CD16 is considered a marker of mature neutrophils with decreasing expression in G-MDSCs (Yungang Wang et al. 2020). Furthermore, a study made in G-CSF-treated donors with several types of cancer described a mature population of CD66b$^+$ CD10$^+$ neutrophils and an immature population of CD66b$^+$ CD10$^-$ neutrophils. By contrast, mature neutrophils showed suppressive activity, whereas immature neutrophils favored T cell activation (Marini et al. 2017). So far, it has been proposed that G-MDSCs arise due to aberrant granulopoiesis (Kramer and Abrams 2020). However, more research is required to achieve a better understanding and a unified nomenclature of these cell subsets.

Under this controversy, the role of G-MDSCs in CRC has been studied in patients and animal models. In CRC patients, serum LOX1 is proposed as a biomarker of poor prognosis associated with the NLR62. Moreover, lysyl oxidase-like 4 (LOXL4) proteins were found exclusively in CRC patients' circulating neutrophils (Palmieri et al. 2020).

The main protumoral mechanism proposed for G-MDSCs is suppressing T cell activation (Yin et al. 2020). Arginase 1, ROS production, and peroxynitrite production were described as the preferred suppression mechanisms of G-MDSCs (Figure 4). In fact, blockade of CCL21 and G-CSF1 reduces the G-MDSCs number in a STAT3-dependent manner, and in consequence, there is a decreased formation of tumors. On the other hand, hypoxia through hypoxia-inducible factor (HIF-1α) induces the secretion of exosomes in G-MDSCs, which can promote CRC cell stemness and CRC progression (Wang et al. 2019). Finally, $CD244^{+}$ G-MDSCs were associated with the peritoneal dissemination of CRC cells in a CRC cell line mouse model (Sugita et al. 2021).

Conclusion

As can be appreciated along with this chapter, the innate immune cells may have a critical role in fueling the growth and establishment of CRC. Nevertheless, innate immune cells also have a protective role in avoiding tumor dispersion. Thus, for researchers and clinicians is very important to know the TME, the innate and adaptive immune cells recruited around the tumor, and, importantly, be aware of their activation profile. If we have access to all these parameters, it is plausible to design specific immunologic strategies to improve the fight against colon tumors. Several immune strategies have been explored to modify the outcome of CRC, and more may be developed in a few years.

Different strategies can target innate immune cells to modify the TME and improve CRC outcomes. On one side, we can boost the activation of innate immune cells, and on the other side, we can eliminate the suppressive activity of the innate cells. In both cases, drugs or immunotherapy can be afforded.

"Vaccination" with DCs was one of the most explored alternatives to improve immune responses against different types of cancers. In CRC it is necessary to find standard, specific, and most immunogenic CRC-antigens to load DCs with these antigens and return them to the host as tumor-loaded DCs, which must be completely mature and correctly activated in order to induce a prominent and effective immune response against the tumor.

Another possibility to improve the effector of the innate immune response against CRC is favoring the cytotoxic activity of neutrophils, which can be expanded by using G-CSF and increasing their Fc receptors to bind to IgG specific antitumor antibodies, increasing its antibody-dependent cellular cytotoxic activity.

One more strategy to modify the activity of innate immunity in CRC may be trying to eliminate the suppressive activity of different populations of innate immune cells, such as M2 macrophages, N2 neutrophils, or MDSCs. For example, it is possible to modify the state of activation of TAMs with an M2 profile by inducing them to an inflammatory and cytotoxic profile by TLR-mediated stimulation or blocking inhibitory signals such as STAT6. Different strategies are expected to be soon available to modulate the innate immune response to CRC, such as specific elimination of N2 neutrophils or inducing maturation on MDSCs to eliminate their potent suppressive activity. Moreover, identifying the selective expression of CD38 on these MDSCs may be possible the development of specific antibodies to eliminate these suppressor cells in CRC patients. Hopefully, some of these strategies will be available soon for patients suffering from CRC.

Acknowledgments

This work was supported by Dirección General de Asuntos del Personal Académico, UNAM, grants PAPIIT IN226519 to LIT; PAPIIT IA209720 to JEO, and by Consejo Nacional de Ciencia y Tecnología, grants A1-S 37879 and 315817 (2021).

References

Al-Haidari, Amr A., Nader Algethami, Mattias Lepsenyi, Milladur Rahman, Ingvar Syk, and Henrik Thorlacius. 2019. "Neutrophil Extracellular Traps Promote Peritoneal Metastasis of Colon Cancer Cells." *Oncotarget* 10 (12). https://doi.org/10.18632/oncotarget.26664.

Ambe, Kenji, Masaki Mori, and Munetomo Enjoji. 1989. "S-100 Protein-Positive Dendritic Cells in Colorectal Adenocarcinomas. Distribution and Relation to the Clinical Prognosis." *Cancer* 63 (3). https://doi.org/10.1002/1097-0142(19890201)63:3496:AID-CNCR2820630318>3.0.CO;2-K.

American Cancer Society. 2021. *Colorectal Cancer Facts & Figures 2020-2022.* 2021. https://www.cancer.org/research/cancer-facts-statistics/colorectal-cancer-facts-figures.html.

Andzinski, Lisa, Nadine Kasnitz, Stephanie Stahnke, Ching-Fang Wu, Marcus Gereke, Maren von Köckritz-Blickwede, Bastian Schilling, Sven Brandau, Siegfried Weiss, and Jadwiga Jablonska. 2016. "Type IIFNs Induce Anti-Tumor Polarization of Tumor Associated Neutrophils in Mice and Human." *International Journal of Cancer* 138 (8). https://doi.org/10.1002/ijc.29945.

Bain, Calum C., and Allan McI. Mowat. 2014. "Macrophages in Intestinal Homeostasis and Inflammation." *Immunological Reviews* 260 (1). https://doi.org/10.1111/imr.12192.

Ballbach, M, A Dannert, A Singh, DM Siegmund, R Handgretinger, L Piali, N Rieber, and D Hartl. 2017. "Expression of Checkpoint Molecules on Myeloid-Derived Suppressor Cells." *Immunology Letters* 192 (December): 1–6. https://doi.org/10.1016/J.IMLET.2017.10.001.

Bergenfelz, Caroline, and Karin Leandersson. 2020. "The Generation and Identity of Human Myeloid-Derived Suppressor Cells." *Frontiers in Oncology* 10 (February). https://doi.org/10.3389/fonc.2020.00109.

Berry, RS, MJ Xiong, A Greenbaum, P Mortaji, RA Nofchissey, F Schultz, C Martinez, L Luo, KT Morris, and JA Hanson. 2017. "High Levels of Tumor-Associated Neutrophils Are Associated with Improved Overall Survival in Patients with Stage II Colorectal Cancer." *PloS One* 12 (12). https://doi.org/10.1371/JOURNAL.PONE.0188799.

Bronte, Vincenzo, Sven Brandau, Shu-Hsia Chen, Mario P. Colombo, Alan B. Frey, Tim F. Greten, Susanna Mandruzzato, et al. 2016. "Recommendations for Myeloid-Derived Suppressor Cell Nomenclature and Characterization Standards." *Nature Communications* 7 (1). https://doi.org/10.1038/ncomms12150.

Callejas, Blanca E., Mónica G. Mendoza-Rodríguez, Olga Villamar-Cruz, Sandy Reyes-Martínez, Cuauhtémoc Angel Sánchez-Barrera, Miriam Rodríguez-Sosa, Norma L. Delgado-Buenrostro, et al. 2019. "Helminth-derived Molecules Inhibit Colitis-associated Colon Cancer Development through NF-κB and STAT3 Regulation." *International Journal of Cancer* 145 (11). https://doi.org/10.1002/ijc.32626.

Chun, Eunyoung, Sydney Lavoie, Monia Michaud, Carey Ann Gallini, Jason Kim, Genevieve Soucy, Robert Odze, Jonathan N. Glickman, and Wendy S. Garrett. 2015. "CCL2 Promotes Colorectal Carcinogenesis by Enhancing Polymorphonuclear Myeloid-Derived Suppressor Cell Population and Function." *Cell Reports* 12 (2): 244–57. https://doi.org/10.1016/J.CELREP.2015.06.024.

Coca, Santiago, Javier Perez-Piqueras, David Martinez, Antonio Colmenarejo, Miguel A. Saez, Carmen Vallejo, Jose A. Martos, and Manuel Moreno. 1997. "The Prognostic Significance of Intratumoral Natural Killer Cells in Patients with Colorectal Carcinoma." *Cancer* 79 (12). https://doi.org/10.1002/(SICI) 1097-0142(19970615) 79:12<2320::AID-CNCR5>3.0.CO;2-P.

Condamine, T, GA Dominguez, JI Youn, AV Kossenkov, S Mony, K Alicea-Torres, E Tcyganov, et al. 2016. "Lectin-Type Oxidized LDL Receptor-1 Distinguishes Population of Human Polymorphonuclear Myeloid-Derived Suppressor Cells in Cancer Patients." *Science Immunology* 1 (2). https://doi.org/10.1126/SCIIMMUNOL.AAF8943.

Corvinus, Florian M, Carina Orth, Richard Moriggl, Svetlana A Tsareva, Stefan Wagner, Edith B Pfitzner, Daniela Baus, et al. 2005. "Persistent STAT3 Activation in Colon Cancer Is Associated with Enhanced Cell Proliferation and Tumor Growth." *Neoplasia (New York, N.Y.)* 7 (6): 545. https://doi.org/10.1593/NEO.04571.

Dadabayev, A. R., M. H. Sandel, A. G. Menon, H. Morreau, C. J. M. Melief, R. Offringa, S. H. van der Burg, et al. 2004. "Dendritic Cells in Colorectal Cancer Correlate with Other Tumor-Infiltrating Immune Cells." *Cancer Immunology, Immunotherapy* 53 (11). https://doi.org/10.1007/s00262-004-0548-2.

Dardalhon, Valerie, Ana C. Anderson, Jozsef Karman, Lionel Apetoh, Rucha Chandwaskar, David H. Lee, Melanie Cornejo, et al. 2010. "Tim-3/Galectin-9 Pathway: Regulation of Th1 Immunity through Promotion of CD11b+Ly-6G+ Myeloid Cells." *Journal of Immunology (Baltimore, Md : 1950)* 185 (3): 1383. https://doi.org/10.4049/JIMMUNOL.0903275.

Delgado-Ramirez, Yael, Itzel Baltazar-Perez, Yamileth Martinez, Blanca E. Callejas, Itzel Medina-Andrade, Jonadab E. Olguín, Norma L. Delgado-Buenrostro, Yolanda I. Chirino, Luis I. Terrazas, and Sonia Leon-Cabrera. 2021. "STAT1 Is Required for Decreasing Accumulation of Granulocytic Cells via IL-17 during Initial Steps of Colitis-Associated Cancer." *International Journal of Molecular Sciences* 22 (14). https://doi.org/10.3390/ijms22147695.

Dell'Aquila, E., C. Cremolini, T. Zeppola, S. Lonardi, F. Bergamo, G. Masi, M. Stellato, et al. 2018. "Prognostic and Predictive Role of Neutrophil/Lymphocytes Ratio in Metastatic Colorectal Cancer: A Retrospective Analysis of the TRIBE Study by GONO." *Annals of Oncology* 29 (4): 924–30. https://doi.org/10.1093/ANNONC/MDY004.

Droeser, Raoul A., Christian Hirt, Serenella Eppenberger-Castori, Inti Zlobec, Carsten T. Viehl, Daniel M. Frey, Christian A. Nebiker, et al. 2013. "High Myeloperoxidase Positive Cell Infiltration in Colorectal Cancer Is an Independent Favorable Prognostic Factor." *PLoS ONE* 8 (5). https://doi.org/10.1371/journal.pone.0064814.

Eaden, J A. 2001. "The Risk of Colorectal Cancer in Ulcerative Colitis: A Meta-Analysis." *Gut* 48 (4). https://doi.org/10.1136/gut.48.4.526.

Fearon, Eric R., and Bert Vogelstein. 1990. "A Genetic Model for Colorectal Tumorigenesis." *Cell* 61 (5). https://doi.org/10.1016/0092-8674(90)90186-I.

Felix, Klaus, and Matthias M. Gaida. 2016. "Neutrophil-Derived Proteases in the Microenvironment of Pancreatic Cancer -Active Players in Tumor Progression." *International Journal of Biological Sciences* 12 (3). https://doi.org/10.7150/ijbs.14996.

Fessler, Evelyn, and Jan Paul Medema. 2016. "Colorectal Cancer Subtypes: Developmental Origin and Microenvironmental Regulation." *Trends in Cancer* 2 (9). https://doi.org/10.1016/j.trecan.2016.07.008.

Fidelle, Marine, Satoru Yonekura, Marion Picard, Alexandria Cogdill, Antoine Hollebecque, Maria Paula Roberti, and Laurence Zitvogel. 2020. "Resolving the Paradox of Colon Cancer Through the Integration of Genetics, Immunology, and the Microbiota." *Frontiers in Immunology* 11 (December). https://doi.org/10.3389/fimmu.2020.600886.

Frick, Adrian, Vineeta Khare, Gregor Paul, Michaela Lang, Franziska Ferk, Siegfried Knasmüller, Andrea Beer, Georg Oberhuber, and Christoph Gasche. 2018. "Overt Increase of Oxidative Stress and DNA Damage in Murine and Human Colitis and Colitis-Associated Neoplasia." *Molecular Cancer Research* 16 (4). https://doi.org/10.1158/1541-7786.MCR-17-0451.

Fridlender, Z. G., and S. M. Albelda. 2012. "Tumor-Associated Neutrophils: Friend or Foe?" *Carcinogenesis* 33 (5). https://doi.org/10.1093/carcin/bgs123.

Fridlender, Zvi G., Jing Sun, Samuel Kim, Veena Kapoor, Guanjun Cheng, Leona Ling, G. Scott Worthen, and Steven M. Albelda. 2009. "Polarization of Tumor-Associated Neutrophil Phenotype by TGF-β: 'N1' versus 'N2' TAN." *Cancer Cell* 16 (3). https://doi.org/10.1016/j.ccr.2009.06.017.

Gabrilovich, Dmitry I. 2017. "Myeloid-Derived Suppressor Cells." *Cancer Immunology Research* 5 (1). https://doi.org/10.1158/2326-6066.CIR-16-0297.

Galdiero, Maria Rosaria, Paolo Bianchi, Fabio Grizzi, Giuseppe di Caro, Gianluca Basso, Andrea Ponzetta, Eduardo Bonavita, et al. 2016. "Occurrence and Significance of Tumor-Associated Neutrophils in Patients with Colorectal Cancer." *International Journal of Cancer* 139 (2). https://doi.org/10.1002/ijc.30076.

Gardner, Alycia, Álvaro de Mingo Pulido, and Brian Ruffell. 2020. "Dendritic Cells and Their Role in Immunotherapy." *Frontiers in Immunology* 11 (May). https://doi.org/10.3389/fimmu.2020.00924.

Germann, Markus, Nadine Zangger, Marc-Olivier Sauvain, Christine Sempoux, Amber D Bowler, Pratyaksha Wirapati, Lana E Kandalaft, et al. 2020. "Neutrophils Suppress Tumor-infiltrating T Cells in Colon Cancer via Matrix

Metalloproteinase-mediated Activation of <scp>TGF</Scp> β." *EMBO Molecular Medicine* 12 (1). https://doi.org/10.15252/emmm.201910681.

Gershkovitz, Maya, Yaki Caspi, Tanya Fainsod-Levi, Ben Katz, Janna Michaeli, Saleh Khawaled, Shaya Lev, et al. 2018. "TRPM2 Mediates Neutrophil Killing of Disseminated Tumor Cells." *Cancer Research* 78 (10): 2680–90. https://doi.org/10.1158/0008-5472.CAN-17-3614.

Gloire, Geoffrey, Sylvie Legrand-Poels, and Jacques Piette. 2006. "NF-KB Activation by Reactive Oxygen Species: Fifteen Years Later." *Biochemical Pharmacology* 72 (11). https://doi.org/10.1016/j.bcp.2006.04.011.

Governa, V, E Trella, V Mele, L Tornillo, F Amicarella, E Cremonesi, MG Muraro, et al. 2017. "The Interplay Between Neutrophils and CD8 + T Cells Improves Survival in Human Colorectal Cancer." *Clinical Cancer Research: An Official Journal of the American Association for Cancer Research* 23 (14): 3847–58. https://doi.org/10.1158/1078-0432.CCR-16-2047.

Grasso, Catherine S., Marios Giannakis, Daniel K. Wells, Tsuyoshi Hamada, Xinmeng Jasmine Mu, Michael Quist, Jonathan A. Nowak, et al. 2018. "Genetic Mechanisms of Immune Evasion in Colorectal Cancer." *Cancer Discovery* 8 (6). https://doi.org/10.1158/2159-8290.CD-17-1327.

Guinney, Justin, Rodrigo Dienstmann, Xin Wang, Aurélien de Reyniès, Andreas Schlicker, Charlotte Soneson, Laetitia Marisa, et al. 2015. "The Consensus Molecular Subtypes of Colorectal Cancer." *Nature Medicine* 21 (11). https://doi.org/10.1038/nm.3967.

Guo, Liang, Chuanlei Wang, Xiang Qiu, Xiaoyu Pu, and Pengyu Chang. 2020. "Colorectal Cancer Immune Infiltrates: Significance in Patient Prognosis and Immunotherapeutic Efficacy." *Frontiers in Immunology* 11 (May). https://doi.org/10.3389/fimmu.2020.01052.

Inagaki, Katsuaki, Shoma Kunisho, Hidehiko Takigawa, Ryo Yuge, Shiro Oka, Shinji Tanaka, Fumio Shimamoto, Kazuaki Chayama, and Yasuhiko Kitadai. 2021. "Role of Tumor-associated Macrophages at the Invasive Front in Human Colorectal Cancer Progression." *Cancer Science* 112 (7). https://doi.org/10.1111/cas.14940.

Ismail, Ayaaz, and Peter Lance. 2004. "Colorectal Adenomas." In *Encyclopedia of Gastroenterology*, edited by Leonard R Johnson, 466–70. New York: Elsevier. https://doi.org/https://doi.org/10.1016/B0-12-386860-2/00154-4.

Itatani, Yoshiro, Kenji Kawada, Yoshiharu Sakai, Susumu Inamoto, Takamasa Yamamoto, and Makoto M. Taketo. 2014. "Myeloid-Derived Suppressor Cells (MDSC) Promotes Colorectal Cancer Invasion through CCL15-CCR1 Chemokine Axis." *Journal of the American College of Surgeons* 219 (4): e177. https://doi.org/10.1016/J.JAMCOLLSURG.2014.07.864.

Itatani, Yoshiro, Takamasa Yamamoto, Cuiling Zhong, Alfredo A. Molinolo, Jane Ruppel, Priti Hegde, M. Mark Taketo, and Napoleone Ferrara. 2020.

"Suppressing Neutrophil-Dependent Angiogenesis Abrogates Resistance to Anti-VEGF Antibody in a Genetic Model of Colorectal Cancer." *Proceedings of the National Academy of Sciences* 117 (35). https://doi.org/10.1073/pnas.2008112117.

Jablonska, J., C.-F. Wu, L. Andzinski, S. Leschner, and S. Weiss. 2014. "CXCR2-Mediated Tumor-Associated Neutrophil Recruitment Is Regulated by IFN-β." *International Journal of Cancer* 134 (6). https://doi.org/10.1002/ijc.28551.

Jakubowska, K, M Koda, W Kisielewski, L Kańczuga-Koda, M Grudzińska, and W Famulski. 2020. "Pre- and Postoperative Neutrophil and Lymphocyte Count and Neutrophil-to-Lymphocyte Ratio in Patients with Colorectal Cancer." *Molecular and Clinical Oncology* 13 (5): 1–10. https://doi.org/10.3892/MCO.2020.2126.

Jasperson, Kory W., Thérèse M. Tuohy, Deborah W. Neklason, and Randall W. Burt. 2010. "Hereditary and Familial Colon Cancer." *Gastroenterology* 138 (6). https://doi.org/10.1053/j.gastro.2010.01.054.

Jayakumar, Asha, and Alfred L.M. Bothwell. 2017. "Stat6 Promotes Intestinal Tumorigenesis in a Mouse Model of Adenomatous Polyposis by Expansion of MDSCs and Inhibition of Cytotoxic CD8 Response." *Neoplasia (New York, N.Y.)* 19 (8): 595. https://doi.org/10.1016/J.NEO.2017.04.006.

Johnson, Constance M., Caimiao Wei, Joe E. Ensor, Derek J. Smolenski, Christopher I. Amos, Bernard Levin, and Donald A. Berry. 2013. "Meta-Analyses of Colorectal Cancer Risk Factors." *Cancer Causes & Control* 24 (6). https://doi.org/10.1007/s10552-013-0201-5.

Karakasheva, TA, GA Dominguez, A Hashimoto, EW Lin, C Chiu, K Sasser, JW Lee, GL Beatty, DI Gabrilovich, and AK Rustgi. 2018. "CD38+ M-MDSC Expansion Characterizes a Subset of Advanced Colorectal Cancer Patients." *JCI Insight* 3 (6). https://doi.org/10.1172/JCI.INSIGHT.97022.

Kim, JM, JS Kim, HC Jung, IS Song, and CY Kim. 2000. "Apoptosis of Human Gastric Epithelial Cells via Caspase-3 Activation in Response to Helicobacter Pylori Infection: Possible Involvement of Neutrophils through Tumor Necrosis Factor Alpha and Soluble Fas Ligands." *Scandinavian Journal of Gastroenterology* 35 (1): 40–48. https://doi.org/10.1080/003655200750024515.

Knaapen, A. M. 2006. "Neutrophils and Respiratory Tract DNA Damage and Mutagenesis: A Review." *Mutagenesis* 21 (4). https://doi.org/10.1093/mutage/gel032.

Koch, Moritz, Philipp Beckhove, Jan op den Winkel, Daniel Autenrieth, Philipp Wagner, Daniel Nummer, Sebastian Specht, et al. 2006. "Tumor Infiltrating T Lymphocytes in Colorectal Cancer." *Annals of Surgery* 244 (6). https://doi.org/10.1097/01.sla.0000247058.43243.7b.

Koga, Y, A Matsuzaki, A Suminoe, H Hattori, and T Hara. 2004. "Neutrophil-Derived TNF-Related Apoptosis-Inducing Ligand (TRAIL): A Novel Mechanism of Antitumor Effect by Neutrophils." *Cancer Research* 64 (3): 1037–43. https://doi.org/10.1158/0008-5472.CAN-03-1808.

Kolaczkowska, Elzbieta, and Paul Kubes. 2013. "Neutrophil Recruitment and Function in Health and Inflammation." *Nature Reviews Immunology* 13 (3). https://doi.org/10.1038/nri3399.

Kramer, Elliot D., and Scott I. Abrams. 2020. "Granulocytic Myeloid-Derived Suppressor Cells as Negative Regulators of Anticancer Immunity." *Frontiers in Immunology* 0 (August): 1963. https://doi.org/10.3389/FIMMU.2020.01963.

Kumar, V, S Patel, E Tcyganov, and DI Gabrilovich. 2016. "The Nature of Myeloid-Derived Suppressor Cells in the Tumor Microenvironment." *Trends in Immunology* 37 (3): 208–20. https://doi.org/10.1016/J.IT.2016.01.004.

Kusmartsev, Sergei, Yulia Nefedova, Daniel Yoder, and Dmitry I. Gabrilovich. 2004. "Antigen-Specific Inhibition of CD8 + T Cell Response by Immature Myeloid Cells in Cancer Is Mediated by Reactive Oxygen Species." *The Journal of Immunology* 172 (2). https://doi.org/10.4049/jimmunol.172.2.989.

Labani-Motlagh, Alireza, Mehrnoush Ashja-Mahdavi, and Angelica Loskog. 2020. "The Tumor Microenvironment: A Milieu Hindering and Obstructing Antitumor Immune Responses." *Frontiers in Immunology* 11 (May). https://doi.org/10.3389/fimmu.2020.00940.

Lanzi, Anastasia, Franck Pagès, Christine Lagorce-Pagès, and Jérôme Galon. 2020. "The Consensus Immunoscore: Toward a New Classification of Colorectal Cancer." *OncoImmunology* 9 (1). https://doi.org/10.1080/2162402X.2020.1789032.

Lasry, Audrey, Adar Zinger, and Yinon Ben-Neriah. 2016. "Inflammatory Networks Underlying Colorectal Cancer." *Nature Immunology* 17 (3). https://doi.org/10.1038/ni.3384.

Leon-Cabrera, SA, E Molina-Guzman, YG Delgado-Ramirez, A Vázquez-Sandoval, Y Ledesma-Soto, CG Pérez-Plasencia, YI Chirino, et al. 2017. "Lack of STAT6 Attenuates Inflammation and Drives Protection against Early Steps of Colitis-Associated Colon Cancer." *Cancer Immunology Research* 5 (5): 385–96. https://doi.org/10.1158/2326-6066.CIR-16-0168.

Leon-Cabrera, Sonia, Armando Vázquez-Sandoval, Emmanuel Molina-Guzman, Yael Delgado-Ramirez, Norma Delgado-Buenrostro, Blanca Callejas, Yolanda Chirino, et al. 2018. "Deficiency in STAT1 Signaling Predisposes Gut Inflammation and Prompts Colorectal Cancer Development." *Cancers* 10 (9). https://doi.org/10.3390/cancers10090341.

Li, Jinyuan, Linhai Li, Yuejin Li, Yaxin Long, Quan Zhao, Yiming Ouyang, Weimin Bao, and Kunmei Gong. 2020. "Tumor-Associated Macrophage

Infiltration and Prognosis in Colorectal Cancer: Systematic Review and Meta-Analysis." *International Journal of Colorectal Disease* 35 (7). https://doi.org/10.1007/s00384-020-03593-z.

Li, Mu-Xing, Xue-Min Liu, Xu-Feng Zhang, Jian-Fei Zhang, Wan-Li Wang, Ying Zhu, Jian Dong, et al. 2014. "Prognostic Role of Neutrophil-to-Lymphocyte Ratio in Colorectal Cancer: A Systematic Review and Meta-Analysis." *International Journal of Cancer* 134 (10): 2403–13. https://doi.org/10.1002/IJC.28536.

Li, W, X Zhang, Y Chen, Y Xie, J Liu, Q Feng, Y Wang, W Yuan, and J Ma. 2016. "G-CSF Is a Key Modulator of MDSC and Could Be a Potential Therapeutic Target in Colitis-Associated Colorectal Cancers." *Protein & Cell* 7 (2): 130–40. https://doi.org/10.1007/S13238-015-0237-2.

Malki, Ahmed, Rasha Abu ElRuz, Ishita Gupta, Asma Allouch, Semir Vranic, and Ala-Eddin al Moustafa. 2020. "Molecular Mechanisms of Colon Cancer Progression and Metastasis: Recent Insights and Advancements." *International Journal of Molecular Sciences* 22 (1). https://doi.org/10.3390/ijms22010130.

Marini, O, S Costa, D Bevilacqua, F Calzetti, N Tamassia, C Spina, D de Sabata, et al. 2017. "Mature CD10 + and Immature CD10 - Neutrophils Present in G-CSF-Treated Donors Display Opposite Effects on T Cells." *Blood* 129 (10): 1343–56. https://doi.org/10.1182/BLOOD-2016-04-713206.

Martin, Amandine, Cédric Seignez, Cindy Racoeur, Nicolas Isambert, Nesrine Mabrouk, Alessandra Scagliarini, Sylvie Reveneau, et al. 2018. "Tumor-Derived Granzyme B-Expressing Neutrophils Acquire Antitumor Potential after Lipid A Treatment." *Oncotarget* 9 (47): 28364. https://doi.org/10.18632/ONCOTARGET.25342.

Martinez, Fernando O., and Siamon Gordon. 2014. "The M1 and M2 Paradigm of Macrophage Activation: Time for Reassessment." *F1000Prime Reports* 6 (March). https://doi.org/10.12703/P6-13.

Matlung, Hanke L., Liane Babes, Xi Wen Zhao, Michel van Houdt, Louise W. Treffers, Dieke J. van Rees, Katka Franke, et al. 2018. "Neutrophils Kill Antibody-Opsonized Cancer Cells by Trogoptosis." *Cell Reports* 23 (13). https://doi.org/10.1016/j.celrep.2018.05.082.

Medina-Andrade, Itzel, Jonadab E. Olguín, Tonathiu Rodríguez, and Luis I. Terrazas. 2021. "The Inflammation during Colorectal Cancer: A Friend or a Foe?" In *Immunotherapy in Resistant Cancer: From the Lab Bench Work to Its Clinical Perspectives*. Elsevier. https://doi.org/10.1016/B978-0-12-822028-3.00003-0.

Meissl, Katrin, Sabine Macho-Maschler, Mathias Müller, and Birgit Strobl. 2017. "The Good and the Bad Faces of STAT1 in Solid Tumours." *Cytokine* 89 (January): 12–20. https://doi.org/10.1016/J.CYTO.2015.11.011.

Mendoza-Rodríguez, Mónica G., C. Ángel Sánchez-Barrera, Blanca E. Callejas, Verónica García-Castillo, Diana L. Beristain-Terrazas, Norma L. Delgado-Buenrostro, Yolanda I. Chirino, et al. 2020. "Use of STAT6 Phosphorylation Inhibitor and Trimethylglycine as New Adjuvant Therapies for 5-Fluorouracil in Colitis-Associated Tumorigenesis." *International Journal of Molecular Sciences* 21 (6). https://doi.org/10.3390/IJMS21062130.

Nakashima-Nakasuga, C, S Hazama, N Suzuki, Y Nakagami, M Xu, S Yoshida, S Tomochika, et al. 2020. "Serum LOX-1 Is a Novel Prognostic Biomarker of Colorectal Cancer." *International Journal of Clinical Oncology* 25 (7): 1308–17. https://doi.org/10.1007/S10147-020-01673-2.

Naszai, Mate, Alina Kurjan, and Timothy S. Maughan. 2021. "The Prognostic Utility of Pre-Treatment Neutrophil-to-Lymphocyte-Ratio (NLR) in Colorectal Cancer: A Systematic Review and Meta-Analysis." *Cancer Medicine* 00 (July): 1–15. https://doi.org/10.1002/CAM4.4143.

National Institutes of Health (NIH). 2021. "Clinical Trials." 2021. https://clinicaltrials.gov/ct2/results?term=dendritic+cell&cond=Colorectal+Cancer&age_v=&gndr=&type=&rslt=&Search=Apply.

Ogawa, R, T Yamamoto, H Hirai, K Hanada, Y Kiyasu, G Nishikawa, R Mizuno, et al. 2019. "Loss of SMAD4 Promotes Colorectal Cancer Progression by Recruiting Tumor-Associated Neutrophils via the CXCL1/8-CXCR2 Axis." *Clinical Cancer Research : An Official Journal of the American Association for Cancer Research* 25 (9): 2887–99. https://doi.org/10.1158/1078-0432.CCR-18-3684.

Øines, Mari, Lise M. Helsingen, Michael Bretthauer, and Louise Emilsson. 2017. "Epidemiology and Risk Factors of Colorectal Polyps." *Best Practice & Research Clinical Gastroenterology* 31 (4). https://doi.org/10.1016/j.bpg.2017.06.004.

Olguín, Jonadab E., Itzel Medina-Andrade, Tonathiu Rodríguez, Miriam Rodríguez-Sosa, and Luis I. Terrazas. 2020. "Relevance of Regulatory T Cells during Colorectal Cancer Development." *Cancers* 12 (7). https://doi.org/10.3390/cancers12071888.

Onuma, Amblessed, Jiayi He, Yujia Xia, Hongji Zhang, Dmitry Genkin, George Tetz, Hai Huang, and Allan Tsung. 2020. "Neutrophil Extracellular Traps Blockade in Combination with PD-1 Inhibition in Treatment of Colorectal Cancer Metastasis." *Journal of Clinical Oncology* 38 (15_suppl). https://doi.org/10.1200/JCO.2020.38.15_suppl.e16002.

Ozdemir, Y, ML Akin, I Sucullu, AZ Balta, and E Yucel. 2014. "Pretreatment Neutrophil/Lymphocyte Ratio as a Prognostic Aid in Colorectal Cancer." *Asian Pacific Journal of Cancer Prevention : APJCP* 15 (6): 2647–50. https://doi.org/10.7314/APJCP.2014.15.6.2647.

Palmieri, V, A Lazaris, TZ Mayer, SK Petrillo, H Alamri, M Rada, G Jarrouj, et al. 2020. "Neutrophils Expressing Lysyl Oxidase-like 4 Protein Are Present in Colorectal Cancer Liver Metastases Resistant to Anti-Angiogenic Therapy." *The Journal of Pathology* 251 (2): 213–23. https://doi.org/10.1002/PATH.5449.

Pernot, Simon. 2014. "Colorectal Cancer and Immunity: What We Know and Perspectives." *World Journal of Gastroenterology* 20 (14). https://doi.org/10.3748/wjg.v20.i14.3738.

Pillay, Janesh, Ineke den Braber, Nienke Vrisekoop, Lydia M. Kwast, Rob J. de Boer, José A. M. Borghans, Kiki Tesselaar, and Leo Koenderman. 2010. "*In Vivo* Labeling with 2H2O Reveals a Human Neutrophil Lifespan of 5.4 Days." *Blood* 116 (4). https://doi.org/10.1182/blood-2010-01-259028.

Pinto, Marta L., Elisabete Rios, Cecília Durães, Ricardo Ribeiro, José C. Machado, Alberto Mantovani, Mário A. Barbosa, Fatima Carneiro, and Maria J. Oliveira. 2019. "The Two Faces of Tumor-Associated Macrophages and Their Clinical Significance in Colorectal Cancer." *Frontiers in Immunology* 10 (August). https://doi.org/10.3389/fimmu.2019.01875.

Pryczynicz, Anna, Dariusz Cepowicz, Konrad Zaręba, Mariusz Gryko, Joanna Hołody-Zaręba, Bogusław Kędra, Andrzej Kemona, and Katarzyna Guzińska-Ustymowicz. 2016. "Dysfunctions in the Mature Dendritic Cells Are Associated with the Presence of Metastases of Colorectal Cancer in the Surrounding Lymph Nodes." *Gastroenterology Research and Practice* 2016. https://doi.org/10.1155/2016/2405437.

Pylaeva, Ekaterina, Stephan Lang, and Jadwiga Jablonska. 2016. "The Essential Role of Type I Interferons in Differentiation and Activation of Tumor-Associated Neutrophils." *Frontiers in Immunology* 7 (December). https://doi.org/10.3389/fimmu.2016.00629.

Queen, Marisa M., Randall E. Ryan, Ryan G. Holzer, Cynthia R. Keller-Peck, and Cheryl L. Jorcyk. 2005. "Breast Cancer Cells Stimulate Neutrophils to Produce Oncostatin M: Potential Implications for Tumor Progression." *Cancer Research* 65 (19). https://doi.org/10.1158/0008-5472.CAN-05-1734.

Rao, HL, JW Chen, M Li, YB Xiao, J Fu, YX Zeng, MY Cai, and D Xie. 2012. "Increased Intratumoral Neutrophil in Colorectal Carcinomas Correlates Closely with Malignant Phenotype and Predicts Patients' Adverse Prognosis." *PloS One* 7 (1). https://doi.org/10.1371/JOURNAL.PONE.0030806.

Rashid, Mohammad H., Thaiz F. Borin, Roxan Ara, Raziye Piranlioglu, Bhagelu R. Achyut, Hasan Korkaya, Yutao Liu, and Ali S. Arbab. 2021. "Critical Immunosuppressive Effect of MDSC-Derived Exosomes in the Tumor Microenvironment." *Oncology Reports* 45 (3): 1171. https://doi.org/10.3892/OR.2021.7936.

Rashtak, Shahrooz, Xiaoyang Ruan, Brooke R. Druliner, Hongfang Liu, Terry Therneau, Mohamad Mouchli, and Lisa A. Boardman. 2017. “Peripheral Neutrophil to Lymphocyte Ratio Improves Prognostication in Colon Cancer.” *Clinical Colorectal Cancer* 16 (2): 115-123.e3. https://doi.org/10.1016/J.CLCC.2017.01.008.

Reid, Fiona S. W., Natasha Egoroff, Peter G. Pockney, and Stephen R. Smith. 2021. “A Systematic Scoping Review on Natural Killer Cell Function in Colorectal Cancer.” *Cancer Immunology, Immunotherapy* 70 (3). https://doi.org/10.1007/s00262-020-02721-6.

Ren, Xiaoli, Jianbiao Xiao, Wanning Zhang, Feifei Wang, Yongrong Yan, Xuehui Wu, Zhicheng Zeng, et al. 2021. “Inhibition of CCL7 Derived from Mo-MDSCs Prevents Metastatic Progression from Latency in Colorectal Cancer.” *Cell Death & Disease 2021 12:5* 12 (5): 1–15. https://doi.org/10.1038/s41419-021-03698-5.

Rex, Douglas K, Dennis J Ahnen, John A Baron, Kenneth P Batts, Carol A Burke, Randall W Burt, John R Goldblum, et al. 2012. “Serrated Lesions of the Colorectum: Review and Recommendations From an Expert Panel.” *American Journal of Gastroenterology* 107 (9). https://doi.org/10.1038/ajg.2012.161.

Richardson, JJR, C Hendrickse, F Gao-Smith, and DR Thickett. 2017. “Characterization of Systemic Neutrophil Function in Patients Undergoing Colorectal Cancer Resection.” *The Journal of Surgical Research* 220 (December): 410-418.e1. https://doi.org/10.1016/J.JSS.2017.07.036.

Rodriguez, Javier, Eduardo Castañón, Jose Luis Perez-Gracia, Inmaculada Rodriguez, Antonio Viudez, Carlos Alfaro, Carmen Oñate, et al. 2018. “A Randomized Phase II Clinical Trial of Dendritic Cell Vaccination Following Complete Resection of Colon Cancer Liver Metastasis.” *Journal for ImmunoTherapy of Cancer* 6 (1). https://doi.org/10.1186/s40425-018-0405-z.

Ruland, J. 2019. “Colon Cancer: Epithelial Notch Signaling Recruits Neutrophils to Drive Metastasis.” *Cancer Cell* 36 (3): 213–14. https://doi.org/10.1016/J.CCELL.2019.08.010.

Sagiv, Jitka Y., Janna Michaeli, Simaan Assi, Inbal Mishalian, Hen Kisos, Liran Levy, Pazzit Damti, et al. 2015. “Phenotypic Diversity and Plasticity in Circulating Neutrophil Subpopulations in Cancer.” *Cell Reports* 10 (4). https://doi.org/10.1016/j.celrep.2014.12.039.

Sandel, Maro H., Alisher R. Dadabayev, Anand G. Menon, Hans Morreau, Cornelis J. M. Melief, Rienk Offringa, Sjoerd H. van der Burg, et al. 2005. “Prognostic Value of Tumor-Infiltrating Dendritic Cells in Colorectal Cancer: Role of Maturation Status and Intratumoral Localization.” *Clinical Cancer Research* 11 (7). https://doi.org/10.1158/1078-0432.CCR-04-1448.

Shang, A, C Gu, C Zhou, Yang. Y, C Chen, B Zeng, J Wu, et al. 2020. "Exosomal KRAS Mutation Promotes the Formation of Tumor-Associated Neutrophil Extracellular Traps and Causes Deterioration of Colorectal Cancer by Inducing IL-8 Expression." *Cell Communication and Signaling: CCS* 18 (1). https://doi.org/10.1186/S12964-020-0517-1.

Shang, K, YP Bai, C Wang, Z Wang, HY Gu, X Du, XY Zhou, et al. 2012. "Crucial Involvement of Tumor-Associated Neutrophils in the Regulation of Chronic Colitis-Associated Carcinogenesis in Mice." *PloS One* 7 (12). https://doi.org/10.1371/JOURNAL.PONE.0051848.

Sharma, Bhawna, Dhananjay M. Nawandar, Kalyan C. Nannuru, Michelle L. Varney, and Rakesh K. Singh. 2013. "Targeting CXCR2 Enhances Chemotherapeutic Response, Inhibits Mammary Tumor Growth, Angiogenesis, and Lung Metastasis." *Molecular Cancer Therapeutics* 12 (5). https://doi.org/10.1158/1535-7163.MCT-12-0529.

Shibutani, M, K Maeda, H Nagahara, E Noda, H Ohtani, Y Nishiguchi, and K Hirakawa. 2013. "A High Preoperative Neutrophil-to-Lymphocyte Ratio Is Associated with Poor Survival in Patients with Colorectal Cancer." *Anticancer Research* 33 (8): 3291–94. https://pubmed.ncbi.nlm.nih.gov/23898094/.

Sieminska, Izabela, and Jarek Baran. 2020. "Myeloid-Derived Suppressor Cells in Colorectal Cancer." *Frontiers in Immunology* 11 (August): 1526. https://doi.org/10.3389/FIMMU.2020.01526.

Sinha, Pratima, Olesya Chornoguz, Virginia K. Clements, Konstantin A. Artemenko, Roman A. Zubarev, and Suzanne Ostrand-Rosenberg. 2011. "Myeloid-Derived Suppressor Cells Express the Death Receptor Fas and Apoptose in Response to T Cell–Expressed FasL." *Blood* 117 (20): 5381. https://doi.org/10.1182/BLOOD-2010-11-321752.

Son, Yong, Yong-Kwan Cheong, Nam-Ho Kim, Hun-Taeg Chung, Dae Gill Kang, and Hyun-Ock Pae. 2011. "Mitogen-Activated Protein Kinases and Reactive Oxygen Species: How Can ROS Activate MAPK Pathways?" *Journal of Signal Transduction* 2011 (February). https://doi.org/10.1155/2011/792639.

Stidham, Ryan, and Peter Higgins. 2018. "Colorectal Cancer in Inflammatory Bowel Disease." *Clinics in Colon and Rectal Surgery* 31 (03). https://doi.org/10.1055/s-0037-1602237.

Sugita, Yutaka, Kimihiro Yamashita, Mitsugu Fujita, Masafumi Saito, Kota Yamada, Kyosuke Agawa, Akihiro Watanabe, et al. 2021. "CD244^{+} Polymorphonuclear Myeloid-derived Suppressor Cells Reflect the Status of Peritoneal Dissemination in a Colon Cancer Mouse Model." *Oncology Reports* 45 (6): 1–13. https://doi.org/10.3892/OR.2021.8057.

Sun, Bingwei, Weiting Qin, Mingming Song, Lu Liu, Yao Yu, Xinxin Qi, and Hui Sun. 2018. "Neutrophil Suppresses Tumor Cell Proliferation via Fas /Fas

Ligand Pathway Mediated Cell Cycle Arrested." *International Journal of Biological Sciences* 14 (14). https://doi.org/10.7150/ijbs.29297.

Sung, Hyuna, Jacques Ferlay, Rebecca L. Siegel, Mathieu Laversanne, Isabelle Soerjomataram, Ahmedin Jemal, and Freddie Bray. 2021. "Global Cancer Statistics 2020: GLOBOCAN Estimates of Incidence and Mortality Worldwide for 36 Cancers in 185 Countries." *CA: A Cancer Journal for Clinicians* 71 (3). https://doi.org/10.3322/caac.21660.

Tallerico, Rossana, Matilde Todaro, Simone di Franco, Cristina Maccalli, Cinzia Garofalo, Rosa Sottile, Camillo Palmieri, et al. 2013. "Human NK Cells Selective Targeting of Colon Cancer–Initiating Cells: A Role for Natural Cytotoxicity Receptors and MHC Class I Molecules." *The Journal of Immunology* 190 (5). https://doi.org/10.4049/jimmunol.1201542.

Tanaka, Y, S Ito, and K Isobe. 2016. "Vancomycin-Sensitive Bacteria Trigger Development of Colitis-Associated Colon Cancer by Attracting Neutrophils." *Scientific Reports* 6 (April). https://doi.org/10.1038/SREP23920.

Tang, Yan-ping, Ming-zhi Xie, Ke-zhi Li, Ji-lin Li, Zheng-min Cai, and Bang-li Hu. 2020. "Prognostic Value of Peripheral Blood Natural Killer Cells in Colorectal Cancer." *BMC Gastroenterology* 20 (1). https://doi.org/10.1186/s12876-020-1177-8.

Tecchio, Cristina, Alessandra Micheletti, and Marco A. Cassatella. 2014. "Neutrophil-Derived Cytokines: Facts Beyond Expression." *Frontiers in Immunology* 5 (October). https://doi.org/10.3389/fimmu.2014.00508.

Terzić, Janoš, Sergei Grivennikov, Eliad Karin, and Michael Karin. 2010. "Inflammation and Colon Cancer." *Gastroenterology* 138 (6). https://doi.org/10.1053/j.gastro.2010.01.058.

Triner, D, SN Devenport, SK Ramakrishnan, X Ma, RA Frieler, JK Greenson, N Inohara, et al. 2019. "Neutrophils Restrict Tumor-Associated Microbiota to Reduce Growth and Invasion of Colon Tumors in Mice." *Gastroenterology* 156 (5): 1467–82. https://doi.org/10.1053/J.GASTRO.2018.12.003.

Umansky, Viktor, Carolin Blattner, Christoffer Gebhardt, and Jochen Utikal. 2016. "The Role of Myeloid-Derived Suppressor Cells (MDSC) in Cancer Progression." *Vaccines* 4 (4). https://doi.org/10.3390/VACCINES4040036.

Uribe-Querol, Eileen, and Carlos Rosales. 2015. "Neutrophils in Cancer: Two Sides of the Same Coin." *Journal of Immunology Research* 2015. https://doi.org/10.1155/2015/983698.

Veglia, Filippo, Emilio Sanseviero, and Dmitry I. Gabrilovich. 2021. "Myeloid-Derived Suppressor Cells in the Era of Increasing Myeloid Cell Diversity." *Nature Reviews Immunology*, February. https://doi.org/10.1038/s41577-020-00490-y.

Veglia, Filippo, Vladimir A. Tyurin, Maria Blasi, Alessandra de Leo, Andrew Kossenkov, Laxminarasimha Donthireddy, To Tsun Ki Jerrick, et al. 2019.

"Fatty Acid Transporter 2 Reprograms Neutrophils in Cancer." *Nature* 569 (7754): 73. https://doi.org/10.1038/S41586-019-1118-2.

Wang, Y, K Wang, GC Han, RX Wang, H Xiao, CM Hou, RF Guo, et al. 2014. "Neutrophil Infiltration Favors Colitis-Associated Tumorigenesis by Activating the Interleukin-1 (IL-1)/IL-6 Axis." *Mucosal Immunology* 7 (5): 1106–15. https://doi.org/10.1038/MI.2013.126.

Wang, Y, K Yin, J Tian, X Xia, J Ma, X Tang, H Xu, and S Wang. 2019. "Granulocytic Myeloid-Derived Suppressor Cells Promote the Stemness of Colorectal Cancer Cells through Exosomal S100A9." *Advanced Science (Weinheim, Baden-Wurttemberg, Germany)* 6 (18). https://doi.org/10.1002/ADVS.201901278.

Wang, Yungang, Yanxia Ding, Yijun Deng, Yu Zheng, and Shengjun Wang. 2020. "Role of Myeloid-Derived Suppressor Cells in the Promotion and Immunotherapy of Colitis-Associated Cancer." *Journal for ImmunoTherapy of Cancer* 8 (2): e000609. https://doi.org/10.1136/JITC-2020-000609.

Watson, Nicholas F.S., Judith M. Ramage, Zahra Madjd, Ian Spendlove, Ian O. Ellis, John H. Scholefield, and Lindy G. Durrant. 2006. "Immunosurveillance Is Active in Colorectal Cancer as Downregulation but Not Complete Loss of MHC Class I Expression Correlates with a Poor Prognosis." *International Journal of Cancer* 118 (1). https://doi.org/10.1002/ijc.21303.

Wei, Chen, Chaogang Yang, Shuyi Wang, Dongdong Shi, Chunxiao Zhang, Xiaobin Lin, Qing Liu, Rongzhang Dou, and Bin Xiong. 2019. "Crosstalk between Cancer Cells and Tumor-Associated Macrophages Is Required for Mesenchymal Circulating Tumor Cell-Mediated Colorectal Cancer Metastasis." *Molecular Cancer* 18 (1). https://doi.org/10.1186/s12943-019-0976-4.

Wei, TT, YT Lin, RY Tseng, CT Shun, YC Lin, MS Wu, JM Fang, and CC Chen. 2016. "Prevention of Colitis and Colitis-Associated Colorectal Cancer by a Novel Polypharmacological Histone Deacetylase Inhibitor." *Clinical Cancer Research: An Official Journal of the American Association for Cancer Research* 22 (16): 4158–69. https://doi.org/10.1158/1078-0432.CCR-15-2379.

Wells, Katerina, and Paul E. Wise. 2017. "Hereditary Colorectal Cancer Syndromes." *Surgical Clinics of North America* 97 (3). https://doi.org/10.1016/j.suc.2017.01.009.

Wen, Chuangyu, Huihui Wang, Xiaobin Wu, Lu He, Qian Zhou, Fang Wang, Siyu Chen, et al. 2019. "ROS-Mediated Inactivation of the PI3K/AKT Pathway Is Involved in the Antigastric Cancer Effects of Thioredoxin Reductase-1 Inhibitor Chaetocin." *Cell Death & Disease* 10 (11). https://doi.org/10.1038/s41419-019-2035-x.

West, Nathan R., Sarah McCuaig, Fanny Franchini, and Fiona Powrie. 2015. "Emerging Cytokine Networks in Colorectal Cancer." *Nature Reviews Immunology* 15 (10). https://doi.org/10.1038/nri3896.

Wickramasinghe, S N, and W N Erber. 2011. "CHAPTER 1 - Normal Blood Cells." In *Blood and Bone Marrow Pathology (Second Edition)*, edited by Anna Porwit, Jeffrey McCullough, and Wendy N Erber, Second Edition, 3–18. Edinburgh: Churchill Livingstone. https://doi.org/10.1016/B978-0-7020-3147-2.00001-8.

Wikberg, Maria L., Agnes Ling, Xingru Li, Åke Öberg, Sofia Edin, and Richard Palmqvist. 2017. "Neutrophil Infiltration Is a Favorable Prognostic Factor in Early Stages of Colon Cancer." *Human Pathology* 68 (October). https://doi.org/10.1016/j.humpath.2017.08.028.

World Health Organization (WHO). 2020. *Global Cancer Observarory 2020: Cancer Today*. 2020. https://gco.iarc.fr/today/home.

Worthley, Daniel L, and Barbara A Leggett. 2010. "Colorectal Cancer: Molecular Features and Clinical Opportunities." *The Clinical Biochemist. Reviews* 31 (2).

Wu, Saxena, Awaji, and Singh. 2019. "Tumor-Associated Neutrophils in Cancer: Going Pro." *Cancers* 11 (4). https://doi.org/10.3390/cancers11040564.

Yamagishi, Hidetsugu, Hajime Kuroda, Yasuo Imai, and Hideyuki Hiraishi. 2016. "Molecular Pathogenesis of Sporadic Colorectal Cancers." *Chinese Journal of Cancer* 35 (1). https://doi.org/10.1186/s40880-015-0066-y.

Yan, Jun, Goetz Kloecker, Chris Fleming, Michael Bousamra, Richard Hansen, Xiaoling Hu, Chuanlin Ding, et al. 2014. "Human Polymorphonuclear Neutrophils Specifically Recognize and Kill Cancerous Cells." *OncoImmunology* 3 (7). https://doi.org/10.4161/15384101.2014.950163.

Yang, Luyu, Lu Liu, Rui Zhang, Jun Hong, Yaping Wang, Jian Wang, Jieliang Zuo, Jubo Zhang, Jinhong Chen, and Hankun Hao. 2020. "IL-8 Mediates a Positive Loop Connecting Increased Neutrophil Extracellular Traps (NETs) and Colorectal Cancer Liver Metastasis." *Journal of Cancer* 11 (15). https://doi.org/10.7150/jca.44215.

Yin, Kai, Xueli Xia, Ke Rui, Tingting Wang, and Shengjun Wang. 2020. "Myeloid-Derived Suppressor Cells: A New and Pivotal Player in Colorectal Cancer Progression." *Frontiers in Oncology* 0 (December): 2795. https://doi.org/10.3389/FONC.2020.610104.

Youn, Je-In, Michelle Collazo, Irina N. Shalova, Subhra K. Biswas, and Dmitry I. Gabrilovich. 2012. "Characterization of the Nature of Granulocytic Myeloid-Derived Suppressor Cells in Tumor-Bearing Mice." *Journal of Leukocyte Biology* 91 (1): 167–81. https://doi.org/10.1189/JLB.0311177.

Youn, JI, and DI Gabrilovich. 2010. "The Biology of Myeloid-Derived Suppressor Cells: The Blessing and the Curse of Morphological and Functional

Heterogeneity." *European Journal of Immunology* 40 (11): 2969–75. https://doi.org/10.1002/EJI.201040895.

Zhang, B, Z Wang, L Wu, M Zhang, Li. W, J Ding, J Zhu, H Wei, and K Zhao. 2013. "Circulating and Tumor-Infiltrating Myeloid-Derived Suppressor Cells in Patients with Colorectal Carcinoma." *PloS One* 8 (2). https://doi.org/10.1371/JOURNAL.PONE.0057114.

Zhang, Jie, Hong-Ying Zhang, Jia Li, Xin-Yu Shao, Chun-Xia Zhang, Jie Zhang, Hong-Ying Zhang, Jia Li, Xin-Yu Shao, and Chun-Xia Zhang. 2017. "The Elevated NLR, PLR and PLT May Predict the Prognosis of Patients with Colorectal Cancer: A Systematic Review and Meta-Analysis." *Oncotarget* 8 (40): 68837–46. https://doi.org/10.18632/ONCOTARGET.18575.

Zhao, Yamei, Xiaoxu Ge, Xiaoming Xu, Shaojun Yu, Jian Wang, and Lifeng Sun. 2019. "Prognostic Value and Clinicopathological Roles of Phenotypes of Tumour-Associated Macrophages in Colorectal Cancer." *Journal of Cancer Research and Clinical Oncology* 145 (12). https://doi.org/10.1007/s00432-019-03041-8.

Zhong, Xiaoming, Bin Chen, and Zhiwen Yang. 2018. "The Role of Tumor-Associated Macrophages in Colorectal Carcinoma Progression." *Cellular Physiology and Biochemistry* 45 (1). https://doi.org/10.1159/000486816.

Zhou, G, K Peng, Y Song, W Yang, W Shu, T Yu, L Yu, et al. 2018. "CD177+ Neutrophils Suppress Epithelial Cell Tumourigenesis in Colitis-Associated Cancer and Predict Good Prognosis in Colorectal Cancer." *Carcinogenesis* 39 (2): 272–82. https://doi.org/10.1093/CARCIN/BGX142.

Zhou, WW, YP Chu, and GY An. 2017. "Significant Difference of Neutrophil-Lymphocyte Ratio between Colorectal Cancer, Adenomatous Polyp and Healthy People." *European Review for Medical and Pharmacological Sciences* 21 (23): 5386–91. https://doi.org/10.26355/EURREV_201712_13924.

Zhou, Xiuling, Xiangqiong Mo, Junlan Qiu, Jingjing Zhao, Shuncong Wang, Cuiling Zhou, Yonghui Su, Zhong Lin, and Haiqing Ma. 2018. "Chemotherapy Combined with Dendritic Cell Vaccine and Cytokine-Induced Killer Cells in the Treatment of Colorectal Carcinoma: A Meta-Analysis." *Cancer Management and Research* Volume 10 (November). https://doi.org/10.2147/CMAR.S173201.

Chapter 5

Plasticity of Natural Killer Cells in Cancer: Opportunities for Therapy?

Mariana Segovia-Mendoza[1]
and Jorge Morales-Montor[2,*]
[1]Departamento de Farmacología, Facultad de Medicina,
Universidad Nacional Autónoma de México,
Ciudad de México, México
[2]Departamento de Inmunología, Instituto de Investigaciones Biomédicas, Universidad Nacional Autónoma de México,
Ciudad de México, México

Abstract

Natural killer (NK) cells are vital components of the innate immune system. NK cells are implicated in pleiotropic processes such as transplant rejection, cell killing after viral infections, inflammatory diseases, and cancer. This kind of cellular immune effector is an important switch for the adaptive immune response mediated by the activation of T lymphocytes. It has been reported that the percentage of NK cells in the tumor microenvironment is low. However, recent works have pointed out that different lineages of NK cells, which are characteristic components of the hyper-secretory profile of cytokines and granzymes, led to the proliferation of malignant cells and the formation of distant metastases. Moreover, the disturbed metabolism of NK cells encouraged by tumor cells can promote the cancer process. Until now,

* Corresponding Author's E-mail: jmontor66@iibiomedicas.unam.mx.

In: The Innate Immune System in Health and Disease
Editor: Jorge Morales-Montor
ISBN: 978-1-68507-507-1

little information about the relationship between different subsets of NK cells, their function, and regulation of cancer promotion is known. Thus, the present chapter focuses on the deep knowledge of the regulation of these cellular mediators in the cancer context to expand the scientific evidence of their possible use as a therapeutic strategy that improves clinical prognosis.

Keywords: innate immune system, NK cells, cancer, therapy

Introduction

Origin and Function of NK Cells

NK cells are originated from a common lymphoid progenitor in the bone marrow; however, they are found in many peripheral blood tissues [1]. Traditionally, NK cells are recognized as important players of the innate immune response with cytotoxic activity against damaged cells due to viral infections or malignant transformation [2, 3]. NK cells express killing receptors (KIRs) and immunoreceptor tyrosine-based inhibitory motifs (ITIMs) that can interact with major histocompatibility complex I (MHC-I). This interaction leads to NK cells recognizing and preventing healthy cells' killing [4-6]. Interleukin (IL)-2, IL-5, and different transcription factors including Nfil3, Id2, GATA-3, EOMES, and T-bet shape the activation development and maturation of NK cells, respectively [7-11]. Therefore, different membrane markers such as neural cell adhesion molecule (NCAM), also known as CD56, are employed to evaluate the state of differentiation of NK cells [12]. Different reports sustain that the differentiation status of NK cells culminates in the peripheral tissues where they reside. It is important to mention that when an NK cell is activated, it can release different proteins, including granzymes and perforins, which are responsible for its cytotoxic effects [13, 14]. To safeguard or release the granzyme and perforin proteins, NK cells need to receive inhibitory or activating signals that impact the function of different receptors including cytokine-binding receptors, integrins, KIRs, receptors recognizing non-self-antigens, among others. Globally, the inhibitory or activating receptors of NK cells can be classified into three types: MHC-I specific, MHC-I related, and MHC-I non-related receptors [15].

On the other hand, that NK cells can activate the adaptive immune response mediated by the activation and migration of T lymphocytes to lymph nodes through the secretion of different cytokines such as interferon-γ (IFN-γ) or tumor necrosis factor (TNF-α) [16]. Moreover, NK cells can also impact the function of other components of the immune system. For instance, NK cells can eradicate or induce the maturation of dendritic cells (DCs). If DCs fail to exert a satisfactory antigen presentation, NK cells will destroy the defective DCs. In contrast, the cellular interaction between NK cells and DCs can enhance the antigen presentation and guarantee in part a successful adaptive immune response [17]. Nowadays, their activity is considered similar to that exerted by cytotoxic T lymphocytes since both kinds of immune cells can induce cell death [18]. Some reports have identified that NK cells can also have and activate an antigen-specific immunologic memory [19]. Thus, new scientific views propose that since NK cells exert sophisticated biological functions, they can be considered a part of innate and adaptive immunity [20].

Subsets of NK Cells

Human NK cells are $CD3^{-}$ $CD16^{+}$ $CD56^{+}$ cells and represent 5% to 20% of circulating lymphocytes. Regarding CD16/CD56 expression, two main subsets have been defined $CD56^{bright}$ $CD16^{-}$ subtype (90-95% of NK cells), which is responsible for the production of diverse cytokines including IFN-γ and TNFα, whereas the $CD56^{dim}CD16^{+}$ (5-10%) NK subpopulation contains a high cytolytic activity through the activation of different killer receptors and releases granules containing perforin and granzymes [21, 22]. Of note, perforin and granzymes release onto the target cell´s surface is a well-described hallmark of the activation of NK cell within the externalization of CD107a, a marker of NK activation and cytotoxic degranulation [23]. We previously mentioned that NK cells can discriminate between normal cells and cells that do not express adequate amounts of major histocompatibility complex (MHC) class I molecules. Additionally, three novel NK-specific triggering surface molecules (NKp46, NKp30, and NKp44) have been identified. They are termed natural cytotoxicity receptors (NCR). In the clinic, monoclonal antibodies (mAbs) or antigens from tumor cells can abrogate lysis through masking the NCRs [24]. Different works have been described activating or inhibitory receptors on NK cells, which together with the expression of other cytokines can determine the phenotype of an immune lineage [25].

NK cells can remember previous exposure to the inflammatory microenvironment; this evokes the secretion of cytokines that were previously released in the first stimulus, generating cell-differentiation from normal NK cells to memory NK (NKm) cells [26, 27]. NKm cells have been identified in the tumor microenvironment; these cells produce high levels of IFN-γ, perforin, and granzyme after restimulation, Figure 1 [28]. However, after the failure of the activation of NK cells, either baseline phenotype or memory phenotype, can promote tumor growth and distant metastasis rather than kill malignant cells [29]. Moreover, it has been reported that the half-life of NK cells is around 1-2 weeks, while NKm cells can live on average twice as long [30]. The above could be a therapeutic strategy to readapt NKm cells to a cytotoxic facet in cancer disease.

On the other hand, different studies have identified resting and activated subtypes of NK cells in the tumor microenvironment [31]. NK cells can divide in NK1 and NK2 phenotypes, which are defined as an analogy with T helper lymphocytes (Th1 and Th2), secreting IFN-γ and IL-5, IL-13, respectively, Figure 1 [32]. A recent well-detailed work of Nersesian et al. reported that the tumor infiltration of NK cells in solid tumors is related to improved overall survival in patients with solid tumors. They found increased infiltration of NK1 cells into the tumors [33].

It is important to mention that besides the cytotoxic activity of NK cells (cNK), they can also acquire different immune functions such as antigen-presenting (AP-NK) cells [34], helper functions NK (NKh) [35], and regulatory NK (NKreg) activity [36]. All these phenotypes are governed by surface molecules, and metabolic fingerprints, which results in specific and individual functions, Figure 1 [36].

NKh cells can primarily act as enhancers of the lithic functions of T cells to maintain tumor rejection. Different reports have identified that members of the heat shock protein (HSP) family promote the helper phenotype of NK cells [37]. Moreover, the NKh phenotype has also been associated with the interplay of IL-18 and prostaglandin (PG) E2, it has been reported that unlike the potentiation of cytotoxic activity promoted by IL-2, the presence of IL-18 in NK cells induces a migratory response through generating CCR7 without enhancing their killer activity. Of note, like DCs, NKh cells can be recruited in the lymph nodes, which might support the development of T helper type 1 (TH1) responses [38].

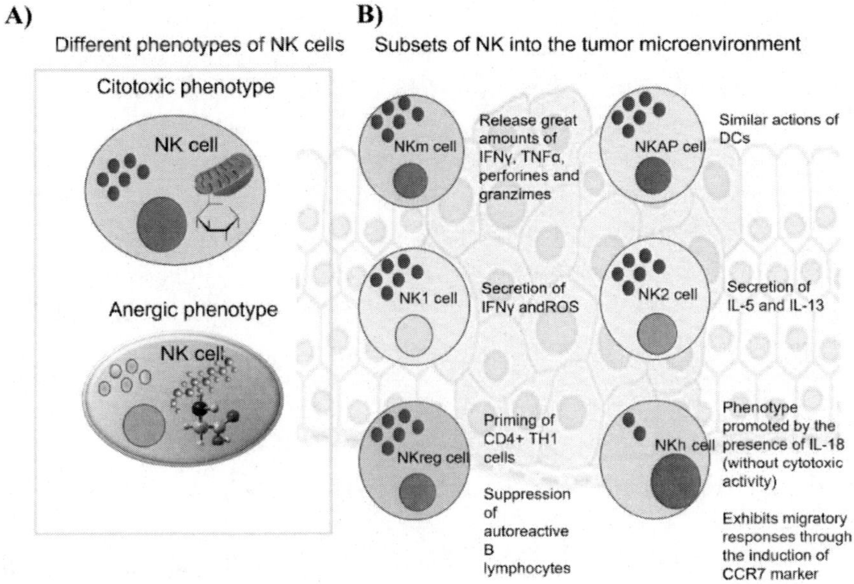

Figure 1. A) Phenotypic state of NK cells. The images display the two states of activation of NK cells, cytotoxic or activated NK cells are mainly dependent on glucose metabolism evoked by the activation of mitochondria; meanwhile, the anergic phenotype receives its energy mainly from amino acid or fatty acid degradation pathways. B) Subsets of NK cells in the tumor microenvironment. Different subsets of NK can shape the tumor microenvironment, these can be NKm (memory), NKAP (antigen-presenting), NK1, NK2, NKreg (regulatory) and NH (helper) cells. The main characteristics of them are named in the image.

NKreg can influence other cell types, such as DCs, T cells, B cells, and endothelial cells. For instance, NKreg can destroy immature DCs, and induce the maturation of DCs [39, 40]. NKreg cells can also promote the priming of CD4+ TH1 cells by secreting IFN-γ, however, NK cells can also kill activated T cells, an activity that is determined by the expression of outstanding amounts of classical or nonclassical MHC class I molecules. Besides, NKreg cells can suppress autoreactive B lymphocytes. Conversely, the depletion of NK cells has been associated with increased autoimmunity. On the other hand, it has been described that NK cells possess adherent molecules such as integrins that serve to attach to endothelial cells; meanwhile, endothelial cells can present different types of injury, in this sense, NKreg cells can induce angiogenesis in a pregnancy context or kill the damaged endothelial cells involved in a virus infection [6].

Delving into the metabolic aspect, the cytotoxicity capacity of NK cells is primarily linked to glucose and glycolysis effector mechanisms. On the contrary, the regulatory functions of NK cells are promoted under low levels of glycolysis and the cells need alternative sources of energy such as fatty acids or amino acids, Figure 1. Memory NK cells exhibit stable mitochondrial activity, mainly denoted by increased respiratory capacity, stable membrane potential, and reduced levels of reactive oxygen species (ROS) [36]. Nevertheless, it has been mentioned that NK cells are functionally dynamic lymphocytes, with up to 30,000 unique phenotypes that comprise the repertoire of each individual [41].

Immune Infiltration of NK Cells at Peripheral and Intratumorally Levels in Patients with Cancer

Like another type of immune cell lineage, NK cells belonging to the tumor microenvironment can be divided into resting or anergic and activated cells. Depending on their state of maturation, they provide a poor or good clinical outcome [42]. Different studies have identified infiltration of NK cells in several types of cancer. For instance, in advanced stages of cancer, the state of activation of NK cells corresponds to an anergic phenotype, and as a final result, they confer poor outcomes [43]. Yang et al. employed a flow cytometry analysis demonstrating that patients with advanced pancreatic cancer displayed a high density of NKs combined with a high neutrophil-to-lymphocyte ratio in the blood of patients with this disease, which was related to a strong inflammatory response and poor overall survival (OS). Related to the above, when comparing the median OS of patients with a high density of NK cells vs. a cohort with a low density of them, it was 4 ± 0.32 months vs 6 ± 0.30 months, respectively. In addition, enzyme-linked immunosorbent assays demonstrated that the systemic response of patients with a high density of NK cells displayed a reduced secretion of IL-2, IFN-γ, and TNF-α, which contributes to the inhibition of NK cell activation. In other words, although the patients with advanced stage of pancreatic cancer contain a high density of NKs, the cells have impaired functions to kill cancer cells [43]. Interestingly, it has been described that NK cells coming from healthy people differ from patients with cancer in the secretory profile of cytokines and in the expression of different receptors [44, 45].

On the other hand, different studies carried out in patients with colorectal cancer have associated the presence of NK cells with a favorable prognostic.

Although low percentages of NK cells have been found in patients with colorectal cancer [46], different studies employing several techniques such as immunohistochemistry and flow cytometry as well as in different tissues including tumor, blood, and lymph nodes have established that the presence of NK cells is feasible in these type of tumors [47-49]. Within the tumor microenvironment, Coca et al. demonstrated that patients with colorectal cancer without adjuvant therapy with little (<50 NK cells), and moderate (50–150 NK cells) NK infiltration showed significantly shorter survival rates than those with extensive (>150 NK cells) infiltration, altogether with a TNM (stands for tumor (T), nodes (N), and metastases) stage I-III. This correlation was carried out by monitoring the patients after 5 years of performing the identification of intratumoral NK cells [47]. It is important to mention that NK cells were found close to neoplastic cells, suggesting that they are exerting a cytotoxic activity. In addition, Okada et al. also support the fact that the presence of NK as well T and B lymphocytes in patients with colorectal cancer is associated with a good prognosis due to these cells were commonly found in larger retrieved lymph nodes (rLNs) of patients with this disease in stage II. The authors found that when NK cells were present in the rLNs of patients with colorectal cancer, the median follow-up period of the patients who survived without recurrence was 118 months. Thus, the authors also attributed that NK cells can be used as a predictor of recurrence [48]. Moreover, it is important to mention that high numbers of lymph nodes (LN) have been associated with good outcomes [50, 51]. Therefore, the association of NK cells and good prognosis in patients with colorectal cancer has been reported in the tumor microenvironment or in LNs and at a systemic level. Tang and colleagues evaluated the presence of NK cells in the blood of patients with colorectal cancer. They found that NK cells could act as an independent predictor of survival; however, NK cells and B lymphocytes could increase the predictive value [49]. This study retrospectively analyzed by flow cytometry the presence of NK cells in the blood of patients with colorectal cancer. It is important to remark that the patients received radical surgery and chemotherapy. Again, a higher percentage of NK cells indicated a longer survival time than a lower percentage. Besides, as was reported by Okada et al. the rates of NK cells positively correlated with the T and B lymphocyte ratio. The association of these immune cells was more evident in patients with early disease [49]. All these studies support the fact that the presence of NK cells in the early stages of cancer disease is vital for getting good clinical outcomes without recurrence. However, they lack a deep identification of a wide range of cell receptors and markers related to the cytotoxic activity of NK cells.

Most of the studies related to the presence of NK cells and the clinical outcome have shown a favorable condition in various types of tumors [52-58], although there are some reports where it is still considered controversial [59-65]. We will delve into this relationship below.

Flow cytometry analyses carried out in the blood of patients with chronic myeloid leukemia who discontinue imatinib (an ATP-competitive inhibitor of the BCR-ABL1 oncoprotein) demonstrated a high number of NK cells, as well as T lymphocytes, was not associated with relapse of the disease. Interestingly, the expression of activating receptors of NK cells including NKG2D, NKp46, NKp30, DNAM-1, and KIR2D was decreased in non-relapsing and relapsing patients compared with healthy individuals. Due to the high percentage of NK cells in patients with chronic myeloid leukemia being associated with a good prognosis, the authors raised that NK cells play a role in controlling leukemia-initiating cells at the origin of relapse after drug cessation. Nevertheless, the number of them should be enough to refund their reduced cytotoxic activity [52]. Besides, additional data on the stage of cancer would be relevant to make an overall prediction. It is important to mention that NK cells in the blood of patients with chronic lymphocytic leukemia without treatment also confers both a favorable prognosis and overall survival [53]. The authors postulated that NK cells could serve as a critical immunological biomarker of a prognostic factor of this disease. It is important to highlight that the authors also evaluated the percentage of NK cells in the two conditions, in individuals where treatment with chemotherapeutic drugs was necessary (fludarabine and cyclophosphamide; fludarabine, cyclophosphamide, and rituximab; high-dose methylprednisolone; cyclophosphamide, vincristine, and prednisone; and chlorambucil with/without rituximab) or when the treatment was unnecessary. They interestingly found that after any scheme of chemotherapy, the count of NK cells decreased, contrary to data to those found in patients naïve to treatment [53]. Similar results about the relationship between the high percentage of peripheral NK cells and better overall survival have also been reported in patients with follicular lymphoma in different clinical stages [54, 66], some patients need treatment based on rituximab or rituximab-containing regimens. According to the above, the authors observed that patients with a low count of NK cells had a lower clinical response to therapy than patients with high levels of these cells [54]. The above findings indicate that the role of peripheral NK cells in disseminated cancer is essential in the optimal response to therapy. Additionally, the importance of NK cells has also been demonstrated in mantle cell lymphoma, a subtype of non-Hodgkin's lymphoma, since a low percentage of NK cells in the peripheral blood of patients with this disease has been associated with an inferior survival [61].

On the other hand, the high presence of NK cells at peripheral and intratumorally levels and good prognosis have also been reported in solid tumors, such as prostate cancer [55, 67]. The study of Pasero et al. was focused on the identification and isolation of NK cells from peripheral blood of patients with metastatic prostate cancer, the authors also characterized different activating receptors of NK cells *ex vivo* by flow cytometry. The results demonstrated that NKp46, NKp30, DNAM-1, and the degranulation marker CD107 were associated with a good prognosis. Interestingly, peripheral NK cells derived from patients with prostate cancer displayed high expression of molecules involved in efficient NK cell cytotoxicity (NKp30, NKp46), co-stimulation (DNAM-1), maturation, (CD57) and degranulation (CD107), which were tested against prostate cancer cell lines [55]. However, despite the fact that the NK cells found at the peripheral level of patients with prostate cancer present both important effector molecules and cytotoxic activity, some hypotheses about the deregulation of their cytotoxic activity in the tumor have been exposed as is the case of exosomes produced by human prostate cancer cells, which promote the expression of ligands for NKG2D, causing a down-regulation activity of NK cells [68].

Like prostate cancer, patients with lung cancer and with a high degree of infiltration of NK cells in the tumor have related with good prognosis [57, 69]. Lung cancer is considered another solid type of neoplasm. Moreover, it is well known that non-small cell lung cancer (NSCLC) is regarded as one of the most aggressive phenotypes of this disease. Still generally, the prognosis of lung-associated malignancies is terrible [70, 71]. Thus, the identification of pharmacological alternatives for the treatment of this disease is crucial for improving clinic results and the quality of life of patients. Jin et al. investigated the differential composition of the immune cell infiltrate, governed by NK cells, in malignant and non-malignant lung tissue areas belonging to patients with NSCLC (including squamous carcinoma, adenocarcinoma, and large cell carcinoma) without chemotherapy, radiotherapy, and biotherapy. The authors focused on the expression of different receptors, functional status, metastasis, and prognosis. First, the authors revealed by immunofluorescence that NK cells were mainly concentrated in the tumor stroma. Interestingly, they found significant differences in $CD56^{+}/CD16^{+}$ NK cell infiltration degree and receptor expression between different pathological types of lung cancer. Of note, they concluded that the more infiltration of NK cells existed, the longer the survival time of patients did. Finally, the authors also employed a mouse lung cancer model to verify the expression of the inhibitory receptor NKG2A. They observed that the level of expression of NKG2A in NK cells was significantly high in metastatic lung cancer cells [57]. Related to the above, the workgroup of Picard et al. found different subsets of NK cells in the blood of patients with NSCLS tumors, all of them were based on the expression of membrane markers ($CD56^{dim}$ $CD16^{+}$, $CD56^{dim}$ $CD16^{-}$ and $CD56^{bright}$). Interestingly, the authors reported that $CD56^{dim}$ CD16 mainly drove the degranulation properties and cytokine production (IFN-γ and TNFα)– NK cell subset in patients with NSCLC; meanwhile, high levels of $NKp46^{+}$ $CD56^{dim}$ $CD16^{+}$ NK subtype has a critical influence on the overall survival. It should be noted that the median overall survival of patients with a high density of the above phenotype was 27 months vs. 16 months for patients with a low number of NK cells [56]. Is also important to consider that the different subtypes of NK cells belonging to patients with NSCLC contain a high expression of co-inhibitory receptors such as TIM3 and PD-1 as compared to healthy donors [56]. Studies employing quantitative real-time polymerase chain reaction (qRT-PCR) have evaluated the expression of NCR1 (which encodes for NKp46) and NCR3 (which encodes for NKp30) in the blood of patients with NSCLC. This study concluded that NK cell receptors have important prognostic features in patients with advanced NSCLC. Although, it is important to consider that both molecules' expression was lower compared with patients with other malignancies or healthy controls. The authors show that NK cell receptors' expression in the blood can yield useful biomarkers for patient stratification [60]. The results above indicate that although there is much similarity between phenotypes and percentages of NK cells at the systemic level between patients with lung cancer and healthy donors, there is a wide range of subtypes with differential cytotoxic functions. Furthermore, the different subtypes of this immune cell lineage vary not only systemically but could also vary within the tumor microenvironment. And surely, this effect applies to different types of cancer. Therefore, the intratumoral release of a drug would be optimal to improve its anticancer efficacy. This assumption is supported due to NK cells that shape the tumor microenvironment having defective degranulation, secretory and cytotoxic functions [72-74].

Another solid tumor widely distributed worldwide is breast cancer [75]. Among the most aggressive types of this disease, there is a type 2 epidermal growth receptor positive phenotype (HER2+) and a triple-negative (TN) one [76]. The third subtype that makes up the variety of breast cancer is the hormonal (which contains the estrogen and progesterone receptor presence) phenotype, which better response to anti-hormonal therapy [77, 78]. However, in the most aggressive phenotypes, predictive biomarkers are needed for personalized breast cancer treatment. In this regard, the studies of Muntasell et al. focused on identifying tumor-infiltrating NK cells and tumor HLA class I pointed out to NK cells as complementary biomarkers with predictive and prognostic value in patients with primary HER2+ breast cancer treated with neo-adjuvant anti-HER2 monoclonal antibodies. They found that the infiltration of NK cells in HER2+ breast tumors has also been linked with disease-free survival and pathologic complete response [58]. The authors observed that NK cells and T lymphocytes were mainly located in tumor stroma areas. Importantly, like reported in databases, the authors found that higher T and NK cell densities were detected mainly in ER-negative compared with ER-positive tumors. The expression of HLA-1 was also performed; the results demonstrated two groups of cancer cells with high and low expression of this molecule. Thus, the authors raised the possibility that patients with low or negative expression of HLA-I in tumors may benefit from treatment enhancing NK cells rather than T-cell antitumor immunity, whereas patients with tumors with high expression of HLA-I can employee treatment based on PD1/PD-L1 checkpoint blockade, enhancing tumor T-cell immunity. In addition, the authors support the fact that NK and T cell-targeted immunotherapy in a personalized manner would benefit outstandingly the treatment of patients with this disease [58]. There is controversy about the infiltration of NK cells in breast tumors, as was reported by Rathore et al., who found that low intratumorally NK count showed a significant and inverse association with tumor grade, stage, and lymph node status. Nevertheless, a low percentage of NK cells had a substantial relationship with response to treatment and good prognosis. The authors discussed that the presence of NK cells is important in the tumor microenvironment, but it could also be an interplay of NK with different immune cells and soluble factors such as cytokines, and stromal-derived factors, among others. However, different limitations of this study including one marker for the immunohistochemistry detection of NK cells was employed. Additionally, markers of cytotoxic activity, estrogen, and HER2 status were also lacking [59].

In liver cancer, a recent report carried out by the workgroup of Sun et al. analyzed the expression of NKG2A and their associated dysfunction of NK cells present in the peripheral blood and intra-or peritumor regions of the liver neoplasm samples taken from blood and fresh tumor tissue during resection surgery. The authors also correlated between the expression of this receptor and the clinical outcome in patients with liver cancer [62]. It is important to remark that NKG2A is considered an inhibitory molecule of the activity of NK cells [79]. The authors found that NKG2A was more expressed in the intratumor samples than peritumoral samples or healthy tissues. The expression of this molecule was associated with a large tumor diameter. Additionally, the expression of other activating and inhibitory receptors was also downregulated and upregulated, respectively, in NK cells coming from tumor samples. Additionally, intratumorally NK cells found have characteristics of functional exhaustion denoted by the reduction of the production of IFN-γ, granzyme B, and perforin. Of Note, intratumorally, liver cells displayed high levels of HLA-E, which is a ligand of NKG2A. All these characteristics corroborate with shorter overall survival (OS). Notably, the authors previously found that high plasma levels of IL-10 were strongly related to the induction of NKG2A expression in patients with liver cancer [80].

Interestingly, the increased expression of IL-10 was also observed in the patients enrolled in this study. The NKG2A was predominantly observed in $CD56^{dim}$ NK cells, rather than $CD56^{bright}$ NK cells belonging to intratumor tissue rather than peripheral blood. With this finding, the authors argued a negative regulatory mechanism in exhausted NK cells through the induction of NKG2A and CD96 in the cancer nests [62, 63].

The dysfunction of NK cells by different molecules has also been observed in digestive cancers. Apparently, the immunological checkpoint of PD-1/PDL-1 is not exclusive to the function of T lymphocytes. Liu et al., analyzed by flow cytometry the expression of PD-1 on peripheral and tumor-infiltrating NK cells from patients with digestive cancers, including esophageal, liver, colorectal, gastric, and biliary cancer. They found a high expression of PD-1 in the NK cells belonging to patients with cancer, which also correlated with the degree of tumor differentiation, tumor lymph nodes metastasis stage. In addition, the expression of PD-1 was related to poor prognosis. On the other hand, PI3K/AKT signaling in NK cells has been related to cytokine stimulation and cytotoxic functions [81]. Notably, the authors found that the activation of the PD-1/PD-L1 axis impairs PI3K phosphorylation in NK cells. On the contrary, the blockade of PD-1 leads to the activation of NK cells [64]. This work supports the fact that NK cells are important in eradicating tumor growth of digestive cancers, but the different therapeutic alternatives based on PD-1 blockade lead to the activation of NK cells. Although it would be necessary to know the mechanisms involved in the upregulation of PD-1 on NK cells.

According to the above, in liver cancer, there also found the high expression of PD-1 and other inhibitory receptors [65]. CD49, an integrin α1 that binds to integrin β1 (CD29) to form the CD49a/CD29 heterodimer, is associated with the retaining lymphocytes in tissues [82]. Of note, the expression of CD49 has been identified in tissue-resident NK cells [83]. A recent report showed that high expression of this molecule in NK cells belonging to the intratumoral and peritumoral regions of samples taken from patients with liver cancer without treatment was associated with an exhausted phenotype [65]. Moreover, transcriptomic analyses were carried out in NK cells with a $CD49^+$ or $CD49^-$ phenotype. The results showed that the expression of CD49 correlated with the presence of other inhibitory molecules including PD-1, NKG2A, CD96, meanwhile, genes related to the secretion of cytokines were drastically downregulated. Interestingly, NK cells $CD49^+$ were associated with poor outcomes [65].

Previous works suggest that the presence of NK cells in the tumor microenvironment and at the peripheral level helps counteract the growth of malignant cells. However, because different phenotypes characterized by the specific presence of molecules that cause an exhausted phenotype, specific identification and personalized and targeted therapy could go a long way in countering tumor growth.

Therapeutic Strategies Based on NK Activity in Cancer

As mentioned in the previous sections, the presence of NK cells in different types of tumors is related to a better prognosis. Therefore, drugs aimed at enhancing their cytotoxic effects or overcoming cell dysfunction would be beneficial to induce the destruction of malignant cells. In this section, we will describe the state of the art of the main therapeutic strategies that have been used for this purpose.

Chimeric Antigen Receptors (CARs)

Like CAR development in T cells, CAR-NK cells resemble the genetic modification of CARs that can recognize a specific antigen preferable expressed or overexpressed in target cells. The signaling of CAR-NK cells is like that in CAR-T cells, typically, they are shaping up the fusion of CD3ζ with one or two costimulatory molecules for the receptors of T cells [84]. The use of CAR- NK strategy against a wide range of tumor antigens has been tested in preclinical studies for hematological malignancies and solid tumors [85, 86].

Killer Cell Engagers

This kind of therapeutic alternative is based on the employee of bi-and tri-specific killer engagers (BiKEs and TriKEs), which are composed of two or three antibody variable fragments of different antigens. Generally, one antibody fragment can recognize the tumor antigen and the other fragment is directed against CD16 on NK cells [87]. In this manner, BiKEs or TrikEs usually engage with CD16 and tumor antigens, inducing immune synapses; that as a final result lead to the antibody-dependent cellular cytotoxicity (ADCC) of NK cells [88]. n preclinical studies, different BiKEs and TriKEs have been designed against different tumor antigens [89-91]. In addition, BiKEs and TriKEs have also been developed with cytokines to further to enhance the functions of NK cells [92].

NK Cell "Priming" Strategies or Cytokine-Induced Human NK Expansion and Activation

This strategy consists of isolating donor-derived fresh NK cells, which have a reduced lytic function. Then, the cells are stimulated with various cytokines to regain the cell cytotoxicity [93]. For this purpose, a mix of IL-2/IL-15/IL-18 generally has been employed in the medium where ex vivo expansion of NK cells is performed. The activation of NK cells with the mix of these soluble factors has offered sustained antitumor effects that resemble the memory type of NK cells [28, 94]. NK cells can also be exposed to cancer cells to enhance *ex vivo* NK cell expansion but also for their cytotoxic maturation [95]. Additionally, the exposure of NK cells to kinase inhibitors alone or in combination with cytokines promotes their maturation, enhancing their cytotoxicity. An example of the above is the inhibition of glycogen synthase 3 (GSK3), which results in the elevated expression of several transcription factors and high production of TNFα and IFNγ [96].

However, different agents such as the tyrosine kinase inhibitor sorafenib have recently promoted the antitumor response of NK cells belonging to liver tumors. Sorafenib causes a positive antitumor interaction between NK cells and tumor-associated macrophages (TAM) since it stimulates the proinflammatory activity of TAMs and this switch on NK cells [97]. On the contrary, histone deacetylase inhibitors (HDACi) such as valproic acid inhibit the STAT3 pathway in human NK cells isolated from the blood and therefore the expression of an important activating receptor (NKG2D), resulting in the decrease of antitumor response [98]. The above findings suggest that the priming of NK cells would have to be studied with various agents to get a great or negative promise for the treating of different types of cancer.

Methods to Enhance the Infiltration and Function of Infused NK Cells

In the section on immune infiltration of NK cells in patients with cancer, different studies denoted that NK cells are considered important effectors for the elimination of cancer cells, however, sometimes, there are not enough percentages of them to eradicate transformed cells. In this regard, different alternatives that direct NK cells towards tumors are also being studied. These therapeutic approaches are based on the fact that tumor infiltration of NK cells is correlated with better prognosis in different types of tumors. Protein-conjugated antibodies (NRPbody) lead the recruitment of NK cells into the tumors [99]. NRPbody has been demonstrated to increase NK-cell infiltration in murine pancreatic ductal adenocarcinoma models, which results in an improved immune response against cancer [99]. In addition, these biological devices can also be used to guide NK cells into tumors and target chemoimmunotherapy to cancer cells, using as an advantage the overexpression of a specific molecule [100, 101]. Other workgroups have also postulated that genetic modifications in the chemokine receptor of NK cells can favor their migration to the tumor microenvironment [102].

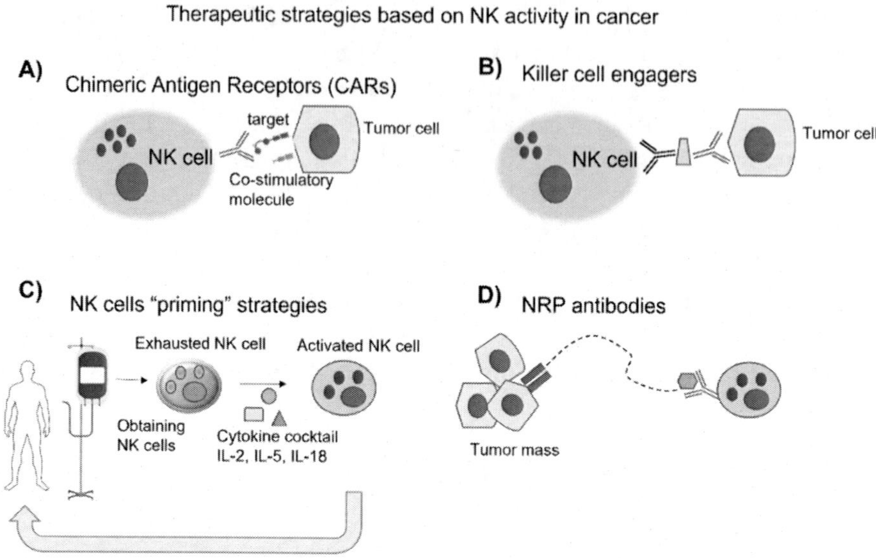

Figure 2. A) Therapeutic strategies based on NK activity in cancer. A) Chimeric antigen receptors (CARs). This kind of therapy is based on the designment of specific antibodies for NK that recognize target molecules of the tumor cells, but also they can bind to costimulatory molecules that enhance the antitumor activity of NK cells. B) Killer cell engager. This kind of therapeutic alternative is based on the employee of bi-and tri-specific killer engagers (BiKEs and TriKEs), which are composed of two or three antibody variable fragments of different antigens. Generally, one antibody fragment can recognize the tumor antigen and the other fragment is directed against CD16 on NK cells. C) NK cell priming strategy. This strategy consists of isolating donor-derived fresh NK cells, which have a reduced lytic function. Then, the cells are stimulated with various cytokines to regain the cell cytotoxicity, then NK cells are re-inoculated into the individual. D) Protein-conjugated antibodies (NRP antibodies), NK-cell–homing protein, named NK-cell–recruiting protein-conjugated antibody, this type of antibody is coupled to a protein that is recruited by tumor cells. *In vivo* models, CXCL16, the NK-cell – recruiting protein component of the NRPbody, resulted in the increased percentage of NK cells in the tumor mass.

Conclusion

Significant advances in medicine, cancer remains one of the leading causes of death, and the number of cancer cases is still increasing worldwide. Recent advances in immuno-oncology have demonstrated the great potential of activating the host's immune system. Here, we exposed the different subsets of NK cells that shape the immune tumor microenvironment. They all serve to understand their biology for the design of therapy to overcome their exhausted phenotype or rise their percentage of the tumor mass. Notably, it is important to remark that the detection of NK cells in different types of tumors depends on the techniques employed, the accuracy of the antibodies, and the molecular targets used.

On the other hand, although immune checkpoint therapy in T lymphocytes has been successful in the clinic, it would be convenient to explore different approaches such as other types of immune populations that are also significantly involved eradicating transformed cells. In addition, there exist different therapeutic strategies that have not been widely used in cancer treatment. Moreover, it would be important to evaluate the combination of NK cell-based therapies with other kinds of drugs to interrupt the co-evolution of the immune system and tumor cells.

References

[1] Bonanni, V., et al., Bone Marrow NK Cells: Origin, Distinctive Features, and Requirements for Tissue Localization. *Frontiers in Immunology,* 2019. 10(1569).

[2] Paul, S. and G. Lal, The Molecular Mechanism of Natural Killer Cells Function and Its Importance in Cancer Immunotherapy. *Frontiers in Immunology,* 2017. 8(1124).

[3] Herberman, R. B., et al., Cell-mediated immunity to leukemia virus- and tumor-associated antigens in mice. *Cancer Res,* 1976. 36(2 pt 2): p. 615-21.

[4] Lanier, L. L. and J. H. Phillips, NK cell recognition of major histocompatibility complex class I molecules. *Semin Immunol,* 1995. 7(2): p. 75-82.

[5] Vivier, E., J. A. Nunes, and F. Vely, Natural killer cell signaling pathways. *Science* (New York, N.Y.), 2004. 306(5701): p. 1517-9.

[6] Vivier, E., et al., Functions of natural killer cells. *Nat Immunol,* 2008. 9(5): p. 503-510.

[7] Huntington, N.D., The unconventional expression of IL-15 and its role in NK cell homeostasis. *Immunol Cell Biol,* 2014. 92(3): p. 210-3.

[8] Wu, Y., Z. Tian, and H. Wei, Developmental and Functional Control of Natural Killer Cells by Cytokines. *Frontiers in Immunology,* 2017. 8: p. 930-930.

[9] Samson, S. I., et al., GATA-3 promotes maturation, IFN-gamma production, and liver-specific homing of NK cells. *Immunity,* 2003. 19(5): p. 701-11.

[10] Townsend, M. J., et al., T-bet regulates the terminal maturation and homeostasis of NK and Valpha14i NKT cells. *Immunity,* 2004. 20(4): p. 477-94.

[11] Fehniger, T. A., et al., CD56bright natural killer cells are present in human lymph nodes and are activated by T cell-derived IL-2: a potential new link between adaptive and innate immunity. *Blood,* 2003. 101(8): p. 3052-7.

[12] Van Acker, H.H., et al., CD56 in the Immune System: More Than a Marker for Cytotoxicity? *Frontiers in Immunology,* 2017. 8(892).

[13] Pardo, J., et al., Granzymes are essential for natural killer cell-mediated and perf-facilitated tumor control. *Eur J Immunol,* 2002. 32(10): p. 2881-7.

[14] Osinska, I., K. Popko, and U. Demkow, Perforin: an important player in immune response. *Cent Eur J Immunol,* 2014. 39(1): p. 109-15.

[15] Wu, S. Y., et al., Natural killer cells in cancer biology and therapy. *Mol Cancer,* 2020. 19(1): p. 120.

[16] He, X. S., et al., T cell-dependent production of IFN-gamma by NK cells in response to influenza A virus. *The Journal of clinical investigation,* 2004. 114(12): p. 1812-1819.

[17] Moretta, A., et al., Early liaisons between cells of the innate immune system in inflamed peripheral tissues. *Trends Immunol,* 2005. 26(12): p. 668-75.

[18] Chiossone, L., et al., Natural killer cells and other innate lymphoid cells in cancer. *Nat Rev Immunol,* 2018. 18(11): p. 671-688.

[19] Paust, S., et al., Critical role for the chemokine receptor CXCR6 in NK cell-mediated antigen-specific memory of haptens and viruses. *Nat Immunol,* 2010. 11(12): p. 1127-35.

[20] Vivier, E., et al., Innate or adaptive immunity? The example of natural killer cells. *Science* (New York, N.Y.), 2011. 331(6013): p. 44-49.

[21] Guillerey, C., N. D. Huntington, and M. J. Smyth, Targeting natural killer cells in cancer immunotherapy. *Nat Immunol,* 2016. 17(9): p. 1025-36.

[22] Hanna, J., et al., Novel insights on human NK cells' immunological modalities revealed by gene expression profiling. *J Immunol*, 2004. 173(11): p. 6547-63.

[23] Bryceson, Y.T., et al., Cytolytic granule polarization and degranulation controlled by different receptors in resting NK cells. *J Exp Med,* 2005. 202(7): p. 1001-12.

[24] Moretta, A., et al., Activating receptors and coreceptors involved in human natural killer cell-mediated cytolysis. *Annu Rev Immunol,* 2001. 19: p. 197-223.

[25] Paul, S. and G. Lal, The Molecular Mechanism of Natural Killer Cells Function and Its Importance in Cancer Immunotherapy. *Frontiers in Immunology,* 2017. 8: p. 1124-1124.

[26] Pal, M., et al., Tumor-priming converts NK cells to memory-like NK cells. *Oncoimmunology,* 2017. 6(6): p. e1317411.

[27] Romee, R., et al., Cytokine activation induces human memory-like NK cells. *Blood,* 2012. 120(24): p. 4751-60.

[28] Ni, J., et al., Sustained effector function of IL-12/15/18-preactivated NK cells against established tumors. *J Exp Med,* 2012. 209(13): p. 2351-65.

[29] Cerwenka, A. and L.L. Lanier, Natural killer cell memory in infection, inflammation and cancer. *Nat Rev Immunol,* 2016. 16(2): p. 112-23.

[30] Romee, R., et al., Cytokine-induced memory-like natural killer cells exhibit enhanced responses against myeloid leukemia. *Sci Transl Med,* 2016. 8(357): p. 357ra123.

[31] Newman, A. M., et al., Robust enumeration of cell subsets from tissue expression profiles. *Nat Methods,* 2015. 12(5): p. 453-7.

[32] Kimura, M. Y. and T. Nakayama, Differentiation of NK1 and NK2 cells. *Crit Rev Immunol,* 2005. 25(5): p. 361-74.

[33] Nersesian, S., et al., NK cell infiltration is associated with improved overall survival in solid cancers: A systematic review and meta-analysis. *Translational Oncology,* 2021. 14(1): p. 100930.

[34] Zeissig, S., et al., Analyzing antigen recognition by Natural Killer T cells. *Methods Mol Biol,* 2013. 960: p. 557-572.

[35] Hanna, J. and O. Mandelboim, When killers become helpers. *Trends Immunol,* 2007. 28(5): p. 201-6.

[36] Poznanski, S. M. and A. A. Ashkar, What Defines NK Cell Functional Fate: Phenotype or Metabolism? *Front Immunol,* 2019. 10: p. 1414.

[37] Sedlacek, A. L., L. B. Kinner-Bibeau, and R. J. Binder, Phenotypically distinct helper NK cells are required for gp96-mediated anti-tumor immunity. *Scientific reports,* 2016. 6: p. 29889-29889.

[38] Mailliard, R. B., et al., IL-18-induced CD83+CCR7+ NK helper cells. *J Exp Med,* 2005. 202(7): p. 941-953.

[39] Ferlazzo, G., et al., Human dendritic cells activate resting natural killer (NK) cells and are recognized via the NKp30 receptor by activated NK cells. *J Exp Med,* 2002. 195(3): p. 343-51.

[40] Moretta, A., Natural killer cells and dendritic cells: rendezvous in abused tissues. *Nat Rev Immunol,* 2002. 2(12): p. 957-64.

[41] Horowitz, A., et al., Genetic and environmental determinants of human NK cell diversity revealed by mass cytometry. *Sci Transl Med,* 2013. 5(208): p. 208ra145.

[42] Ali, H. R., et al., Patterns of Immune Infiltration in Breast Cancer and Their Clinical Implications: A Gene-Expression-Based Retrospective Study. *PLOS Medicine,* 2016. 13(12): p. e1002194.

[43] Yang, C., et al., Anergic natural killer cells educated by tumor cells are associated with a poor prognosis in patients with advanced pancreatic ductal adenocarcinoma. *Cancer Immunol Immunother,* 2018. 67(12): p. 1815-1823.

[44] Pesce, S., et al., Identification of a subset of human natural killer cells expressing high levels of programmed death 1: A phenotypic and functional characterization. *J Allergy Clin Immunol,* 2017. 139(1): p. 335-346 e3.

[45] Ndhlovu, L. C., et al., Tim-3 marks human natural killer cell maturation and suppresses cell-mediated cytotoxicity. *Blood,* 2012. 119(16): p. 3734-43.

[46] Sconocchia, G., et al., NK cells and T cells cooperate during the clinical course of colorectal cancer. *Oncoimmunology,* 2014. 3(8): p. e952197.

[47] Coca, S., et al., The prognostic significance of intratumoral natural killer cells in patients with colorectal carcinoma. *Cancer,* 1997. 79(12): p. 2320-8.

[48] Okada, K., et al., The Number of Natural Killer Cells in the Largest Diameter Lymph Nodes Is Associated with the Number of Retrieved Lymph Nodes and Lymph Node Size, and Is an Independent Prognostic Factor in Patients with Stage II Colon Cancer. *Oncology,* 2018. 95(5): p. 288-296.

[49] Tang, Y. P., et al., Prognostic value of peripheral blood natural killer cells in colorectal cancer. *BMC Gastroenterology,* 2020. 20(1): p. 31.

[50] Sarli, L., et al., Number of lymph nodes examined and prognosis of TNM stage II colorectal cancer. *Eur J Cancer,* 2005. 41(2): p. 272-9.

[51] Okada, K., et al., The size of retrieved lymph nodes correlates with the number of retrieved lymph nodes and is an independent prognostic factor in patients with stage II colon cancer. *Int J Colorectal Dis,* 2015. 30(12): p. 1685-93.

[52] Rea, D., et al., Natural killer-cell counts are associated with molecular relapse-free survival after imatinib discontinuation in chronic myeloid leukemia: the IMMUNOSTIM study. *Haematologica,* 2017. 102(8): p. 1368-1377.

[53] Wang, W. T., et al., Elevated absolute NK cell counts in peripheral blood predict good prognosis in chronic lymphocytic leukemia. *J Cancer Res Clin Oncol,* 2018. 144(3): p. 449-457.

[54] Shafer, D., et al., Low NK cell counts in peripheral blood are associated with inferior overall survival in patients with follicular lymphoma. *Leuk Res,* 2013. 37(10): p. 1213-5.

[55] Pasero, C., et al., Highly effective NK cells are associated with good prognosis in patients with metastatic prostate cancer. *Oncotarget,* 2015. 6(16): p. 14360-73.

[56] Picard, E., et al., Circulating NKp46(+) Natural Killer cells have a potential regulatory property and predict distinct survival in Non-Small Cell Lung Cancer. *Oncoimmunology,* 2019. 8(2): p. e1527498.

[57] Jin, S., et al., NK cell phenotypic modulation in lung cancer environment. *PLoS One,* 2014. 9(10): p. e109976.

[58] Muntasell, A., et al., NK Cell Infiltrates and HLA Class I Expression in Primary HER2(+) Breast Cancer Predict and Uncouple Pathological Response and Disease-free Survival. *Clin Cancer Res,* 2019. 25(5): p. 1535-1545.

[59] Rathore, A. S., et al., Is the tumor infiltrating natural killer cell (NK-TILs) count in infiltrating ductal carcinoma of breast prognostically significant? *Asian Pac J Cancer Prev,* 2014. 15(8): p. 3757-61.

[60] Fend, L., et al., Prognostic impact of the expression of NCR1 and NCR3 NK cell receptors and PD-L1 on advanced non-small cell lung cancer. *Oncoimmunology,* 2017. 6(1): p. e1163456.

[61] Zhou, X. H., et al., Low absolute NK cell counts in peripheral blood are associated with inferior survival in patients with mantle cell lymphoma. *Cancer Biomark,* 2019. 24(4): p. 439-447.

[62] Sun, C., et al., High NKG2A expression contributes to NK cell exhaustion and predicts a poor prognosis of patients with liver cancer. *Oncoimmunology,* 2017. 6(1): p. e1264562.

[63] Sun, H., et al., Human CD96 Correlates to Natural Killer Cell Exhaustion and Predicts the Prognosis of Human Hepatocellular Carcinoma. *Hepatology,* 2019. 70(1): p. 168-183.

[64] Liu, Y., et al., Increased expression of programmed cell death protein 1 on NK cells inhibits NK-cell-mediated anti-tumor function and indicates poor prognosis in digestive cancers. *Oncogene,* 2017. 36(44): p. 6143-6153.

[65] Sun, H., et al., Accumulation of Tumor-Infiltrating CD49a + NK Cells Correlates with Poor Prognosis for Human Hepatocellular Carcinoma. *Cancer Immunology Research,* 2019. 7(9): p. 1535.

[66] Klanova, M., et al., Prognostic Impact of Natural Killer Cell Count in Follicular Lymphoma and Diffuse Large B-cell Lymphoma Patients

Treated with Immunochemotherapy. *Clin Cancer Res,* 2019. 25(15): p. 4634-4643.

[67] Gannon, P. O., et al., Characterization of the intra-prostatic immune cell infiltration in androgen-deprived prostate cancer patients. *J Immunol Methods,* 2009. 348(1-2): p. 9-17.

[68] Lundholm, M., et al., Prostate tumor-derived exosomes down-regulate NKG2D expression on natural killer cells and CD8+ T cells: mechanism of immune evasion. *PLoS One,* 2014. 9(9): p. e108925.

[69] Villegas, F. R., et al., Prognostic significance of tumor infiltrating natural killer cells subset CD57 in patients with squamous cell lung cancer. *Lung Cancer,* 2002. 35(1): p. 23-8.

[70] Herbst, R. S., J. V. Heymach, and S. M. Lippman, Lung cancer. *N Engl J Med,* 2008. 359(13): p. 1367-80.

[71] Siegel, R. L., K. D. Miller, and A. Jemal, Cancer statistics, 2020. *CA Cancer J Clin,* 2020. 70(1): p. 7-30.

[72] Platonova, S., et al., Profound coordinated alterations of intratumoral NK cell phenotype and function in lung carcinoma. *Cancer Res,* 2011. 71(16): p. 5412-22.

[73] Le Maux Chansac, B., et al., NK cells infiltrating a MHC class I-deficient lung adenocarcinoma display impaired cytotoxic activity toward autologous tumor cells associated with altered NK cell-triggering receptors. *J Immunol,* 2005. 175(9): p. 5790-8.

[74] Carrega, P., et al., Natural killer cells infiltrating human nonsmall-cell lung cancer are enriched in CD56 bright CD16(-) cells and display an impaired capability to kill tumor cells. *Cancer,* 2008. 112(4): p. 863-75.

[75] Sung, H., et al., Global Cancer Statistics 2020: GLOBOCAN Estimates of Incidence and Mortality Worldwide for 36 Cancers in 185 Countries. *CA Cancer J Clin,* 2021. 71(3): p. 209-249.

[76] Slamon, D. J., et al., Human breast cancer: correlation of relapse and survival with amplification of the HER-2/neu oncogene. *Science* (New York, N.Y.), 1987. 235(4785): p. 177-82.

[77] Cho, N., Molecular subtypes and imaging phenotypes of breast cancer. *Ultrasonography,* 2016. 35(4): p. 281-8.

[78] Sorlie, T., et al., Gene expression patterns of breast carcinomas distinguish tumor subclasses with clinical implications. *Proc Natl Acad Sci U S A,* 2001. 98(19): p. 10869-74.

[79] Sun, C., et al., NK cell receptor imbalance and NK cell dysfunction in HBV infection and hepatocellular carcinoma. *Cell Mol Immunol,* 2015. 12(3): p. 292-302.

[80] Li, F., et al., Blocking the natural killer cell inhibitory receptor NKG2A increases activity of human natural killer cells and clears hepatitis B virus infection in mice. *Gastroenterology,* 2013. 144(2): p. 392-401.

[81] Ali, A. K., N. Nandagopal, and S. H. Lee, IL-15-PI3K-AKT-mTOR: A Critical Pathway in the Life Journey of Natural Killer Cells. *Front Immunol,* 2015. 6: p. 355.

[82] Ray, S. J., et al., The collagen binding alpha1beta1 integrin VLA-1 regulates CD8 T cell-mediated immune protection against heterologous influenza infection. *Immunity,* 2004. 20(2): p. 167-79.

[83] Peng, H., et al., Liver-resident NK cells confer adaptive immunity in skin-contact inflammation. *The Journal of clinical investigation,* 2013. 123(4): p. 1444-56.

[84] Xie, G., et al., CAR-NK cells: A promising cellular immunotherapy for cancer. *EBioMedicine,* 2020. 59: p. 102975.

[85] Liu, S., et al., NK cell-based cancer immunotherapy: from basic biology to clinical development. *Journal of Hematology & Oncology,* 2021. 14(1): p. 7.

[86] Yilmaz, A., et al., Chimeric antigen receptor-engineered natural killer cells for cancer immunotherapy. *Journal of Hematology & Oncology,* 2020. 13(1): p. 168-168.

[87] Tay, S. S., H. Carol, and M. Biro, TriKEs and BiKEs join CARs on the cancer immunotherapy highway. *Human vaccines & immunotherapeutics,* 2016. 12(11): p. 2790-2796.

[88] Felices, M., et al., Generation of BiKEs and TriKEs to Improve NK Cell-Mediated Targeting of Tumor Cells. *Methods Mol Biol,* 2016. 1441: p. 333-46.

[89] Wiernik, A., et al., Targeting natural killer cells to acute myeloid leukemia in vitro with a CD16 x 33 bispecific killer cell engager and ADAM17 inhibition. *Clin Cancer Res,* 2013. 19(14): p. 3844-55.

[90] Gleason, M. K., et al., Bispecific and trispecific killer cell engagers directly activate human NK cells through CD16 signaling and induce cytotoxicity and cytokine production. *Mol Cancer Ther,* 2012. 11(12): p. 2674-84.

[91] Davis, Z. B., et al., Natural killer cells unleashed: Checkpoint receptor blockade and BiKE/TriKE utilization in NK-mediated anti-tumor immunotherapy. *Semin Immunol,* 2017. 31: p. 64-75.

[92] Vallera, D. A., et al., IL15 Trispecific Killer Engagers (TriKE) Make Natural Killer Cells Specific to CD33+ Targets While Also Inducing Persistence, In Vivo Expansion, and Enhanced Function. *Clin Cancer Res,* 2016. 22(14): p. 3440-50.

[93] Bryceson, Y. T., et al., Synergy among receptors on resting NK cells for the activation of natural cytotoxicity and cytokine secretion. *Blood,* 2006. 107(1): p. 159-66.

[94] Miller, J. S., et al., Successful adoptive transfer and in vivo expansion of human haploidentical NK cells in patients with cancer. *Blood,* 2005. 105(8): p. 3051-7.

[95] Phillips, J. H. and L. L. Lanier, A model for the differentiation of human natural killer cells. Studies on the in vitro activation of Leu-11+ granular lymphocytes with a natural killer-sensitive tumor cell, K562. *J Exp Med,* 1985. 161(6): p. 1464-1482.

[96] Cichocki, F., et al., GSK3 Inhibition Drives Maturation of NK Cells and Enhances Their Antitumor Activity. *Cancer Res,* 2017. 77(20): p. 5664-5675.

[97] Sprinzl, M. F., et al., Sorafenib perpetuates cellular anticancer effector functions by modulating the crosstalk between macrophages and natural killer cells. *Hepatology,* 2013. 57(6): p. 2358-68.

[98] Ni, L., et al., The histone deacetylase inhibitor valproic acid inhibits NKG2D expression in natural killer cells through suppression of STAT3 and HDAC3. *Scientific reports,* 2017. 7: p. 45266-45266.

[99] Lee, J., et al., An Antibody Designed to Improve Adoptive NK-Cell Therapy Inhibits Pancreatic Cancer Progression in a Murine Model. *Cancer Immunol Res,* 2019. 7(2): p. 219-229.

[100] Au, K. M., S. I. Park, and A. Z. Wang, Trispecific natural killer cell nanoengagers for targeted chemoimmunotherapy. *Science advances,* 2020. 6(27): p. eaba8564-eaba8564.

[101] Carlsten, M., et al., Efficient mRNA-Based Genetic Engineering of Human NK Cells with High-Affinity CD16 and CCR7 Augments Rituximab-Induced ADCC against Lymphoma and Targets NK Cell Migration toward the Lymph Node-Associated Chemokine CCL19. *Front Immunol,* 2016. 7: p. 105.

[102] Levi, I., et al., Characterization of tumor infiltrating natural killer cell subset. *Oncotarget,* 2015. 6(15): p. 13835-43.

[103] Miyazato, K. and Y. Hayakawa, Pharmacological targeting of natural killer cells for cancer immunotherapy. *Cancer Sci,* 2020. 111(6): p. 1869-1875.

[104] Dianat-Moghadam, H., et al., Natural killer cell-based immunotherapy: From transplantation toward targeting cancer stem cells. *J Cell Physiol,* 2018. 234(1): p. 259-273.

[105] Luna, J. I., et al., Targeting Cancer Stem Cells with Natural Killer Cell Immunotherapy. *Expert Opin Biol Ther,* 2017. 17(3): p. 313-324.

Chapter 6

The Dual Role of Neutrophils in Cancer Development

Eileen Uribe-Querol[1] and Carlos Rosales[2,*]
[1]Laboratorio de Biología del Desarrollo,
División de Estudios de Posgrado e Investigación,
Facultad de Odontología, Universidad Nacional Autónoma de México,
Mexico City, Mexico
[2]Departamento de Inmunología,
Instituto de Investigaciones Biomédicas,
Universidad Nacional Autónoma de México,
Mexico City, Mexico

Abstract

Neutrophils, the most abundant leukocytes in blood, are the first cells to be recruited to sites of inflammation and infection. At these sites, neutrophils display an array of anti-microbial functions, including phagocytosis, degranulation, production of reactive oxygen species, and formation of neutrophil extracellular traps (NET). In addition, neutrophils are also found infiltrating many types of tumors. Neutrophils seem to have an important role in progression of malignant disease. In fact, increasing clinical evidence suggests that elevated neutrophil counts in blood correlate with poor prognosis. The tumor microenvironment controls neutrophil recruitment and in turn tumor-associated neutrophils (TAN) help tumor progression. However, it is well established that neutrophils also possess potent anti-tumor effector functions. Hence,

* Corresponding Author's E-mail: carosal@unam.mx.

In: The Innate Immune System in Health and Disease
Editor: Jorge Morales-Montor
ISBN: 978-1-68507-507-1

TAN can be detrimental or beneficial to the host. In this chapter, we describe these two aspects of neutrophil biology during cancer progression. We discuss recent studies that provide insight into the mechanisms for neutrophil migration to the tumor, for neutrophil promotion of tumor progression, and for neutrophil activation to enhance their anti-tumor functions.

Keywords: cancer, inflammation, metastasis, tumor-associated neutrophil, elastase, neutrophil extracellular traps, immunotherapy

Introduction

Neutrophils are the most abundant leukocytes in blood and are an essential component of resistance against infections as part of innate immunity (Borregaard, 2010). Neutrophils are also a hallmark of acute inflammation, responding to invading microorganisms and other tissue insults. Neutrophils are recruited from the circulation into affected tissues, where they display important anti-microbial functions (Naussef, 2007; Gazendam et al., 2016), including phagocytosis (Rosales and Uribe-Querol, 2017; Uribe-Querol and Rosales, 2021a), degranulation (Cowland and Borregaard, 2016), production of reactive oxygen species (ROS) (El-Benna et al., 2016; Winterbourn et al., 2016), and formation of neutrophil extracellular traps (NET) (Deniset and Kubes, 2016; Brinkmann, 2018). Also, neutrophils express an extensive repertoire of cytokines, immunosuppressive and stimulatory molecules, which can induce chronic inflammation, as well as to modulate the adaptive immune response (Yang et al., 2017; Rosales, 2020).

In addition to the role of neutrophils in innate and adaptive immunity, neutrophils are also now appreciated as an important component of cancer. The tumor microenvironment is characterized by an inflammatory response (Mantovani et al., 2008; Hanahan and Weinberg, 2011). Consequently, inflammatory cells such as neutrophils are key players in cancer development. In fact, several lines of evidence in clinical observations and laboratory studies have shown that the presence of neutrophils in tumors correlates with poor prognosis (Uribe-Querol and Rosales, 2015; Coffelt et al., 2016). Neutrophils infiltrate solid tumors to a variable extent. However, these tumor-associated neutrophils (TAN) were in many cases considered as simple bystander cells because it was hard to imagine that neutrophils being short-lived cells, could have an effect on progressive diseases such as cancer. However, recently it is clear that TAN have relevant roles in cancer (Shaul and Fridlender, 2019). The tumor microenvironment promotes neutrophil recruitment and in turn TAN help tumor progression. TAN are different from circulating neutrophils, and in many cases, TAN can display a pro-tumorigenic phenotype. Mechanisms leading to this phenotype are just beginning to be elucidated, as discussed later. Neutrophils displaying a pro-tumor phenotype have been documented in several cancer types including melanoma (Schmidt et al., 2005), renal carcinoma (Jensen et al., 2009), and head and neck squamous cell carcinoma (HNSCC) (Trellakis et al., 2011). However, neutrophils may also be potent anti-tumor effector cells (Gregory and Houghton, 2011), since the various antimicrobial mechanisms of neutrophils can also kill malignant cells. Additionally, cytokines and chemokines secreted by neutrophils can activate other immune cells such as natural-killer (NK) cells and cytotoxic lymphocytes (CTL), to destroy tumors (Tecchio et al., 2013). Hence, neutrophils can be beneficial or detrimental to the host (Uribe-Querol and Rosales, 2015; Ustyanovska Avtenyuk et al., 2020). In mouse cancer models, these two functions of neutrophils have been described in tumors. TAN with either one of these functions were named N1 or N2 (Fridlender et al., 2009) in a similar manner as anti-tumor or pro-tumor macrophages (Galdiero et al., 2013).

In this chapter, we review the pro-tumor and anti-tumor functions of neutrophils. We describe recent studies related to the mechanisms for neutrophil recruitment to tumors, for neutrophil promotion of tumor growth, and for neutrophil activation to enhance their anti-tumor functions. Finally, we discuss the potential manipulation of neutrophils to improve cancer immunotherapy in the future.

Neutrophils in Cancer

In many patients with cancer, elevated counts of neutrophils in blood are found. How tumors induce neutrophilia is uncertain, but altered hematopoiesis is usually observed. Several tumors can produce granulocyte-macrophage colony-stimulating factor (GM-CSF) (McGary et al., 1995), which is central for controlling the production and release of neutrophils from the bone marrow (Wirths et al., 2014; Yvan-Charvet and Ng, 2019). Also, in the tumor microenvironment other cytokines such as granulocyte colony stimulating factor (G-CSF), interleukin (IL)-1, IL-6, and IL-17 are produced. All these cytokines can contribute to mobilize more neutrophils into the blood (Lechner et al., 2010; Coffelt et al., 2015; Gorczynski, 2020). High neutrophil numbers in circulation are usually associated with poor prognosis in several types of cancer (Atzpodien and Reitz, 2008; Coffelt et al., 2016). Because neutrophilia is also associated with inflammatory responses, elevated neutrophil counts in blood further support the idea that cancer-related inflammation induces tumor progression (Hanahan and Weinberg, 2011).

The Neutrophil to Lymphocyte Ratio (NLR)

The relation of neutrophil numbers in blood to other leukocyte counts has been suggested to serve as a prognostic factor for cancer. Thus, the neutrophil to lymphocyte ratio (NLR) was introduced as a prognostic factor for colorectal cancer (Walsh et al., 2005). Later, a high NLR was found to be associated with bad prognosis in a wide spectrum of tumors (for example, melanoma, as well as colorectal, breast, prostate, and lung cancers) (Peng et al., 2015; Faria et al., 2016). Due to its simplicity NLR represents an accessible and inexpensive biomarker for many types of tumors. In general, NLR is elevated in patients with more advanced or aggressive disease, as indicated by enlarged tumors and more metastatic lesions (Ocana et al., 2017). Also, a high NLR correlates with an adverse overall survival of patients with various solid tumors (Paramanathan et al., 2014; Templeton et al., 2014). The prognostic value of NLR was confirmed in meta-analysis of one hundred studies comprising 40559 patients with 22 different types of tumors (Templeton et al., 2014). However, the significance of NLR in the clinic is still not confirmed (Vano et al., 2018; Ivars Rubio et al., 2019). As an example, in patients with metastatic breast cancer, an elevated NLR was indeed associated with disease stage and presence of metastases. However, the prognostic value of an elevated NLR was lost in multivariate analysis (Ivars Rubio et al., 2019). This means that further studies are required in particular types of cancer in order to validate optimal cut-off values of NLR to adequately support the prognosis of the disease (Vano et al., 2018).

Recruitment of Neutrophils to Tumors

Solid tumors are composed of several cell types, including tumor cells and stromal cells. The tumor stroma contains fibroblasts, endothelial cells of blood vessels, and in many cases immune cells, such as neutrophils (Reid et al.,

2011; Trellakis et al., 2011). Neutrophils are recruited to tumors by the action of neutrophil-attracting chemokines that are found in the tumor microenvironment (Figure 1). Tumors can produce growth factors (G- CSF and GM- CSF) that stimulate granulopoiesis in the bone marrow (Lawrence et al., 2018). This leads to release into the circulation of fresh neutrophils, including immature neutrophils (band cells) that could have immunosuppressive functions, as discussed below (Gabrilovich et al., 2012; Coffelt et al., 2016; Shaul and Fridlender, 2019). Many tumor cells can also produce chemokines for neutrophils, such as CXCL1, CXCL2, CXCL5, CXCL6, CXCL7 (Lazennec and Richmond, 2010), and CXCL8 (IL-8) (Sparmann and Bar-Sagi, 2004; Ji et al., 2006). In addition to these molecules, tumor-derived oxysterols (Raccosta et al., 2013) and the complement component anaphylatoxin C5a (Roumenina et al., 2019) can also participate in neutrophil recruitment into mouse tumors (Figure 1).

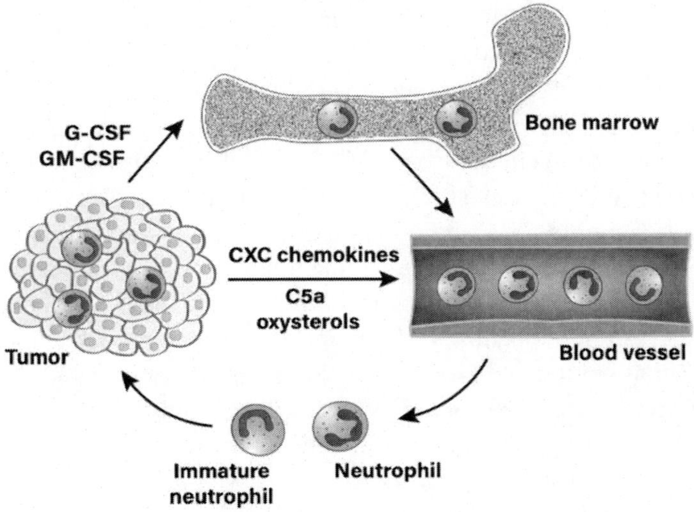

Figure 1. *Mechanisms of neutrophil recruitment to tumors*. Tumor cells produce granulocyte colony stimulating factor (G-CSF) or granulocyte-macrophage colony-stimulating factor (GM-CSF), which are central for controlling the production and release of both immature and mature neutrophils from the bone marrow. Also, in the tumor microenvironment, many CXC chemokines, as well as complement products such as C5a, and oxysterols are produced. These substances are important chemoattractants for neutrophils, which then migrate out of the blood circulation into the tumor.

Neutrophil Phenotypes

Neutrophil Subsets in Blood

Under homeostatic conditions, neutrophils in blood and tissues display considerable diversity, with a wide range of different phenotypes and functions. Neutrophil heterogeneity is the result of maturation and aging processes as well as signals from other cells in the tissues (Ng et al., 2019). Mature neutrophils freshly released from the bone marrow into the circulation are characterized by high expression of CD62L (L-selectin) and CXCR2 and low expression of CXCR4 (Figure 2A). These neutrophils, in mice are released during the night and the early morning, when animals are more active (Casanova-Acebes et al., 2013). Neutrophils then undergo an aging process that follows a circadian cycle. Over a period of 6–8 h, expression of CD62L is reduced and now the aged neutrophils, characterized by high expression of CXCR4, are smaller, have a multilobullar nucleus, and contain fewer granules (Casanova-Acebes et al., 2013; Adrover et al., 2019) (Figure 2A). These phenotypic changes of aging in circulation suggest a mechanism to facilitate neutrophil migration into tissues and indicate that neutrophil functions in inflammation and in immunity are not constant over time and may change during the circadian cycle. This idea is supported by the observation that aged neutrophils migrate less than mature neutrophils into inflamed tissues (Adrover et al., 2019). The changes neutrophils undergo during aging also modify the neutrophil proteome, including chemokine receptors, adhesion receptors, pattern recognition receptors, the inflammasome, and proteins involved in migration and NET production (Adrover et al., 2020).

In normal homeostasis, neutrophils are also found in many tissues, where they display different phenotypes and perform specialized functions (Casanova-Acebes et al., 2018). For example, in the lung some neutrophils are kept in the vascular lumen through a CXCR4-dependent mechanism (Devi et al., 2013). In the spleen, neutrophils display a B-helper function. These B-helper neutrophils concentrate in the marginal zone (Deniset et al., 2017; Scapini and Cassatella, 2017). In lymph nodes, some neutrophils interact with T lymphocytes (Hampton and Chtanova, 2016), carrying antigens to mediate T cell activation (Duffy et al., 2012; Arokiasamy et al., 2017). Another subset of neutrophils (approximately 45–60% in blood) express the molecule CD177 (Hu et al., 2014). These CD177$^+$ neutrophils have increased antimicrobial activity (Zhou et al., 2018) and promote restoration of epithelial cells in colitis (Zindl et al., 2013). Thus, it is thought that CD177$^+$ neutrophils are anti-inflammatory and protective during inflammatory bowel disease (IBD) (Zhou et al., 2018).

Clearly, the existence of many "types" of neutrophils is nowadays well accepted (Figure 2A). However, it is still not clear whether these different phenotypes represent truly different cell lineages or are just examples of multiple activation and functional states (Rosales, 2018).

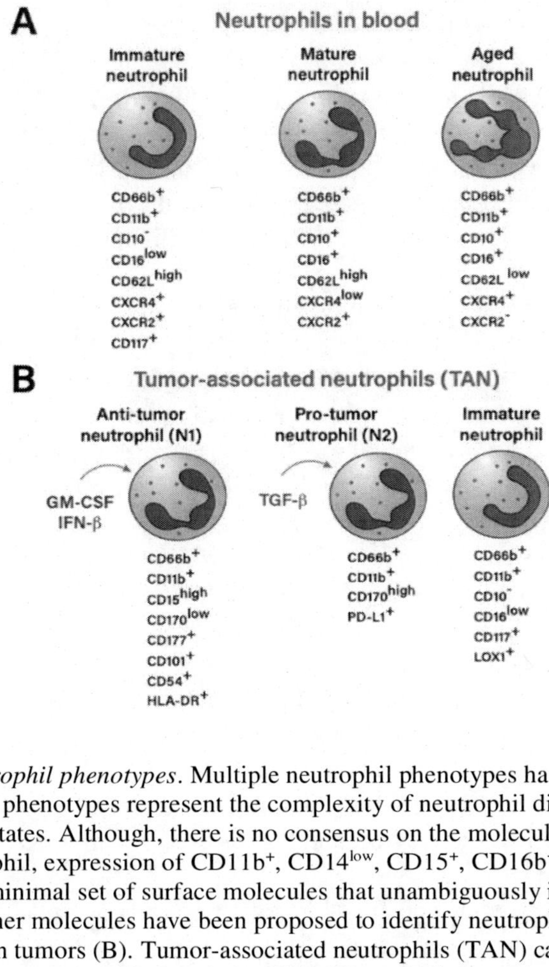

Figure 2. *Neutrophil phenotypes*. Multiple neutrophil phenotypes have been reported. These phenotypes represent the complexity of neutrophil differentiation and activation states. Although, there is no consensus on the molecular markers that define a neutrophil, expression of $CD11b^+$, $CD14^{low}$, $CD15^+$, $CD16b^+$, and $CD66b^+$ represents the minimal set of surface molecules that unambiguously identifies neutrophils. Other molecules have been proposed to identify neutrophil subsets in blood (A) and in tumors (B). Tumor-associated neutrophils (TAN) can display anti-tumor (N1) of pro-tumor (N2) functions. N1 TAN can be induced by granulocyte–macrophage colony-stimulating factor (GM-CSF), or interferon beta (IFN-β), while N2 TAN can be promoted by transforming growth factor-beta (TGF- β).

Neutrophil Subsets in Tumors

In cancer, granulopoiesis is dysregulated causing alterations in leukocyte production and function. Recently, different neutrophil subsets that participate in tumor progression have been identified in the bone marrow. One such subset is a committed proliferative neutrophil precursor (PreNeu) population that retains proliferative capacity and expands in the bone marrow and spleen of tumor bearing mice (Evrard et al., 2018). These PreNeu cells expand under tumoral stress and can differentiate into immature and mature neutrophils. The immature neutrophils are preferentially recruited to growing tumors (Evrard et al., 2018). Also, another subset is a committed unipotent early-stage neutrophil progenitor (NeP) that has been found in adult mouse and human bone marrow. These NeP cells generate only neutrophils and, when transferred into humanized mice, stimulated tumor growth by inhibiting T cell activation (Zhu et al., 2018).

Using mice tumor models, another subset of neutrophils was found by density centrifugation. Normally, leukocytes are purified with a buoyancy density gradient (Böyum, 1968; García-García et al., 2013). In this process, neutrophils go to the bottom of the tube, while mononuclear cells, which for having more buoyancy separate to the upper part of the tube (low-density fraction). Among the mononuclear cells, cells with neutrophil morphology are also found and they have been named low-density neutrophils (LDN) (Sagiv et al., 2015). A small number of LDN has also been reported in the blood on healthy individuals (Blanco-Camarillo et al., 2021). In several reports, the name high-density neutrophils (HDN) has been used to describe "normal" neutrophils which appear at the bottom (denser part) of the density gradient. The term HDN is not adequate because it implies that neutrophils have higher than normal density. We also think, as it has been proposed (Cassatella and Scapini, 2020), that it should not be used while referring to neutrophils. Therefore, we use the term "neutrophils" when referring to typical polymorphonuclear leukocytes, and the term "LDN" for the neutrophils with higher buoyancy.

The characteristics and functions of LDN are not well described; but in cancer, LDN increase in blood and have been associated with augmented tumor growth and metastasis (Sagiv et al., 2015; Hsu et al., 2019). The LDN subset comprises both mature and immature neutrophils. These immature LDN seem to be similar to the granulocytic myeloid-derived suppressor cells (G-MDSC) induced in tumor-bearing animals (Waight et al., 2011). These G-MDSC possess inhibitory activity on T-cell helper responses, thus they can promote tumor development (Veglia et al., 2021). In contrast neutrophils seem to be anti-tumor (Sagiv et al., 2015). However, neutrophils from tumor-bearing mice are capable of becoming LDNs upon treatment with transforming growth factor-beta (TGF-β) (Sagiv et al., 2015). Together all these reports suggest that three subsets of neutrophils are present in cancer: normal neutrophils, immature LDN (G-MDSC), and large mature LDN. However, these cell types present diversity in function and plasticity, changing under the influence of chemokines and cytokines in the tumor microenvironment (Sionov et al., 2015). Further characterization of the heterogeneous LDN and G-MDSC populations is required to determine their precise origin and role in cancer development (Veglia et al., 2021).

Neutrophils infiltrating tumors are also named TAN (tumor-associated neutrophils), and the presence of these cells seems to be an indicator of poor outcome (Sionov et al., 2015). In tumor-bearing mice, TAN have been classified as N1 and N2 (Fridlender et al., 2009), depending on their anti-tumorigenic, or pro-tumorigenic phenotype, respectively (Fridlender et al., 2009) (Figure 2B). N1 cells display anti-tumorigenic activity by releasing cytokines that recruit and activate $CD8^+$ T cells (Ocana et al., 2017). Based on a murine model of lung carcinoma it seems that TAN present an N1 phenotype at the early stage of tumor, but during tumor progression they change into an N2 phenotype (Mishalian et al., 2013). Such a phenotypic change may be induced by factors in the tumor microenvironment, such as TGF-β (Fridlender et al., 2009; Sionov et al., 2015). In contrast, type I interferons induced TAN with a N1 phenotype (Pylaeva et al., 2016). Therefore, murine TAN can have an N1 or N2 phenotype depending on the tumor microenvironment (Sionov et al., 2015) (Figure 2B). Despite this classification, the nature and function of TAN in the cancer microenvironment remains largely unknown, particularly with human tumors. TAN were isolated from early stage lung tumors. These TAN had an activated phenotype, released proinflammatory cytokines, and functioned as antigen presenting cells (Eruslanov et al., 2014; Eruslanov, 2017). Thus, these TAN displayed an N1 phenotype. In contrast, in colorectal cancer tumors TAN were described as $CD45^+/Lin^-/HLADR^-/CD11b^+/CD33^+/CD66b^+$ neutrophils producing more arginase 1 and ROS than autologous neutrophils. These TAN inhibited proliferation of activated T cells and thus displayed an N2 phenotype (Wu et al., 2014). Clearly, neutrophils can have both pro-tumor or anti-tumor functions depending on the tumor microenvironment and the stage of cancer progression. We will describe those functions, next.

Role of Neutrophils in Tumor Progression

Clinical evidence suggests that neutrophils participate in cancer development and tumor progression. Increased numbers of blood neutrophils and TAN are associated with poor prognosis for cancer patients. This negative association has been reported for many types of solid tumors (Dumitru et al., 2012; Brandau et al., 2013). Neutrophils can mediate tumor progression by several mechanisms, including stimulation of proliferation, angiogenesis, metastasis, and immunesuppression (Uribe-Querol and Rosales, 2015) (Figure 3).

Proliferation

Neutrophil elastase (NE) (a protein of azurophilic granules) (Korkmaz et al., 2008) is reported to have pro-tumor effects both *in vivo* and *in vitro* (Sato et al., 2006). NE directly induced lung carcinoma cell proliferation *in vitro* when murine neutrophils were cultured with tumor cells (Houghton et al., 2010). NE effect on proliferation was dependent on phosphatidylinositol 3-kinase (PI-3K) (Houghton et al., 2010). First, NE entered tumor cells via clathrin-coated pits (Gregory et al., 2012). Then, NE degraded insulin receptor substrate-1 (IRS-1), which binds to the regulatory unit of PI-3K. In consequence, IRS-1 degradation resulted in more PI-3K available to activate proliferation (Gabrilovich et al., 2007). A similar effect has been reported on other tumor cells, such as esophageal cancer (Wada et al., 2007), gastric cancer (Wada et al., 2006), and breast cancer (Nawa et al., 2012).

Matrix metalloproteinase (MMP)-9 (gelatinase B) is released from secondary (specific) granules of neutrophils. MMP-9 was reported to stimulate keratinocyte tumor proliferation in a skin carcinogenesis model (Coussens et al., 2000). Also, MMP-9 was reported to prevent apoptosis of lung tumor cells (Acuff et al., 2006). Thus, MMP-9 can promote tumor growth via both increased proliferation and reduced apoptosis of tumor cells.

Neutrophils produce ROS for destruction of microorganisms. However, ROS can also directly induce genotoxicity on tumor cells in circumstances when ROS do not kill the cells. ROS-mediated genotoxicity involves two pathways: oxidative DNA damage and myeloperoxidase (MPO)-catalyzed activation of chemical carcinogens (Knaapen et al., 2006). These pathways then indirectly induce tumor proliferation.

Neutrophils can also produce cytokines or growth factors, which promote proliferation of cancer cells (Tecchio et al., 2013). For example, breast cancer cells could stimulate neutrophils to release Oncostatin-M (Grenier, 2001; Cross et al., 2004; Goren et al., 2006), which in turn stimulated breast cancer cells to secrete vascular endothelial growth factor (VEGF) (Queen et al., 2005). Also, hepatocellular carcinoma cells induced neutrophils to release hepatocyte growth factor (HGF), which in turn, stimulated tumor cell proliferation (Imai et al., 2005) (Figure 3).

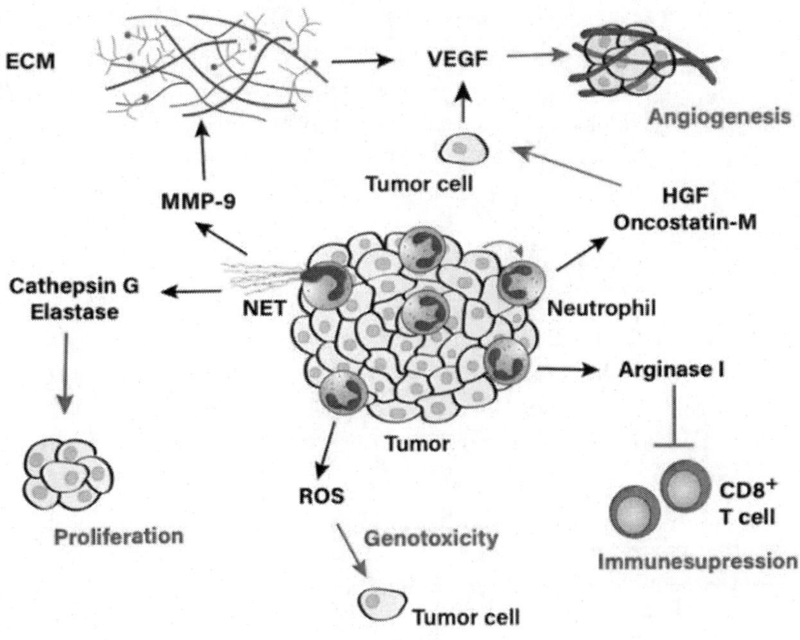

Figure 3. *Pro-tumor activity of neutrophils.* Neutrophils help tumor progression in several ways. Neutrophils can secrete matrix metalloproteinase-9 (MMP-9) that releases vascular endothelial growth factor (VEGF) from the extracellular matrix (ECM) to promote angiogenesis. Certain tumors induce neutrophils to produce hepatocyte growth factor (HGF) or Oncostatin-M, an IL-6-like cytokine, which then stimulates tumor cells to secrete VEGF to promote angiogenesis. Neutrophils can secrete arginase 1, which inhibits CD8$^+$ T cells, creating an immunosuppressive state. Neutrophils also produce reactive oxygen species (ROS) that can damage DNA, inducing genotoxic effects on tumor cells. Many proteins from neutrophil granules are present on neutrophil extracellular traps (NET), which are DNA fibers released by activated neutrophils in the tumor microenvironment. Serine proteases, such as elastase and cathepsin G, from neutrophil granules and exposed on NET, seem to have a direct effect on tumor cells for inducing proliferation.

Angiogenesis

Angiogenesis, the formation of new blood vessels, is an important factor for tumor development. MMP-9 from neutrophils can also induce angiogenesis by releasing the VEGF sequestered in the extracellular matrix (ECM) (Hawinkels et al., 2008; Ebrahem et al., 2010; Wilson et al., 2010; Kuang et al., 2011). This angiogenic effect of neutrophils via MMP-9 has been observed in several cancer models (Bergers et al., 2000; Van Coillie et al., 2001; Nozawa et al., 2006; Jablonska et al., 2010; Deryugina and Quigley, 2015). Together these reports confirm that neutrophil-derived MMP-9 can promote angiogenesis by releasing VEGF from the ECM in many types of tumors (Figure 3).

Metastasis

Neutrophils can also influence the metastatic potential of cancer cells. The mechanisms neutrophils use to increase the migratory capacity of tumor cells are only partially understood. One way involves the action of NE. This enzyme could reduce the surface expression of E-cadherin on pancreatic ductal adenocarcinoma cells, leading to separation of the cells from the culture monolayer (Gaida et al., 2012). Similarly, NE also enhanced the migratory capacity of esophageal cancer cells (Wada et al., 2007). Cathepsin G (a peptidase from azurophilic granules) can also promote tumor cell migration (Wilson et al., 2010; Morimoto-Kamata et al., 2012; Yui et al., 2014). Cathepsin G degrades ECM molecules, leading to E-cadherin-mediated homotypic cell-cell adhesion (Yui et al., 2014). These tumor cell aggregates allow tumor cells to disseminate via the circulation to distant sites and establish new metastases (Wilson et al., 2008; Wilson et al., 2009). Tumors can induce activation of neutrophils to release inflammatory factors that promote tumor migration. Several tumor cells release hyaluronan, which can then activate neutrophils via toll-like receptor (TLR)-4. In response, neutrophils induce migration of the tumor cells (Wu et al., 2011).

Once in circulation, neutrophils can also help tumor cells to survive by inducing tumor cell adherence. Tumor cells have been observed in close association with neutrophils (Crissman et al., 1985), and this association is dependent on β2 integrins on neutrophils. In this way, neutrophils promote tumor cell retention and in consequence induce more metastasis (Huh et al.,

2010; Spicer et al., 2012). Clearly, the mechanisms neutrophils use to promote tumor cell migration and metastasis are diverse and complex.

Immunesuppression

Arginase 1 is an enzyme released from neutrophil granules, that degrades extracellular arginine. Because arginine is an indispensable amino acid for T cell activation, degranulation of neutrophils can inhibit T cells and produce an immunosuppressive effect in the tumor microenvironment (Dumitru et al., 2012). Supporting this view is the fact that elimination of TAN in tumor-bearing animals increased activated $CD8^+$ T cells and resulted in smaller tumors (Fridlender et al., 2009) (Figure 3). Also, ROS produced after neutrophil activation can lead to immune suppression (Aarts et al., 2019). The hydrogen peroxide (H_2O_2) produced by activated neutrophils reduced expression of the T cell receptor (TCR) CD3 ζ chain and decreased cytokine production from T cells (Schmielau and Finn, 2001; Szuster-Ciesielska et al., 2004). These T cells also presented defective chemotaxis, which was caused by oxidation of the actin-remodeling protein cofilin (Klemke et al., 2008). In addition, H_2O_2 can induce T cell inhibition by preventing NF-κB (nuclear factor kappa-light-chain-enhancer of activated B cells) activation (Malmberg et al., 2001).

T cells can also be inhibited by direct interaction with neutrophils expressing the immune checkpoint molecule PD-L1 (programmed death-ligand 1). The ligation of PD-1 (programmed death cell protein 1) on T-cells with PD-L1 leads to an arrest of T cell proliferation (Freeman et al., 2000). Expression of PD-L1 on neutrophils can be induced by IFN-γ (interferon gamma), GM-CSF, or hypoxic conditions (Bankey et al., 2010; de Kleijn et al., 2013; Noman et al., 2014). Inhibition of T cell function by "supressive" neutrophils expressing PD-L1was demonstrated in patients with hepatocellular carcinoma (He et al., 2015). The interaction between PD-1 and PD-L1 leading to T cell suppression is an important novel therapeutic approach in cancer (Dermani et al., 2019).

Other mechanisms that can impair T cell function by direct interaction between neutrophils and lymphocytes include the induction of apoptosis in the T cells (Michaeli et al., 2017), and also direct damage to the lymphocyte membrane by neutrophils, which can take little pieces of the T cell membrane (Aarts et al., 2019) in a process known as trogocytosis (Bettadapur et al., 2020; Uribe-Querol and Rosales, 2021b).

In addition, to their direct mechanisms (cell-cell contact or release of granule proteins) to inhibit T cells, neutrophils can also promote tumor growth by affecting other immune cells. In tumor-bearing mice, neutrophils were found to also block NK cell activation (Li et al., 2009). NK cells did not perform cytotoxicity of tumor cells, which resulted in enhanced metastasis (Sceneay et al., 2012; Spiegel et al., 2016). Finally, neutrophils can also induce the development of regulatory T cells (Treg) (Chaudhary and Elkord, 2016). Treg augmented in the tumor microenvironment because they are less susceptible to the oxidative stress generated by neutrophils (Mougiakakos et al., 2009; Mougiakakos et al., 2011).

Neutrophil Extracellular Traps

As indicated in the introduction, NET constitute a recently described anti-microbial mechanism of neutrophils. NET are fibers of chromatin decorated with proteins from the neutrophil granules (Brinkmann and Zychlinsky, 2012). NET form a mesh-like structure where microorganisms get trapped and are either directly killed in some cases, or more often subsequently phagocytosed by other neutrophils (Papayannopoulos, 2018).

However, NET have been shown to play an important role in cancer, since these fibers can promote tumor growth. In an initial study, tumorous NET were found in Ewing sarcoma tumors (Berger-Achituv et al., 2013). In cancer models of chronic myelogenous leukemia, mammary and lung carcinoma, neutrophils were more susceptible to release NET (Demers et al., 2012). Also, as mentioned earlier, many tumors infiltrated with TAN are associated with poor prognosis. In these tumors NET have been found. In initial tumors TAN were found in the periphery of the tumor, but in mature tumors TAN and free DNA were within the tumor (Mishalian et al., 2013). Thus, NET are also associated with tumor progression (Cedervall and Olsson, 2016).

Recent reports have provided important clues on the mechanisms by which NET within primary tumors can promote tumor progression. In the tumor microenvironment, changes such as coagulation and enhanced thrombosis have been associated to increased tumor growth. NET were found to be a link between these changes and tumorigenesis. Neutrophils stimulated with lipopolysaccharides (LPS) showed an upregulation of complement C3a receptor. Activation of complement led to increased NET formation and induction of coagulation.

At the same time, neutrophils underwent polarization to an N2 phenotype, which stimulated tumorigenesis (Guglietta et al., 2016). In addition, it has also been reported that immature neutrophils preferentially respond to tumor-derived C3a inducing chemotaxis to metastatic sites (Hsu et al., 2019). Similarly, breast cancer tumors that produced high levels of G-CSF and IL-1β had elevated neutrophil counts and more NET formation, which led to tumor-associated thrombosis (Gomes et al., 2019). Another way in which NET can induce tumor cell proliferation is through the actions of NE, which activated TLR-4 on cancer cells, leading to increased mitochondrial biogenesis, and accelerated tumor growth (Yazdani et al., 2019). Together these data support a model for primary tumor development. Neutrophils would migrate to the new tumor, where TAN would produce NETs, which would promote tumor growth.

Neutrophil Functions Against Tumors

Recently, it is more and more recognized that neutrophils have a negative role in many types of cancer. However, neutrophils can also display anti-tumor functions. As mentioned before, neutrophils first entering murine tumors display an anti-tumor (N1) phenotype (Fridlender et al., 2009). In fact, very early reports recognized the anti-tumor capacity of neutrophils both *in vitro* (Gerrard et al., 1981) and *in vivo* (Katano and Torisu, 1982). Also, the importance of tumor-infiltrating neutrophils on anti-tumor responses is underscored by the fact that depletion of murine neutrophils results in enhanced tumor growth (Suttmann et al., 2006; Kousis et al., 2007; Fridlender et al., 2009). In addition, there are reports showing that neutrophils can prevent metastasis to the lung of breast cancer (Granot et al., 2011) and renal carcinoma (López-Lago et al., 2013). In both models, neutrophils prevented metastasis. In the breast cancer model, tumor cells produced CCL2 that induced neutrophil ROS production (Granot et al., 2011), while in the renal carcinoma model, tumor-derived IL-8 recruited tumor cytotoxic neutrophils (López-Lago et al., 2013). These findings emphasize the potential of neutrophils to display anti-tumor functions, and suggest that neutrophils could be induced to enhance their participation in the fight against cancer. The known mechanisms for neutrophil-mediated cell killing are presented next.

Cytotoxicity

There is no doubt that neutrophils can directly kill tumor cells. However, the mechanisms of killing are not completely described, but they seem to be different from the classic cytotoxic mechanism used by NK cells (Bodduluru et al., 2015). Neutrophil cytotoxicity involves some of the same antimicrobial and immune functions of neutrophils. It has been reported that ROS produced through the respiratory burst participate in direct killing of tumor cells (Zivkovic et al., 2007; Yan et al., 2014; Gershkovitz et al., 2019). Neutrophils can deliver hypochlorous acid (HOCl) directly at the cell membrane (Dallegri et al., 1991) and eliminate tumor cells at the early phase of tumor development (Zivkovic et al., 2005) (Figure 4). More recently, it was found that H_2O_2 produced by neutrophils activates the TRPM2 (transient receptor potential melastatin 2) calcium channel (Syed Mortadza et al., 2015; Kühn, 2020) on tumor cells leading to an influx of Ca^{2+} that results in tumor cell killing (Gershkovitz et al., 2018a). TRPM2 is upregulated in tumor cells after epithelial-to-mesenchymal transition (EMT) and tumor cells that experienced EMT were killed more efficiently by neutrophils (Gershkovitz et al., 2018b). Another way that murine neutrophils killed tumors in a H_2O_2-dependent manner involved the interaction of cathepsin G with RAGE (receptor for advanced glycation end products) expressed on tumor cells (Sionov et al., 2019). The capacity of neutrophils to produce ROS is also influenced by the tumor microenvironment. In hypoxic conditions, the ROS-dependent cytotoxicity on tumor cells by murine neutrophils was greatly reduced (Mahiddine et al., 2020). This in part, may explain the observation that newly arrived neutrophils to the tumor site may lose their anti-tumor potential during tumor progression.

Neutrophils can also mediate direct killing of tumor cells by inducing apoptosis. This effect is mediated by the tumor necrosis factor-related apoptosis inducing ligand (TRAIL) (Figure 4). Neutrophils from patients with carcinoma of the bladder who received BCG (*Mycobacterium bovis* bacillus Calmette–Guérin) immunotherapy expressed high levels of TRAIL (Ludwig et al., 2004), and these neutrophils could induce apoptosis of tumor cells (Fulda, 2014). In addition, neutrophils isolated from healthy individuals were capable of blocking proliferation of tumor cells, after direct cell-cell contact, through Fas ligand/Fas interaction (Sun et al., 2018). Fas ligand/Fas regulates the expression of cell cycle checkpoint proteins, leading to early cell cycle arrest (Sun et al., 2018).

Antibody-Dependent Cell Killing

Immunotherapy using monoclonal antibodies (mAbs) has become an important tool for treatment of cancer and has offered significant benefits for many cancer patients (Eccles, 2001; Mellman et al., 2011). Therapeutic mAbs can induce direct tumor cell killing by binding to specific antigens and altering intrinsic function of the tumor cell, or by binding, via their Fc domain to the C1q component of the complement system and inducing complement-dependent cytotoxicity (CDC) (Gelderman et al., 2004). More often however, mAbs can also bind to activating Fc receptors on cells of the immune system, which in turn stimulate indirect killing of opsonized cells via antibody-dependent cell-mediated cytotoxicity (ADCC) mechanisms (Chan and Carter, 2010; Mellman et al., 2011; van Egmond and Bakema, 2013; Bakema and van Egmond, 2014; Heemskerk and van Egmond, 2018). ADCC efficiently eliminates various types of tumors.

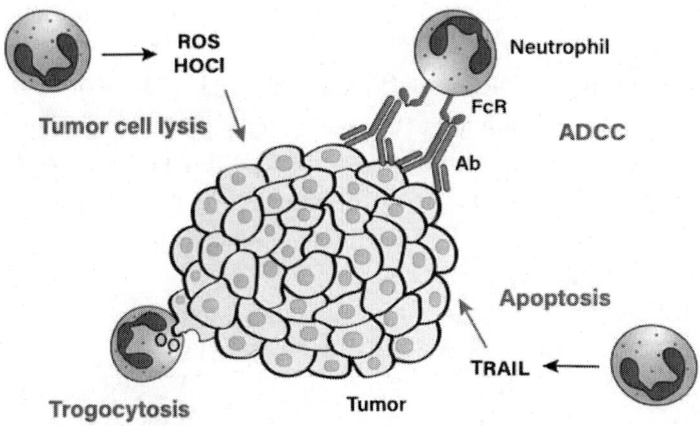

Figure 4. *Anti-tumor activity of neutrophils.* Neutrophils produce reactive oxygen species (ROS) and hypochlorous acid (HOCl) that can directly damage and destroy tumor cells. The most effective anti-tumor mechanism is antibody-dependent cell-mediated cytotoxicity (ADCC). Antibody molecules (Ab) that bind to tumor antigens are recognized by Fc receptors (FcR) on neutrophils. This binding activates a cytotoxic response against the tumor cell. By direct contact or by release of TRAIL, neutrophils can also induce apoptosis of certain tumor cells. Neutrophils can also kill tumor cells by trogocytosis, which is the process of taking little pieces of the tumor cell membrane.

NK cells are particularly efficient in performing ADCC (Wang et al., 2015), via the enzymes perforin and granzyme (Bodduluru et al., 2015). Neutrophils also present efficient ADCC against cells that have been marked by antibodies (van Egmond and Bakema, 2013). In fact, neutrophil ADCC was recognized more than 30 years ago as an efficient mechanism to eliminate antibody-opsonized tumor cells (Ohara et al., 1983) (Figure 4).

Fc Receptors on Neutrophils

All approved mAbs used in cancer therapy are IgG, particularly IgG1 and IgG2 (Vidarsson et al., 2014). This is relevant because there are several Fcγ receptors that are expressed differentially in multiple immune cells (Rosales and Uribe-Querol, 2013a), and each class of IgG can bind to these receptors with distinct affinities (Rosales and Uribe-Querol, 2013b). As a result, each cell activates diverse and highly regulated responses (Bruhns et al., 2009; Nimmerjahn and Ravetch, 2011; Rosales, 2017). Six Fcγ receptors can be expressed by human innate immune cells: FcγRI (CD64), FcγRIIa (CD32a), FcγRIIb (CD32b), FcγRIIc (CD32c), FcγRIIIa (CD16a), and FcγRIIIb (CD16b). From these, human neutrophils constitutively express FcγRIIa and FcγRIIIb (a glycosylphosphatidylinositol (GPI)-linked receptor). Neutrophils under stimulated conditions (mainly with IFN-γ) also express FcγRI. It is noteworthy that FcγRIIa and FcγRIIIb are unique Fcγ receptors not present in other species (Bruhns and Jönsson, 2015).

FcγRI and FcγRIIa are activating receptors, that transduce signals via an immunoreceptor tyrosine-based activation motif (ITAM) located either in the associated common FcRγ chain (for FcγRI) or in the cytoplasmic tail of the Fcγ receptor itself (for FcγRIIa). Upon activation, tyrosine residues within the ITAM get phosphorylated and then Syk tyrosine kinases can bind to them. This initiates downstream signaling cascades that can activate multiple functions including phagocytosis, degranulation, and ADCC (Rosales, 2007). The GPI-linked FcγRIIIb, instead, lacks both transmembrane and cytoplasmic domains. However, it can still signal the cell by a mechanism not yet described (Rosales, 2017). Human neutrophils present more efficient ADCC when they engage FcγRIIa (Ottonello et al., 2001; Schneider-Merck et al., 2010). Nevertheless, FcγRI also seems capable of activating ADCC against tumors (Valerius et al., 1993), and in particular with squamous head and neck cancer (Su et al., 2006). However, in other studies, it was shown that immature neutrophils with high expression of FcγRI had reduced ADCC activity via this receptor (Otten et al., 2007b).

In addition to IgG, other classes of antibodies can also activate ADCC in neutrophils. In particular IgA antibodies, which specifically bind the IgA Fcα receptor I (FcαRI or CD89) (Breedveld and van Egmond, 2019). On neutrophils, IgA antibodies have been found to be more potent inducers of ADCC than IgG antibodies (Bakema and van Egmond, 2014; Heemskerk and van Egmond, 2018; Brandsma et al., 2019). This effect has been reported for several tumor-associated antigens such as HER2 in breast cancer, EGFR in colorectal cancer, and CD20 in lymphoma, among others (Lohse et al., 2011; Borrok et al., 2015; Lohse et al., 2018; Evers et al., 2020). Interestingly, immature neutrophils released from the bone marrow after G-CSF treatment were capable of a more efficient ADCC response against tumor cells *in vitro* when stimulated with IgA antibodies than when stimulated with IgG antibodies (Otten et al., 2005; Otten et al., 2007a). The stronger ADCC response induced by IgA seems to be due to the higher avidity of FcαRI, that binds bivalently to IgA. This results in recruitment of more ITAM sequences when the receptors are aggregated and hence induce a more robust signaling for cell activation (Borrok et al., 2015).

Mechanisms of ADCC

The mechanism for tumor cytotoxicity from neutrophils is not completely known. Initially, a ROS-dependent mechanism was proposed for neutrophil ADCC (Dallegri et al., 1984). In this case, cell-cell contact was required for direct release of HOCl to the tumor cell (Dallegri et al., 1991). However, neutrophils from chronic granulomatous disease (CGD) patients and neutrophils in the presence of ROS scavengers can still kill tumor cells. These results suggest that ROS are not as important for ADCC (Kushner and Cheung, 1991). Hence, ROS-independent mechanisms for ADCC have also been proposed. One of them involves direct cell lysis via perforin and granzyme (Bodduluru et al., 2015). However, expression of these enzymes in neutrophils remains controversial (Metkar and Froelich, 2004; Wagner et al., 2004). Thus, this type of cytotoxicity does not seem to have a role in neutrophil ADCC. Recently, another way for neutrophils inducing cell death has been described. Neutrophils can perform trogocytosis on antibody-opsonized tumor cells leading to their destruction (Valgardsdottir et al., 2017; Matlung et al., 2018) (Figure 4). Originally, trogocytosis was identified as an active mechanism in amoebas that involves the transfer of small pieces "bites" of plasma membrane from one cell to another (Bettadapur et al., 2020; Uribe-Querol and Rosales, 2021b). Neutrophils have been shown to be able to kill the parasite *Trichomonas vaginalis* using trogocytosis (Mercer et al., 2018), as well as the excess sperm cells in the vaginal track (Olivera-Valle et al., 2020). Thus, neutrophils seem to employ trogocytosis to eliminate various types of cells. In the case of antibody-mediated cytotoxicity, neutrophils engage Fc receptors leading to trogocytosis for tumor cell killing (Ustyanovska Avtenyuk et al., 2020).

Therapeutic Strategies

Even though neutrophils have a negative influence on cancer disease when they infiltrate tumors, neutrophils clearly also have the capacity to eliminate tumor cells. Different strategies to modify the behavior on neutrophils during cancer are under consideration. On one hand, these therapeutic strategies involve reducing the access of neutrophils into tumors, and on the other increasing the anti-tumor functions of neutrophils. These therapeutic strategies are briefly described next.

Inhibition of Neutrophil Recruitment

Since elevated numbers of neutrophils are associated with poor outcomes, one therapeutic tactic is to inhibit infiltration of neutrophils into tumors. Many tumors produce chemokines, for example IL-8, that recruit neutrophils to the tumor. Blocking IL-8 with a specific mAb (ABX-IL8) reduced neutrophil chemotaxis, tumor growth, and also metastasis of melanoma (Huang et al., 2002) and lung cancer (Keane et al., 2004). Because other chemokines also interact with the receptors CXCR1 and CXCR2 (Lazennec and Richmond, 2010), another way to inhibit neutrophil recruitment is by blocking these receptors (Mollica Poeta et al., 2019). Antagonists of these receptors have been produced with the intention of preventing neutrophil infiltration and reducing tumor growth (Walters et al., 2008). For example, GSK135756 is an antagonist for CXCR2 that could have antitumor potential (Quigley and Deryugina, 2012). Similarly, pepducin another antagonist of CXCR2 was able to inhibit inflammation and tumorigenesis (Jamieson et al., 2012), as well as metastases of pancreatic ductal adenocarcinoma (Steele et al., 2016).

Reparixin is an antagonist for CXCR1 that has shown to be effective at inhibiting neutrophil recruitment into tissues, and to selectively target human breast cancer stem cells in xenograft models in mice (Ginestier et al., 2010). This chemokine receptor antagonist is now entering clinical evaluation for breast cancer (Schott et al., 2017). As mentioned before, HGF produced by tumor cells recruits neutrophils to the tumor and elevated levels of HGF are associated with poor prognosis in human cancer (Wislez et al., 2003). Neutrophils recognize HGF with the receptor cMET, which is a receptor tyrosine kinase that activates chemotaxis. Thus, blocking cMET resulted in reduced recruitment of neutrophils into tumors and draining lymph nodes (Glodde et al., 2017). Antagonists of the HGF/cMET pathway have been developed and are now being considered for treatment of various types of cancer (De Silva et al., 2017).

Inhibition of Pro-Tumor Activity

In addition to preventing neutrophil recruitment, inhibition of the pro-tumor activity of neutrophils is another strategy being explored. As indicated, several neutrophil enzymes are known to facilitate tumor growth, and inhibition of them may have important benefits in cancer therapy. Inhibition of NE reduced tumor growth of lung adenocarcinomas in a mouse model (Houghton et al., 2010). Also, blocking the activity of NE with the specific inhibitor sivelestat has proven useful in reducing tumor growth in murine models of colorectal and prostate cancer (Ho et al., 2014; Lerman et al., 2017). In addition, sivelestat improved the condition after surgery of patients with carcinoma of the thoracic esophagus (Suda et al., 2007).

The metalloprotease activity of MMP-9 has also been targeted to prevent angiogenesis. Using bisphosphonate zoledronic acid, a strong MMP inhibitor, resulted in reduced angiogenesis and cervical cancer burden (Giraudo et al., 2004). Unfortunately, in other models and clinical trials inhibition of MMP-9 was not effective at reducing tumor growth (Deryugina et al., 2005; Fingleton, 2008). In addition to NE and MMP-9, other components of neutrophils that could favor tumor proliferation are found on NET. As mentioned, the presence of NET in the tumor microenvironment is also associated with bad prognosis. Thus, elimination of NET with administration of enzymes that degrade DNA fibers (e.g., DNAse I), is another approach to control tumor development (Hawes et al., 2015).

Normally, circulating neutrophils are anti-tumorigenic, but may acquire a suppressor phenotype in the tumor microenvironment. This was clearly observed on neutrophils that were exposed to ascites supernatant from patients with epithelial ovarian cancer (Singel et al., 2019). This change was mediated by complement, since blocking complement receptor 3 (CR3) signaling abrogated the neutrophil suppressor phenotype (Singel et al., 2019). Thus, preventing complement activation may be a way to avoid the development of pro-tumor neutrophils.

Activation of Anti-Tumor Activity

Because, normally neutrophils in blood have an activated phenotype that keeps tumors in check, it has been proposed that increasing the number of neutrophils could be a good strategy against cancer. In fact, a single dose of G-CSF increased the number of blood neutrophils with excellent cytotoxic activity against alemtuzumab-opsonized $CD52^+$ tumor cells (Siders et al., 2010). Also, tumor cells modified to express G-CSF attracted neutrophils capable of inhibiting tumor growth (Colombo et al., 1992). Moreover, direct activation of neutrophils with G-CSF and IFN-β produced cells with an anti-tumor phenotype (Jablonska et al., 2010). However, in other cases G-CSF treatment released from the bone marrow immature neutrophils with an immunosuppressive phenotype, which favor tumor progression (Raychaudhuri et al., 2011; Marini et al., 2017). These reports indicate that the use of G-SCF to mobilize neutrophils in cancer must be regarded with extreme caution. More importantly, due to the fundamental role of neutrophils in antimicrobial responses, a general increase and activation of these cells may not be a good therapeutic approach since activated neutrophils without target specificity could cause excessive tissue damage.

Another strategy against cancer is based on the fact that neutrophil TRAIL induces apoptosis of various tumor cells but not normal cells (Brincks et al., 2013; Amarante-Mendes and Griffith, 2015). Thus, TRAIL is considered a candidate molecule against cancer (Fulda, 2014), and recombinant forms of TRAIL or TRAIL receptor agonists are being developed as novel therapeutic tools (Amarante-Mendes and Griffith, 2015).

Another way for further increasing the anti-tumor capacity of neutrophils is by blocking the signaling from inhibitory receptors on their membrane (Feng et al., 2019). The inhibitory receptor signal regulatory protein-α (SIRPα) is an immune checkpoint protein (on neutrophils) that engages the "don't eat me" signal CD47 expressed on tumor and normal cells (Zhao et al., 2011). Thus, by blocking the interaction of SIRPα with CD47, the cytotoxic activity of neutrophils could be enhanced (Zhang et al., 2020). Several reports showed the efficacy of interfering with the interaction between CD47 and SIRPα on neutrophils, leading to enhanced ADCC activity against both solid and hematological tumors (Horrigan and Reproducibility Project: Cancer Biology, 2017; Ring et al., 2017). Based on this, several antibodies have been developed against either CD47 or SIRPα. These antibodies were found to be effective and with minimal toxicity side effects (Sikic et al., 2019; Sim et al., 2019; Voets et al., 2019). One such mAb, Hu5F9-G4, against CD47 was already tried in a phase 1b clinical trial for the treatment of Non-Hodgkin's lymphoma (Advani et al., 2018). Thus, interfering with the CD47 or SIRPα is a promising novel therapeutic approach to involve neutrophils in the fight against cancer.

A more promising strategy to improve the anti-tumor potential of neutrophils is immunotherapy using mAbs against tumor antigens, to activate ADCC (Heemskerk and van Egmond, 2018). Nowadays, most mAbs used in the clinic belong to the IgG1 class, and they can efficiently activate NK cells via FcγRIIIa (Wang et al., 2015). In contrast, neutrophils preferentially engage IgG2 class antibodies via FcγRIIa to activate ADCC (Schneider-Merck et al., 2010). This IgG2-mediated ADCC was influenced by the functional FcγRIIa-R131H polymorphism. ADCC was induced more effectively on neutrophils (from homozygous donors) expressing the receptor FcγRIIa-131H than on neutrophils expressing the receptor FcγRIIa-131R (Schneider-Merck et al., 2010). Therefore, there is a big interest in developing new improved antibodies through Fc engineering technologies in order to potentiate FcγR-mediated functions (Derer et al., 2012). Based on this technology, it was possible to change the Fc portion of an anti-EGFR antibody so that it could bind better to FcγRIIa on neutrophils that to FcγRIIIa on NK cells, resulting in effective neutrophil ADCC activity against EGFR-expressing tumors (Derer et al., 2014).

However, it seems that FcαRI (CD89) is a more potent inducer of ADCC by neutrophils (Brandsma et al., 2019; Breedveld and van Egmond, 2019). Thus, it has been proposed that a new generation of cancer therapeutic mAbs should include IgA class antibodies to fully take advantage of the cytotoxic potential of neutrophils (Bakema and van Egmond, 2011). Indeed, this idea is supported by a new IgA2 anti-EGFR antibody derived from the IgG anti EGFR mAb cetuximab. IgA2 EGFR was more effective than cetuximab *in vivo* against EGFR-transfected Ba/F3 target cells (Boross et al., 2013). It was also shown that the combination of IgG and IgA mAbs to two different tumor targets (EGFR and HER2) led to enhanced cytotoxicity compared to each isotype alone (Brandsma et al., 2015). Very recently, it was also confirmed that IgA mAbs against CD20 were more effective at activating ADCC towards malignant B cell lines than IgG antibodies (Evers et al., 2020). Moreover, this IgA-mediated neutrophil cytotoxicity could be enhanced further by inhibition of the CD47-SIRPα checkpoint pathway (Treffers et al., 2020). Together these reports indicate that targeting CD47-SIRPα interactions in combination with IgA therapeutic antibodies has the potential to improve immunotherapy efficacy in cancer patients.

Conclusion

Tumor development is influenced by the tumor microenvironment and in many cases, neutrophils infiltrate the tumors. Neutrophils then, on one hand, can mediate tumor progression by inducing proliferation, angiogenesis, and metastasis. In these processes, the release of granule components and NET formation are important. In addition, neutrophils can also function as immune suppressor cells by blocking T cell activation, affecting NK cells and inducing Treg. On the other hand, neutrophils can also display strong anti-tumor activity. Tumor cells can be directly damaged by ROS and can be induced to enter apoptosis by TRAIL and Fas. The more potent anti-tumor function of neutrophils is ADCC, which seems to involve Fc receptor-mediated trogocytosis. New therapeutic strategies to activate the anti-tumor functions of neutrophils are being investigated. Promising results have been found with the use of therapeutic antibodies, which are strong activators of ADCC, in particular IgA class antibodies binding to FcαRI. New antibodies are being designed so that they have better affinity for particular Fc receptors and induce stronger anti-tumor responses. Future research will certainly provide us with new clues on how to direct neutrophils to fight cancer.

Acknowledgments

Research in the authors' laboratory was supported by grants PAPIIT IN222120 (to EUQ) and IN202520 (to CR) from Dirección General de

Asuntos del Personal Académico, Universidad Nacional Autónoma de México, Mexico.

Conflict of Interest

The authors declare not having any commercial or financial relationships that could be interpreted as a potential conflict of interest

References

Aarts, C. E. M., Hiemstra, I. H., Béguin, E. P., Hoogendijk, A. J., Bouchmal, S., van Houdt, M., et al., (2019). Activated neutrophils exert myeloid-derived suppressor cell activity damaging T cells beyond repair. *Blood Adv.* 3: 3562-3574. doi: 10.1182/bloodadvances.2019031609.

Acuff, H., Carter, K., Fingleton, B., Gorden, D., and Matrisian, L. (2006). Matrix metalloproteinase-9 from bone marrow-derived cells contributes to survival but not growth of tumor cells in the lung microenvironment. *Cancer Res.* 66: 259-266. doi: 10.1158/0008-5472.CAN-05-2502.

Adrover, J. M., Aroca-Crevillén, A., Crainiciuc, G., Ostos, F., Rojas-Vega, Y., Rubio-Ponce, A., et al., (2020). Programmed 'disarming' of the neutrophil proteome reduces the magnitude of inflammation. *Nat. Immunol.* 21: 135-144. doi: 10.1038/s41590-019-0571-2.

Adrover, J. M., Del Fresno, C., Crainiciuc, G., Cuartero, M.I., Casanova-Acebes, M., Weiss, L. A., et al., (2019). A neutrophil timer coordinates immune defense and vascular protection. *Immunity* 50: 390-402. doi: 10.1016/j.immuni.2019.01.002.

Advani, R., Flinn, I., Popplewell, L., Forero, A., Bartlett, N. L., Ghosh, N., et al., (2018). CD47 blockade by Hu5F9-G4 and Rituximab in non-Hodgkin's lymphoma. *N. Engl. J. Med.* 379: 1711-1721. doi: 10.1056/NEJMoa1807315.

Amarante-Mendes, G. P., and Griffith, T. S. (2015). Therapeutic applications of TRAIL receptor agonists in cancer and beyond. *Pharmacol. Ther.* 155:117-131. doi: 10.1016/j.pharmthera.2015.09.001.

Arokiasamy, S., Zakian, C., Dilliway, J., Wang, W., Nourshargh, S., and Voisin, M. B. (2017). Endogenous TNFα orchestrates the trafficking of neutrophils into and within lymphatic vessels during acute inflammation. *Sci. Rep.* 13: 44189. doi: 10.1038/srep44189.

Atzpodien, J., and Reitz, M. (2008). Peripheral blood neutrophils as independent immunologic predictor of response and long-term survival upon immunotherapy in metastatic renal-cell carcinoma. *Cancer Biother. Radiopharm.* 23: 129-134. doi: 10.1089/cbr.2007.0429.

Bakema, J. E., and van Egmond, M. (2011). Immunoglobulin A: A next generation of therapeutic antibodies? *MAbs*. 3: 352-361.

Bakema, J. E., and van Egmond, M. (2014). Fc receptor-dependent mechanisms of monoclonal antibody therapy of cancer. *Curr. Top. Microbiol. Immunol.* 382:373-392. doi: 10.1007/978-3-319-07911-0_17.

Bankey, P. E., Banerjee, S., Zucchiatti, A., De, M., Sleem, R. W., Lin, C. F., et al., (2010). Cytokine induced expression of programmed death ligands in human neutrophils. *Immunol. Lett.* 129: 100-107. doi: 10.1016/j.imlet.2010.01.006.

Berger-Achituv, S., Brinkmann, V., Abed, U. A., Kühn, L. I., Ben-Ezra, J., Elhasid, R., et al., (2013). A proposed role for neutrophil extracellular traps in cancer immunoediting. *Front. Immunol.* 4: 48. doi: 10.3389/fimmu.2013.00048.

Bergers, G., Brekken, R., McMahon, G., Vu, T., Itoh, T., Tamaki, K., et al., (2000). Matrix metalloproteinase-9 triggers the angiogenic switch during carcinogenesis. *Nat. Cell. Biol.* 2: 737-744. doi: 10.1038/35036374.

Bettadapur, A., Miller, H. W., and Ralston, K. S. (2020). Biting off what can be chewed: Trogocytosis in health, infection, and disease. *Infect. Immun.* 88: e00930-00919. doi: 10.1128/IAI.00930-19.

Blanco-Camarillo, C., Alemán, O. R., and Rosales, C. (2021). Low-density neutrophils in healthy individuals display a mature primed phenotype. *Front. Immunol.* 12: 672520. doi: 10.3389/fimmu.2021.672520.

Bodduluru, L. N., Kasala, E. R., Madhana, R. M., and Sriram, C. S. (2015). Natural killer cells: the journey from puzzles in biology to treatment of cancer. *Cancer Lett.* 357: 454-467. doi: 10.1016/j.canlet.2014.12.020.

Boross, P., Lohse, S., Nederend, M., Jansen, J. H., van Tetering, G., Dechant, M., et al., (2013). IgA EGFR antibodies mediate tumour killing *in vivo*. *EMBO Mol. Med.* 5: 1213-1226. doi: 10.1002/emmm.201201929.

Borregaard, N. (2010). Neutrophils, from marrow to microbes. *Immunity* 33: 657-670.

Borrok, M. J., Luheshi, N. M., Beyaz, N., Davies, G. C., Legg, J. W., Wu, H., et al., (2015). Enhancement of antibody-dependent cell-mediated cytotoxicity by endowing IgG with FcαRI (CD89) binding. *MAbs* 7: 743-751. doi: 10.1080/19420862.2015.1047570.

Böyum, A. (1968). Isolation of mononuclear cells and granulocytes from human blood. Isolation of monuclear cells by one centrifugation, and of granulocytes by combining centrifugation and sedimentation at 1 g. *Scand. J. Clin. Lab. Invest. Suppl.* 97: 77-89.

Brandau, S., Dumitru, C. A., and Lang, S. (2013). Protumor and antitumor functions of neutrophil granulocytes. *Semin. Immunopathol.* 35: 163-176. doi: 10.1007/s00281-012-0344-6.

Brandsma, A. M., Bondza, S., Evers, M., Koutstaal, R., Nederend, M., Jansen, J. H. M., et al., (2019). Potent Fc receptor signaling by IgA leads to superior killing of cancer cells by neutrophils compared to IgG. *Front. Immunol.* 10: 704. doi: 10.3389/fimmu.2019.00704.

Brandsma, A. M., Ten Broeke, T., Nederend, M., Meulenbroek, L. A., van Tetering, G., Meyer, S., et al., (2015). Simultaneous targeting of FcγRs and FcαRI enhances tumor cell killing. *Cancer Immunol. Res.* 3: 1316-1324. doi: 10.1158/2326-6066.CIR-15-0099-T.

Breedveld, A., and van Egmond, M. (2019). IgA and FcαRI: Pathological roles and therapeutic opportunities. *Front. Immunol.* 10: 553. doi: 10.3389/fimmu.2019.00553.

Brincks, E. L., Risk, M. C., and Griffith, T. S. (2013). PMN and anti-tumor immunity--the case of bladder cancer immunotherapy. *Semin. Cancer Biol.* 23: 183-189. doi: 10.1016/j.semcancer.2013.02.002.

Brinkmann, V. (2018). Neutrophil extracellular traps in the second decade. *J. Innate Immun.* 10: 414-421. doi: 10.1159/000489829.

Brinkmann, V., and Zychlinsky, A. (2012). Neutrophil extracellular traps: is immunity the second function of chromatin? *J. Cell Biol.* 198: 773-783. doi: 10.1083/jcb.201203170.

Bruhns, P., Iannascoli, B., England, P., Mancardi, D., Fernandez, N., Jorieux, S., et al., (2009). Specificity and affinity of human Fcγ receptors and their polymorphic variants for human IgG subclasses. *Blood* 113: 3716-3725. doi: 10.1182/blood-2008-09-179754.

Bruhns, P., and Jönsson, F. (2015). Mouse and human FcR effector functions. *Immunol. Rev.* 268: 25-51. doi: 10.1111/imr.12350.

Casanova-Acebes, M., C, P., Weiss, L. A., Nombela-Arrieta, C., Chèvre, R., A-González, N., et al., (2013). Rhythmic modulation of the hematopoietic niche through neutrophil clearance. *Cell* 153: 1025-1035. doi: 10.1016/j.cell.2013.04.040.

Casanova-Acebes, M., Nicolás-Ávila, J. A., Li, J. L., García-Silva, S., Balachander, A., Rubio-Ponce, A., et al., (2018). Neutrophils instruct homeostatic and pathological states in naive tissues. *J. Exp. Med.* 215: 2778-2795. doi: 10.1084/jem.20181468.

Cassatella, M. A., and Scapini, P. (2020). On the improper use of the term high-density neutrophils. *Trends Immunol.* 41: 1059-1061. doi: 10.1016/j.it.2020.10.008.

Cedervall, J., and Olsson, A. K. (2016). Immunity gone astray - NETs in cancer. *Trends Cancer.* 2: 633-634. doi: 10.1016/j.trecan.2016.10.012.

Chan, A. C., and Carter, P. J. (2010). Therapeutic antibodies for autoimmunity and inflammation. *Nat. Rev. Immunol.* 10: 301-316. doi: 10.1038/nri2761.

Chaudhary, B., and Elkord, E. (2016). Regulatory T cells in the tumor microenvironment and cancer progression: Role and therapeutic targeting. *Vaccines (Basel)* 4: 28. doi: 10.3390/vaccines4030028.

Coffelt, S. B., Kersten, K., Doornebal, C. W., Weiden, J., Vrijland, K., Hau, C. S., et al., (2015). IL-17-producing γδ T cells and neutrophils conspire to promote breast cancer metastasis. *Nature* 522: 345-348. doi: 10.1038/nature14282.

Coffelt, S. B., Wellenstein, M. D., and de Visser, K. E. (2016). Neutrophils in cancer: neutral no more. *Nat. Rev. Cancer.* 16: 431-446. doi: 10.1038/nrc.2016.52.

Colombo, M. P., Lombardi, L., Stoppacciaro, A., Melani, C., Parenza, M., Bottazzi, B., et al., (1992). Granulocyte colony-stimulating factor (G-CSF) gene transduction in murine adenocarcinoma drives neutrophil-mediated tumor inhibition in vivo. Neutrophils discriminate between G-CSF-producing and G-CSF-nonproducing tumor cells. *J. Immunol.* 149: 113-119.

Coussens, L., Tinkle, C., Hanahan, D., and Werb, Z. (2000). MMP-9 supplied by bone marrow-derived cells contributes to skin carcinogenesis. *Cell* 103: 481-490. doi: 10.1016/S0092-8674(00)00139-2.

Cowland, J. B., and Borregaard, N. (2016). Granulopoiesis and granules of human neutrophils. *Immunol. Rev.* 273: 11-28. doi: 10.1111/imr.12440.

Crissman, J. D., Hatfield, J., Schaldenbrand, M., Sloane, B. F., and Honn, K. V. (1985). Arrest and extravasation of B16 amelanotic melanoma in murine lungs. A light and electron microscopic study. *Lab. Invest.* 53: 470-478.

Cross, A., Edwards, S. W., Bucknall, R. C., and Moots, R. J. (2004). Secretion of oncostatin M by neutrophils in rheumatoid arthritis. *Arthritis Rheum.* 50: 1430-1436.

Dallegri, F., Ottonello, L., Ballestrero, A., Dapino, P., Ferrando, F., Patrone, F., et al., (1991). Tumor cell lysis by activated human neutrophils: analysis of neutrophil-delivered oxidative attack and role of leukocyte function-associated antigen 1. *Inflammation* 15: 15-30.

Dallegri, F., Patrone, F., Frumento, G., and Sacchetti, C. (1984). Antibody-dependent killing of tumor cells by polymorphonuclear leukocytes. Involvement of oxidative and nonoxidative mechanisms. *J. Natl. Cancer Inst.* 73: 331-339.

de Kleijn, S., Langereis, J. D., Leentjens, J., Kox, M., Netea, M. G., Koenderman, L., et al., (2013). IFN-γ-stimulated neutrophils suppress lymphocyte proliferation through expression of PD-L1. *PLoS One* 8: e72249. doi: 10.1371/journal.pone.0072249.

De Silva, D. M., Roy, A., Kato, T., Cecchi, F., Lee, Y. H., Matsumoto, K., et al., (2017). Targeting the hepatocyte growth factor/Met pathway in cancer. *Biochem. Soc. Trans.* 45: 855-870. doi: 10.1042/BST20160132.

Demers, M., Krause, D. S., Schatzberg, D., Martinod, K., Voorhees, J. R., Fuchs, T. A., et al., (2012). Cancers predispose neutrophils to release extracellular DNA traps that contribute to cancer-associated thrombosis. *Proc. Natl. Acad. Sci. USA.* 109: 13076-13081. doi: 10.1073/pnas.1200419109.

Deniset, J. F., and Kubes, P. (2016). Recent advances in understanding neutrophils. *F1000Res.* 5: 2912. doi: 10.12688/f1000research.9691.1.

Deniset, J. F., Surewaard, B. G., Lee, W. Y., and Kubes, P. (2017). Splenic Ly6G^{high} mature and Ly6G^{int} immature neutrophils contribute to eradication of *S. pneumoniae*. *J. Exp. Med.* 214: 1333-1350. doi: 10.1084/jem.20161621.

Derer, S., Glorius, P., Schlaeth, M., Lohse, S., Klausz, K., Muchhal, U., et al., (2014). Increasing FcγRIIa affinity of an FcγRIII-optimized anti-EGFR antibody restores neutrophil-mediated cytotoxicity. *MAbs.* 6: 409-421. doi: 10.4161/mabs.27457.

Derer, S., Kellner, C., Berger, S., Valerius, T., and Peipp, M. (2012). Fc engineering: design, expression, and functional characterization of antibody variants with improved effector function. *Methods Mol. Biol.* 907: 519-536. doi: 10.1007/978-1-61779-974-7_30.

Dermani, F. K., Samadi, P., Rahmani, G., Kohlan, A. K., and Najafi, R. (2019). PD-1/PD-L1 immune checkpoint: Potential target for cancer therapy. *J. Cell. Physiol.* 234: 1313-1325. doi: 10.1002/jcp.27172.

Deryugina, E., Zijlstra, A., Partridge, J., Kupriyanova, T., Madsen, M., Papagiannakopoulos, T., et al., (2005). Unexpected effect of matrix metalloproteinase down-regulation on vascular intravasation and metastasis of human fibrosarcoma cells selected in vivo for high rates of dissemination. *Cancer Res.* 65: 10959-10969. doi: 10.1158/0008-5472.CAN-05-2228

Deryugina, E. I., and Quigley, J. P. (2015). Tumor angiogenesis: MMP-mediated induction of intravasation- and metastasis-sustaining neovasculature. *Matrix Biol.* 44-46: 94-112. doi: 10.1016/j.matbio.2015.04.004.

Devi, S., Wang, Y., Chew, W. K., Lima, R., A-González, N., Mattar, C. N., et al., (2013). Neutrophil mobilization via plerixafor-mediated CXCR4 inhibition arises from lung demargination and blockade of neutrophil homing to the bone marrow. *J. Exp. Med.* 210: 2321-2336. doi: 10.1084/jem.20130056.

Duffy, D., Perrin, H., Abadie, V., Benhabiles, N., Boissonnas, A., Liard, C., et al., (2012). Neutrophils transport antigen from the dermis to the bone marrow, initiating a source of memory CD8$^+$ T cells. *Immunity* 37: 917-929. 10.1016/j.immuni.2012.07.015.

Dumitru, C., Moses, K., Trellakis, S., Lang, S., and Brandau, S. (2012). Neutrophils and granulocytic myeloid-derived suppressor cells:

immunophenotyping, cell biology and clinical relevance in human oncology. *Cancer Immunol. Immunother.* 61: 1155-1167. doi: 10.1007/s00262-012-1294-5.

Ebrahem, Q., Chaurasia, S. S., Vasanji, A., Qi, J. H., Klenotic, P. A., Cutler, A., et al., (2010). Cross-talk between vascular endothelial growth factor and matrix metalloproteinases in the induction of neovascularization in vivo. *Am. J. Pathol.* 176: 496-503. doi: 10.2353/ajpath.2010.080642.

Eccles, S. A. (2001). Monoclonal antibodies targeting cancer: 'magic bullets' or just the trigger? *Breast Cancer Res.* 3: 86-90.

El-Benna, J., Hurtado-Nedelec, M., Marzaioli, V., Marie, J. C., Gougerot-Pocidalo, M. A., and Dang, P. M. (2016). Priming of the neutrophil respiratory burst: role in host defense and inflammation. *Immunol. Rev.* 273: 180-193. doi: 10.1111/imr.12447.

Eruslanov, E. B. (2017). Phenotype and function of tumor-associated neutrophils and their subsets in early-stage human lung cancer. *Cancer Immunol. Immunother.* 66: 997-1006. doi: 10.1007/s00262-017-1976-0.

Eruslanov, E. B., Bhojnagarwala, P. S., Quatromoni, J. G., Stephen, T. L., Ranganathan, A., Deshpande, C., et al., (2014). Tumor-associated neutrophils stimulate T cell responses in early-stage human lung cancer. *J. Clin. Invest.* 124: 5466-5480. doi: 10.1172/JCI77053.

Evers, M., Ten Broeke, T., Jansen, J. H. M., Nederend, M., Hamdan, F., Reiding, K. R., et al., (2020). Novel chimerized IgA CD20 antibodies: Improving neutrophil activation against CD20-positive malignancies. *MAbs* 12: 1795505. doi: 10.1080/19420862.2020.1795505.

Evrard, M., Kwok, I. W. H., Chong, S. Z., Teng, K. W. W., Becht, E., Chen, J., et al., (2018). Developmental analysis of bone marrow neutrophils reveals populations specialized in expansion, trafficking, and effector functions. *Immunity* 48: 364-379. doi: 10.1016/j.immuni.2018.02.002.

Faria, S. S., Fernandes, P. C., Jr, Silva, M. J., Lima, V. C., Fontes, W., Freitas-Junior, R., et al., (2016). The neutrophil-to-lymphocyte ratio: A narrative review. *Ecancer medical science.* 10: 702. doi: 10.3332/ecancer.2016.702.

Feng, M., Jiang, W., Kim, B. Y. S., Zhang, C. C., Fu, Y. X., and Weissman, I. L. (2019). Phagocytosis checkpoints as new targets for cancer immunotherapy. *Nat. Rev. Cancer* 19: 568-586. doi: 10.1038/s41568-019-0183-z.

Fingleton, B. (2008). MMPs as therapeutic targets--still a viable option?. *Semin. Cell. Dev. Biol.* 19: 61-68. doi:10.1016/j.semcdb.2007.06.006.

Freeman, G. J., Long, A. J., Iwai, Y., Bourque, K., Chernova, T., Nishimura, H., et al., (2000). Engagement of the PD-1 immunoinhibitory receptor by a novel B7 family member leads to negative regulation of lymphocyte activation. *J. Exp. Med.* 192: 1027-1034. doi: 10.1084/jem.192.7.1027.

Fridlender, Z., Sun, J., Kim, S., Kapoor, V., Cheng, G., Ling, L., et al., (2009). Polarization of tumor-associated neutrophil phenotype by TGF-beta: "N1" versus "N2" TAN. *Cancer Cell* 16: 183-194. doi: 10.1016/j.ccr.2009.06.017.

Fulda, S. (2014). Tumor-necrosis-factor-related apoptosis-inducing ligand (TRAIL). *Adv. Exp. Med. Biol.* 818: 167-180. doi: 10.1007/978-1-4471-6458-6_8.

Gabrilovich, D., Bronte, V., Chen, S., Colombo, M., Ochoa, A., Ostrand-Rosenberg, S., et al., (2007). The terminology issue for myeloid-derived suppressor cells. *Cancer Res.* 67: 425. doi: 10.1158/0008-5472.CAN-06-3037

Gabrilovich, D., Ostrand-Rosenberg, S., and Bronte, V. (2012). Coordinated regulation of myeloid cells by tumours. *Nat. Rev. Immunol.* 12: 253-268. doi: 10.1038/nri3175.

Gaida, M. M., Steffen, T. G., Günther, F., Tschaharganeh, D. F., Felix, K., Bergmann, F., et al., (2012). Polymorphonuclear neutrophils promote dyshesion of tumor cells and elastase-mediated degradation of E-cadherin in pancreatic tumors. *Eur. J. Immunol.* 42: 3369-3380. doi: 10.1002/eji.201242628.

Galdiero, M. R., Garlanda, C., Jaillon, S., Marone, G., and Mantovani, A. (2013). Tumor associated macrophages and neutrophils in tumor progression. *J. Cell. Physiol.* 228: 1404-1412. doi: 10.1002/jcp.24260.

García-García, E., Uribe-Querol, E., and Rosales, C. (2013). A simple and efficient method to detect nuclear factor activation in human neutrophils by flow cytometry. *J. Vis. Exp.* 74: e50410. doi:10.3791/50410.

Gazendam, R. P., van de Geer, A., Roos, D., van den Berg, T. K., and Kuijpers, T. W. (2016). How neutrophils kill fungi. *Immunol Rev.* 273: 299-311. doi: 10.1111/imr.12454.

Gelderman, K. A., Tomlinson, S., Ross, G. D., and Gorter, A. (2004). Complement function in mAb-mediated cancer immunotherapy. *Trends Immunol.* 25: 158-164. doi: 10.1016/j.it.2004.01.008.

Gerrard, T. L., Cohen, D. J., and Kaplan, A. M. (1981). Human neutrophil-mediated cytotoxicity to tumor cells. *J. Natl. Cancer Inst.* 66: 483-488.

Gershkovitz, M., Caspi, Y., Fainsod-Levi, T., Katz, B., Michaeli, J., Khawaled, S., et al., (2018a). TRPM2 mediates neutrophil killing of disseminated tumor cells. *Cancer Res.* 78: 2680-2690. doi: 10.1158/0008-5472.CAN-17-3614.

Gershkovitz, M., Fainsod-Levi, T., Khawaled, S., Shaul, M. E., Sionov, R. V., Cohen-Daniel, L., et al., (2018b). Microenvironmental cues determine tumor cell susceptibility to neutrophil cytotoxicity. *Cancer Res.* 78: 5050-5059. doi: 10.1158/0008-5472.CAN-18-0540.

Gershkovitz, M., Fainsod-Levi, T., Zelter, T., Sionov, R. V., and Granot, Z. (2019). TRPM2 modulates neutrophil attraction to murine tumor cells by regulating CXCL2 expression. *Cancer Immunol. Immunother.* 68: 33-43. doi: 10.1007/s00262-018-2249-2.

Ginestier, C., Liu, S., Diebel, M. E., Korkaya, H., Luo, M., Brown, M., et al., (2010). CXCR1 blockade selectively targets human breast cancer stem cells in vitro and in xenografts. *J. Clin. Invest.* 120: 485-497. doi: 10.1172/JCI39397.

Giraudo, E., Inoue, M., and Hanahan, D. (2004). An amino-bisphosphonate targets MMP-9-expressing macrophages and angiogenesis to impair cervical carcinogenesis. *J. Clin. Invest.* 114: 623-633.

Glodde, N., Bald, T., van den Boorn-Konijnenberg, D., Nakamura, K., O'Donnell, J. S., Szczepanski, S., et al., (2017). Reactive neutrophil responses dependent on the receptor tyrosine kinase c-MET limit cancer immunotherapy. *Immunity* 47: 789-802. doi: 10.1016/j.immuni.2017.09.012.

Gomes, T., Várady, C. B. S., Lourenço, A. L., Mizurini, D. M., Rondon, A. M. R., Leal, A. C., et al., (2019). IL-1β blockade attenuates thrombosis in a neutrophil extracellular trap-dependent breast cancer model. *Front. Immunol.* 10: 2088. doi: 10.3389/fimmu.2019.02088.

Gorczynski, R. M. (2020). IL-17 signaling in the tumor microenvironment. *Adv. Exp. Med. Biol.* 1240: 47-58. doi: 10.1007/978-3-030-38315-2_4.

Goren, I., Kämpfer, H., Müller, E., Schiefelbein, D., Pfeilschifter, J., and Frank, S. (2006). Oncostatin M expression is functionally connected to neutrophils in the early inflammatory phase of skin repair: implications for normal and diabetes-impaired wounds. *J. Invest. Dermatol.* 126: 628-637.

Granot, Z., Henke, E., Comen, E., King, T., Norton, L., and Benezra, R. (2011). Tumor entrained neutrophils inhibit seeding in the premetastatic lung. *Cancer Cell* 20: 300-314. doi:10.1016/j.ccr.2011.08.012.

Gregory, A., and Houghton, A. (2011). Tumor-associated neutrophils: new targets for cancer therapy. *Cancer Res.* 71: 2411-2416. doi: 10.1158/0008-5472.CAN-10-2583.

Gregory, A. D., Hale, P., Perlmutter, D. H., and Houghton, A. M. (2012). Clathrin pit-mediated endocytosis of neutrophil elastase and cathepsin G by cancer cells. *J. Biol. Chem.* 287: 35341-35350. doi: 10.1074/jbc.M112.385617.

Grenier, A. (2001). Oncostatin M production by blood and alveolar neutrophils during acute lung injury. *Lab. Invest.* 81: 133-141.

Guglietta, S., Chiavelli, A., Zagato, E., Krieg, C., Gandini, S., Ravenda, P. S., et al., (2016). Coagulation induced by C3aR-dependent NETosis drives protumorigenic neutrophils during small intestinal tumorigenesis. *Nat. Commun.* 7: 11037. doi: 10.1038/ncomms11037.

Hampton, H. R., and Chtanova, T. (2016). The lymph node neutrophil. *Semin. Immunol.* 28: 129-136. doi: 10.1016/j.smim.2016.03.008.

Hanahan, D., and Weinberg, R. (2011). Hallmarks of cancer: the next generation. *Cell* 144: 646-674. doi: 10.1016/j.cell.2011.02.013.

Hawes, M. C., Wen, F., and Elquza, E. (2015). Extracellular DNA: A bridge to cancer. *Cancer Res.* 75: 4260-4264. doi: 10.1158/0008-5472.CAN-15-1546.

Hawinkels, L. J., Zuidwijk, K., Verspaget, H. W., de Jonge-Muller, E. S., van Duijn, W., Ferreira, V., et al., (2008). VEGF release by MMP-9 mediated heparan sulphate cleavage induces colorectal cancer angiogenesis. *Eur. J. Cancer.* 44: 1904-1913. doi: 10.1016/j.ejca.2008.06.031.

He, G., Zhang, H., Zhou, J., Wang, B., Chen, Y., Kong, Y., et al., (2015). Peritumoural neutrophils negatively regulate adaptive immunity via the PD-L1/PD-1 signalling pathway in hepatocellular carcinoma. *J. Exp. Clin. Cancer Res.* 34: 141. doi: 10.1186/s13046-015-0256-0.

Heemskerk, N., and van Egmond, M. (2018). Monoclonal antibody-mediated killing of tumour cells by neutrophils. *Eur. J. Clin. Invest.* 48: e12962. doi: 10.1111/eci.12962.

Ho, A. S., Chen, C. H., Cheng, C. C., Wang, C. C., Lin, H. C., Luo, T. Y., et al., (2014). Neutrophil elastase as a diagnostic marker and therapeutic target in colorectal cancers. *Oncotarget* 5: 473-480. doi: 10.18632/oncotarget.1631.

Horrigan, S. K., and Reproducibility Project: Cancer Biology (2017). Replication study: The CD47-signal regulatory protein alpha (SIRPα) interaction is a therapeutic target for human solid tumors. *Elife* 6: e18173. doi: 10.7554/eLife.18173.

Houghton, A., Rzymkiewicz, D., Ji, H., Gregory, A., Egea, E., Metz, H., et al., (2010). Neutrophil elastase-mediated degradation of IRS-1 accelerates lung tumor growth. *Nat. Med.* 16: 219-223. doi: 10.1038/nm.2084.

Hsu, B. E., Tabariès, S., Johnson, R. M., Andrzejewski, S., Senecal, J., Lehuédé, C., et al., (2019). Immature low-density neutrophils exhibit metabolic flexibility that facilitates breast cancer liver metastasis. *Cell Rep.* 27: 3902-3915. doi: 10.1016/j.celrep.2019.05.091.

Hu, N., Mora-Jensen, H., Theilgaard-Mönch, K., Doornbos-van der Meer, B., Huitema, M. G., Stegeman, C. A., et al., (2014). Differential expression of granulopoiesis related genes in neutrophil subsets distinguished by membrane expression of CD177. *PLoS One* 9: e99671. doi: 10.1371/journal.pone.0099671.

Huang, S., Mills, L., Mian, B., Tellez, C., McCarty, M., Yang, X., et al., (2002). Fully humanized neutralizing antibodies to interleukin-8 (ABX-IL8) inhibit angiogenesis, tumor growth, and metastasis of human melanoma. *Am. J. Pathol.* 161: 125-134.

Huh, S., Liang, S., Sharma, A., Dong, C., and Robertson, G. (2010). Transiently entrapped circulating tumor cells interact with neutrophils to facilitate lung metastasis development. *Cancer Res.* 70: 6071-6082. doi: 10.1158/0008-5472.CAN-09-4442.

Imai, Y., Kubota, Y., Yamamoto, S., Tsuji, K., Shimatani, M., Shibatani, N., et al., (2005). Neutrophils enhance invasion activity of human cholangiocellular carcinoma and hepatocellular carcinoma cells: an in vitro study. *J. Gastroenterol. Hepatol.* 20: 287-293. doi: 10.1111/j.1440-1746.2004.03575.x.

Ivars Rubio, A., Yufera, J. C., de la Morena, P., Fernández Sánchez, A., Navarro Manzano, E., García Garre, E., et al., (2019). Neutrophil-lymphocyte ratio in metastatic breast cancer is not an independent predictor of survival, but depends on other variables. *Sci. Rep.* 9: 16979. doi: 10.1038/s41598-019-53606-3.

Jablonska, J., Leschner, S., Westphal, K., Lienenklaus, S., and Weiss, S. (2010). Neutrophils responsive to endogenous IFN-beta regulate tumor angiogenesis and growth in a mouse tumor model. *J. Clin. Invest.* 120: 1151-1164. doi: 10.1172/JCI37223.

Jamieson, T., Clarke, M., Steele, C. W., Samuel, M. S., Neumann, J., Jung, A., et al., (2012). Inhibition of CXCR2 profoundly suppresses inflammation-driven and spontaneous tumorigenesis. *J. Clin. Invest.* 122: 3127-3144. doi: 10.1172/JCI61067.

Jensen, H., Donskov, F., Marcussen, N., Nordsmark, M., Lundbeck, F., and von der Maase, H. (2009). Presence of intratumoral neutrophils is an independent prognostic factor in localized renal cell carcinoma. *J. Clin. Oncol.* 27: 4709-4717. doi: 10.1200/JCO.2008.18.9498.

Ji, H., Houghton, A., Mariani, T., Perera, S., Kim, C., Padera, R., et al., (2006). K-ras activation generates an inflammatory response in lung tumors. *Oncogene* 25: 2105-2112.

Katano, M., and Torisu, M. (1982). Neutrophil-mediated tumor cell destruction in cancer ascites. *Cancer* 50: 62-68.

Keane, M., Belperio, J., Xue, Y., Burdick, M., and Strieter, R. (2004). Depletion of CXCR2 inhibits tumor growth and angiogenesis in a murine model of lung cancer. *J. Immunol.* 172: 2853-2860. doi: 10.4049/jimmunol.172.5.2853.

Klemke, M., Wabnitz, G. H., Funke, F., Funk, B., Kirchgessner, H., and Samstag, Y. (2008). Oxidation of cofilin mediates T cell hyporesponsiveness under oxidative stress conditions. *Immunity* 29: 404-413. doi: 10.1016/j.immuni.2008.06.016.

Knaapen, A. M., Güngör, N., Schins, R.P., Borm, P. J., and Van Schooten, F. J. (2006). Neutrophils and respiratory tract DNA damage and mutagenesis: a review. *Mutagenesis.* 21: 225-236.

Korkmaz, B., Moreau, T., and Gauthier, F. (2008). Neutrophil elastase, proteinase 3 and cathepsin G: physicochemical properties, activity and physiopathological functions. *Biochimie*. 90: 227-242.

Kousis, P., Henderson, B., Maier, P., and Gollnick, S. (2007). Photodynamic therapy enhancement of antitumor immunity is regulated by neutrophils. *Cancer Res.* 67: 10501-10510. doi: 10.1158/0008-5472.CAN-07-1778.

Kuang, D., Zhao, Q., Wu, Y., Peng, C., Wang, J., Xu, Z., et al., (2011). Peritumoral neutrophils link inflammatory response to disease progression by fostering angiogenesis in hepatocellular carcinoma. *J. Hepatol.* 54: 948-955. doi: 10.1016/j.jhep.2010.08.041.

Kühn, F. J. P. (2020). Structure-function relationship of TRPM2: Recent advances, contradictions, and open questions. *Int. J. Mol. Sci.* 21: 6481. doi: 10.3390/ijms21186481.

Kushner, B. H., and Cheung, N. K. (1991). Clinically effective monoclonal antibody 3F8 mediates nonoxidative lysis of human neuroectodermal tumor cells by polymorphonuclear leukocytes. *Cancer Res.* 51: 4865-4870.

Lawrence, S. M., Corriden, R., and Nizet, V. (2018). The ontogeny of a neutrophil: Mechanisms of granulopoiesis and homeostasis. *Microbiol. Mol. Biol. Rev.* 82: e00057-00017. doi: 10.1128/MMBR.00057-17.

Lazennec, G., and Richmond, A. (2010). Chemokines and chemokine receptors: new insights into cancer-related inflammation. *Trends Mol. Med.* 16: 133-144. doi: 10.1016/j.molmed.2010.01.003.

Lechner, M., Liebertz, D., and Epstein, A. (2010). Characterization of cytokine-induced myeloid-derived suppressor cells from normal human peripheral blood mononuclear cells. *J. Immunol.* 185: 2273-2284. doi: 10.4049/jimmunol.1000901

Lerman, I., Garcia-Hernandez, M. L., Rangel-Moreno, J., Chiriboga, L., Pan, C., Nastiuk, K. L., et al., (2017). Infiltrating myeloid cells exert protumorigenic actions via neutrophil elastase. *Mol. Cancer Res.* 15: 1138-1152. doi: 10.1158/1541-7786.MCR-17-0003.

Li, H., Han, Y., Guo, Q., Zhang, M., and Cao, X. (2009). Cancer-expanded myeloid-derived suppressor cells induce anergy of NK cells through membrane-bound TGF-beta 1. *J. Immunol.* 182: 240-249. doi: 10.4049/jimmunol.182.1.240.

Lohse, S., Derer, S., Beyer, T., Klausz, K., Peipp, M., Leusen, J. H., et al., (2011). Recombinant dimeric IgA antibodies against the epidermal growth factor receptor mediate effective tumor cell killing. *J. Immunol.* 186: 3770-3778. doi: 10.4049/jimmunol.1003082.

Lohse, S., Loew, S., Kretschmer, A., Jansen, J. H. M., Meyer, S., Ten Broeke, T., et al., (2018). Effector mechanisms of IgA antibodies against CD20 include recruitment of myeloid cells for antibody-dependent cell-mediated cytotoxicity and complement-dependent cytotoxicity. *Br. J. Haematol.* 181: 413-417. doi: 10.1111/bjh.14624.

López-Lago, M., Posner, S., Thodima, V., Molina, A., Motzer, R., and Chaganti, R. (2013). Neutrophil chemokines secreted by tumor cells mount a lung antimetastatic response during renal cell carcinoma progression. *Oncogene* 32: 1752-1760. doi: 10.1038/onc.2012.201.

Ludwig, A. T., Moore, J. M., Luo, Y., Chen, X., Saltsgaver, N. A., O'Donnell, M. A., et al., (2004). Tumor necrosis factor-related apoptosis-inducing ligand: a novel mechanism for Bacillus Calmette-Guérin-induced antitumor activit. *Cancer Res.* 64: 3386-3390.

Mahiddine, K., Blaisdell, A., Ma, S., Créquer-Grandhomme, A., Lowell, C. A., and Erlebacher, A. (2020). Relief of tumor hypoxia unleashes the tumoricidal potential of neutrophils. *J. Clin. Invest.* 130: 389-403. doi: 10.1172/JCI130952.

Malmberg, K. J., Arulampalam, V., Ichihara, F., Petersson, M., Seki, K., Andersson, T., et al., (2001). Inhibition of activated/memory ($CD45RO^+$) T cells by oxidative stress associated with block of NF-κB activation. *J. Immunol.* 167: 2595-2601.

Mantovani, A., Allavena, P., Sica, A., and Balkwill, F. (2008). Cancer-related inflammation. *Nature* 454: 436-444. doi: 10.1038/nature07205.

Marini, O., Costa, S., Bevilacqua, D., Calzetti, F., Tamassia, N., Spina, C., et al., (2017). Mature $CD10^+$ and immature $CD10^-$ neutrophils present in G-CSF-treated donors display opposite effects on T cells. *Blood* 129: 1343-1356. doi: 10.1182/blood-2016-04-713206.

Matlung, H. L., Babes, L., Zhao, X. W., van Houdt, M., Treffers, L. W., van Rees, D. J., et al., (2018). Neutrophils kill antibody-opsonized cancer cells by trogoptosis. *Cell Rep.* 23: 3946-3959. doi: 10.1016/j.celrep.2018.05.082.

McGary, C. T., Miele, M. E., and Welch, D. R. (1995). Highly metastatic 13762NF rat mammary adenocarcinoma cell clones stimulate bone marrow by secretion of granulocyte-macrophage colony-stimulating factor /interleukin-3 activity. *Am. J. Pathol.* 147: 1668-1681.

Mellman, I., Coukos, G., and Dranoff, G. (2011). Cancer immunotherapy comes of age. *Nature* 480: 480-489. doi: 10.1038/nature10673.

Mercer, F., Ng, S. H., Brown, T. M., Boatman, G., and Johnson, P. J. (2018). Neutrophils kill the parasite *Trichomonas vaginalis* using trogocytosis. *PLoS Biol.* 16: e2003885. doi: 10.1371/journal.pbio.2003885.

Metkar, S., and Froelich, C. (2004). Human neutrophils lack granzyme A, granzyme B, and perforin. *Blood* 104: 905-906.

Michaeli, J., Shaul, M. E., Mishalian, I., Hovav, A. H., Levy, L., Zolotriov, L., et al., (2017). Tumor-associated neutrophils induce apoptosis of non-activated CD8 T-cells in a TNFα and NO-dependent mechanism, promoting a tumor-supportive environment. *Oncoimmunology* 6: e1356965. doi: 10.1080/2162402X.2017.1356965.

Mishalian, I., Bayuh, R., Levy, L., Zolotarov, L., Michaeli, J., and Fridlender, Z.G. (2013). Tumor-associated neutrophils (TAN) develop pro-tumorigenic properties during tumor progression. *Cancer Immunol. Immunother.* 62: 1745-1756. doi: 10.1007/s00262-013-1476-9.

Mollica Poeta, V., Massara, M., Capucetti, A., and Bonecchi, R. (2019). Chemokines and chemokine receptors: New targets for cancer immunotherapy. *Front. Immunol.* 10: 379. doi: 10.3389/fimmu.2019.00379.

Morimoto-Kamata, R., Mizoguchi, S., Ichisugi, T., and Yui, S. (2012). Cathepsin G induces cell aggregation of human breast cancer MCF-7 cells via a 2-step mechanism: catalytic site-independent binding to the cell surface and enzymatic activity-dependent induction of the cell aggregation. *Mediators Inflamm.* 2012: 456462. doi: 10.1155/2012/456462.

Mougiakakos, D., Johansson, C. C., Jitschin, R., Böttcher, M., and Kiessling, R. (2011). Increased thioredoxin-1 production in human naturally occurring regulatory T cells confers enhanced tolerance to oxidative stress. *Blood* 117: 857-861. doi: 10.1182/blood-2010-09-307041.

Mougiakakos, D., Johansson, C. C., and Kiessling, R. (2009). Naturally occurring regulatory T cells show reduced sensitivity toward oxidative stress-induced cell death. *Blood* 113: 3542-3545. doi: 10.1182/blood-2008-09-181040.

Naussef, W. M. (2007). How human neutrophils kill and degrade microbes. An integrated view. *Immunol. Rev.* 219: 88-102.

Nawa, M., Osada, S., Morimitsu, K., Nonaka, K., Futamura, M., Kawaguchi, Y., et al., (2012). Growth effect of neutrophil elastase on breast cancer: favorable action of sivelestat and application to anti-HER2 therapy. *Anticancer Res.* 32: 13-19.

Ng, L. G., Ostuni, R., and Hidalgo, A. (2019). Heterogeneity of neutrophils. *Nat. Rev. Immunol.* 19: 255-265. doi: 10.1038/s41577-019-0141-8.

Nimmerjahn, F., and Ravetch, J. V. (2011). FcγRs in health and disease. *Curr. Top. Microbiol. Immunol.* 350: 105-125. doi: 10.1007/82_2010_86.

Noman, M. Z., Desantis, G., Janji, B., Hasmim, M., Karray, S., Dessen, P., et al., (2014). PD-L1 is a novel direct target of HIF-1α, and its blockade under hypoxia enhanced MDSC-mediated T cell activation. *J. Exp. Med.* 211: 781-790. doi: 10.1084/jem.20131916.

Nozawa, H., Chiu, C., and Hanahan, D. (2006). Infiltrating neutrophils mediate the initial angiogenic switch in a mouse model of multistage carcinogenesis. *Proc. Natl. Acad. Sci. USA.* 103: 12493-12498. doi: 10.1073/pnas.0601807103.

Ocana, A., Nieto-Jiménez, C., Pandiella, A., and Templeton, A. J. (2017). Neutrophils in cancer: prognostic role and therapeutic strategies. *Mol. Cancer.* 16: 137. doi: 10.1186/s12943-017-0707-7.

Ohara, M., Shirado, M., Miyata, M., Sakuma, H., Nanamiya, M., Yoshida, N., et al., (1983). Natural and antibody-dependent cellular cytotoxicity of polymorphonuclear leukocytes. *Tohoku J. Exp. Med.* 140: 59-66. doi: 10.1620/tjem.140.59.

Olivera-Valle, I., Latorre, M. C., Calvo, M., Gaspar, B., Gómez-Oro, C., Collazos, A., et al., (2020). Vaginal neutrophils eliminate sperm by trogocytosis. *Hum. Reprod.* 35: 2567-2578. doi: 10.1093/humrep/deaa198.

Otten, M. A., Leusen, J. H., Rudolph, E., van der Linden, J.A., Beelen, R. H., van de Winkel, J. G., et al., (2007a). FcR gamma-chain dependent signaling in immature neutrophils is mediated by FcalphaRI, but not by FcgammaRI. *J. Immunol.* 179: 2918-2924. doi: 10.4049/jimmunol.179.5.2918.

Otten, M. A., Leusen, J. H., Rudolph, E., van der Linden, J.A., Beelen, R. H., van de Winkel, J. G., et al., (2007b). FcR γ-chain dependent signaling in immature neutrophils is mediated by FcαRI, but not by FcγRI. *J. Immunol.* 179: 2918-2924.

Otten, M. A., Rudolph, E., Dechant, M., Tuk, C. W., Reijmers, R. M., Beelen, R. H., et al., (2005). Immature neutrophils mediate tumor cell killing via IgA but not IgG Fc receptors. *J. Immunol.* 174: 5472-5480. doi: 10.4049/jimmunol.174.9.5472.

Ottonello, L., Epstein, A., Mancini, M., Tortolina, G., Dapino, P., and Dallegri, F. (2001). Chimaeric Lym-1 monoclonal antibody-mediated cytolysis by neutrophils from G-CSF-treated patients: stimulation by GM-CSF and role of Fcγ-receptors. *Br. J. Cancer* 85: 463-469. doi: 10.1054/bjoc.2000.1940.

Papayannopoulos, V. (2018). Neutrophil extracellular traps in immunity and disease. *Nat. Rev. Immunol.* 18: 134-147. doi:10.1038/nri.2017.105.

Paramanathan, A., Saxena, A., and Morris, D. L. (2014). A systematic review and meta-analysis on the impact of pre-operative neutrophil lymphocyte ratio on long term outcomes after curative intent resection of solid tumours. *Surg. Oncol.* 23: 31-39. doi: 10.1016/j.suronc.2013.12.001.

Peng, B., Wang, Y. H., Liu, Y. M., and Ma, L. X. (2015). Prognostic significance of the neutrophil to lymphocyte ratio in patients with non-small cell lung cancer: a systemic review and meta-analysis. *Int. J. Clin. Exp. Med.* 8: 3098-3106.

Pylaeva, E., Lang, S., and Jablonska, J. (2016). The essential role of type I interferons in differentiation and activation of tumor-associated neutrophils. *Front. Immunol.* 7: 629. doi: 10.3389/fimmu.2016.00629.

Queen, M., Ryan, R., Holzer, R., Keller-Peck, C., and Jorcyk, C. (2005). Breast cancer cells stimulate neutrophils to produce oncostatin M: potential implications for tumor progression. *Cancer Res.* 65: 8896-8904. doi: 10.1158/0008-5472.CAN-05-1734

Quigley, J. P., and Deryugina, E.I. (2012). Combating angiogenesis early: potential of targeting tumor-recruited neutrophils in cancer therapy. *Future Oncol.* 8: 5-8. doi: 10.2217/fon.11.133.

Raccosta, L., Fontana, R., Maggioni, D., Lanterna, C., Villablanca, E. J., Paniccia, A., et al., (2013). The oxysterol–CXCR2 axis plays a key role in the recruitment of tumor-promoting neutrophils. *J. Exp. Med.* 210: 1711-1728. doi: 10.1084/jem.20130440.

Raychaudhuri, B., Rayman, P., Ireland, J., Ko, J., Rini, B., Borden, E. C., et al., (2011). Myeloid-derived suppressor cell accumulation and function in patients with newly diagnosed glioblastoma. *Neuro Oncol.* 13: 591-599. doi: 10.1093/neuonc/nor042.

Reid, M., Basturk, O., Thirabanjasak, D., Hruban, R., Klimstra, D., Bagci, P., et al., (2011). Tumor-infiltrating neutrophils in pancreatic neoplasia. *Mod. Pathol.* 24: 1612-1619. doi: 10.1038/modpathol.2011.113.

Ring, N. G., Herndler-Brandstetter, D., Weiskopf, K., Shan, L., Volkmer, J. P., George, B. M., et al., (2017). Anti-SIRPα antibody immunotherapy enhances neutrophil and macrophage antitumor activity. *Proc. Natl. Acad. Sci. USA.* 114: E10578-E10585. doi: 10.1073/pnas.1710877114.

Rosales, C. (2007). Fc receptor and integrin signaling in phagocytes. *Signal Transduction* 7: 386-401. doi: 10.1002/sita.200700141.

Rosales, C. (2017). Fcγ receptor heterogeneity in leukocyte functional responses. *Front. Immunol.* 8: 280. doi: 10.3389/fimmu.2017.00280.

Rosales, C. (2018). Neutrophil: A cell with many roles in inflammation or several cell types? *Front. Physiol.* 9: 113. doi: 10.3389/fphys.2018.00113.

Rosales, C. (2020). Neutrophils at the crossroads of innate and adaptive immunity. *J. Leukoc. Biol.* 108: 377-396. doi: 10.1002/JLB.4MIR0220-574RR.

Rosales, C., and Uribe-Querol, E. (2013a). Antibody - Fc receptor interactions in antimicrobial functions. *Curr. Immunol. Rev.* 9: 44-55. doi: 10.2174/1573395511309010006.

Rosales, C., and Uribe-Querol, E. (2013b). Fc receptors: cell activators of antibody functions. *Adv. Biosci. Biotech.* 4: 21-33. doi:10.4236/abb.2013.44A004.

Rosales, C., and Uribe-Querol, E. (2017). Phagocytosis: A fundamental process in immunity. *Biomed. Res. Int.* 2017: 9042851. doi: 10.1155/2017/9042851.

Roumenina, L. T., Daugan, M. V., Petitprez, F., Sautès-Fridman, C., and Fridman, W. H. (2019). Context-dependent roles of complement in cancer. *Nat. Rev. Cancer* 19: 698-715. doi: 10.1038/s41568-019-0210-0.

Sagiv, J. Y., Michaeli, J., Assi, S., Mishalian, I., Kisos, H., Levy, L., et al., (2015). Phenotypic diversity and plasticity in circulating neutrophil subpopulations in cancer. *Cell Rep.* 10: 562-573. doi: 10.1016/j.celrep.2014.12.039.

Sato, T., Takahashi, S., Mizumoto, T., Harao, M., Akizuki, M., Takasugi, M., et al., (2006). Neutrophil elastase and cancer. *Surg. Oncol.* 15: 217-222. doi:10.1016/j.suronc.2007.01.003.

Scapini, P., and Cassatella, M. A. (2017). Location in the spleen dictates the function of murine neutrophils. *J. Exp. Med.* 214: 1207-1209. doi: 10.1084/jem.20170655.

Sceneay, J., Chow, M. T., Chen, A., Halse, H. M., Wong, C. S., Andrews, D. M., et al., (2012). Primary tumor hypoxia recruits CD11b+/Ly6Cmed/ Ly6G+immune suppressor cells and compromises NK cell cytotoxicity in the premetastatic niche. *Cancer Res.* 72: 3906-3911. doi: 10.1158/0008-5472.CAN-11-3873.

Schmidt, H., Bastholt, L., Geertsen, P., Christensen, I. J., Larsen, S., Gehl, J., et al., (2005). Elevated neutrophil and monocyte counts in peripheral blood are associated with poor survival in patients with metastatic melanoma: a prognostic model. *Br. J. Cancer.* 93: 273-278.

Schmielau, J., and Finn, O. J. (2001). Activated granulocytes and granulocyte-derived hydrogen peroxide are the underlying mechanism of suppression of T-cell function in advanced cancer patients. *Cancer Res.* 61: 4756-4760.

Schneider-Merck, T., Lammerts van Bueren, J. J., Berger, S., Rossen, K., van Berkel, P. H., Derer, S., et al., (2010). Human IgG2 antibodies against epidermal growth factor receptor effectively trigger antibody-dependent cellular cytotoxicity but, in contrast to IgG1, only by cells of myeloid lineage. *J. Immunol.* 184: 512-520. doi: 10.4049/jimmunol.0900847.

Schott, A. F., Goldstein, L. J., Cristofanilli, M., Ruffini, P. A., McCanna, S., Reuben, J. M., et al., (2017). Phase Ib pilot study to evaluate reparixin in combination with weekly paclitaxel in patients with HER-2-negative metastatic breast cancer. *Clin. Cancer Res.* 23: 5358-5365. doi: 10.1158/1078-0432.CCR-16-2748.

Shaul, M. E., and Fridlender, Z. G. (2019). Tumour-associated neutrophils in patients with cancer. *Nat. Rev. Clin. Oncol.* 16: 601-620. doi: 10.1038/s41571-019-0222-4.

Siders, W. M., Shields, J., Garron, C., Hu, Y., Boutin, P., Shankara, S., et al., (2010). Involvement of neutrophils and natural killer cells in the anti-tumor activity of alemtuzumab in xenograft tumor models. *Leuk. Lymphoma* 51: 1293-1304. doi: 10.3109/10428191003777963.

Sikic, B. I., Lakhani, N., Patnaik, A., Shah, S. A., Chandana, S. R., Rasco, D., et al., (2019). First-in-human, first-in-class phase I trial of the anti-CD47 antibody Hu5F9-G4 in patients with advanced cancers. *J. Clin. Oncol.* 37: 946-953. doi: 10.1200/JCO.18.02018.

Sim, J., Sockolosky, J. T., Sangalang, E., Izquierdo, S., Pedersen, D., Harriman, W., et al., (2019). Discovery of high affinity, pan-allelic, and pan-mammalian reactive antibodies against the myeloid checkpoint receptor SIRPα. *MAbs* 11: 1036-1052. doi: 10.1080/19420862.2019.1624123.

Singel, K. L., Emmons, T. R., Khan, A. N. H., Mayor, P. C., Shen, S., Wong, J. T., et al., (2019). Mature neutrophils suppress T cell immunity in ovarian cancer microenvironment. *JCI Insight* 4: e122311. doi: 10.1172/jci.insight.122311.

Sionov, R. V., Fainsod-Levi, T., Zelter, T., Polyansky, L., Pham, C. T., and Granot, Z. (2019). Neutrophil cathepsin G and tumor cell RAGE facilitate neutrophil anti-tumor cytotoxicity. *Oncoimmunology* 8: e1624129. doi: 10.1080/2162402X.2019.1624129.

Sionov, R. V., Fridlender, Z. G., and Granot, Z. (2015). The multifaceted roles neutrophils play in the tumor microenvironment. *Cancer Microenviron.* 8: 125-158. doi: 10.1007/s12307-014-0147-5.

Sparmann, A., and Bar-Sagi, D. (2004). Ras-induced interleukin-8 expression plays a critical role in tumor growth and angiogenesis. *Cancer Cell* 6: 447-458. doi: 10.1016/j.ccr.2004.09.028

Spicer, J., McDonald, B., Cools-Lartigue, J., Chow, S., Giannias, B., Kubes, P., et al., (2012). Neutrophils promote liver metastasis via Mac-1-mediated interactions with circulating tumor cells. *Cancer Res.* 72: 3919-3927. doi: 10.1158/0008-5472.CAN-11-2393.

Spiegel, A., Brooks, M.W., Houshyar, S., Reinhardt, F., Ardolino, M., Fessler, E., et al., (2016). Neutrophils suppress intraluminal NK cell-mediated tumor cell clearance and enhance extravasation of disseminated carcinoma cells. *Cancer Discov.* 6: 630-649. doi: 10.1158/2159-8290.CD-15-1157.

Steele, C. W., Karim, S. A., Leach, J. D. G., Bailey, P., Upstill-Goddard, R., Rishi, L., et al., (2016). CXCR2 inhibition profoundly suppresses metastases and augments immunotherapy in pancreatic ductal adenocarcinoma. *Cancer Cell* 29: 832-845. doi: 10.1016/j.ccell.2016.04.014.

Su, Y., Vickers, A., Zelefsky, M., Kraus, D., Shaha, A., Shah, J., et al., (2006). Double-blind, placebo-controlled, randomized trial of granulocyte-colony stimulating factor during postoperative radiotherapy for squamous head and neck cancer. *Cancer J.* 12: 182-188.

Suda, K., Kitagawa, Y., Ozawa, S., Miyasho, T., Okamoto, M., Saikawa, Y., et al., (2007). Neutrophil elastase inhibitor improves postoperative clinical courses after thoracic esophagectomy. *Dis. Esophagus* 20: 478-486. doi: 10.1111/j.1442-2050.2007.00699.x.

Sun, B., Qin, W., Song, M., Liu, L., Yu, Y., Qi, X., et al., (2018). Neutrophil suppresses tumor cell proliferation via Fas /Fas ligand pathway mediated cell cycle arrested. *Int. J. Biol. Sci.* 14: 2103-2113. doi: 10.7150/ijbs.29297.

Suttmann, H., Riemensberger, J., Bentien, G., Schmaltz, D., Stöckle, M., Jocham, D., et al., (2006). Neutrophil granulocytes are required for effective Bacillus Calmette-Guérin immunotherapy of bladder cancer and orchestrate local immune responses. *Cancer Res.* 66: 8250-8257. doi: 10.1158/0008-5472.CAN-06-1416.

Syed Mortadza, S. A., Wang, L., Li, D., and Jiang, L. H. (2015). TRPM2 channel–mediated ROS-sensitive Ca^{2+} signaling mechanisms in immune cells. *Front. Immunol.* 6: 407. doi: 10.3389/fimmu.2015.00407.

Szuster-Ciesielska, A., Hryciuk-Umer, E., Stepulak, A., Kupisz, K., and Kandefer-Szerszeń, M. (2004). Reactive oxygen species production by blood neutrophils of patients with laryngeal carcinoma and antioxidative enzyme activity in their blood. *Acta Oncol.* 43: 252-258. doi: 10.1080/02841860410029708.

Tecchio, C., Scapini, P., Pizzolo, G., and Cassatella, M. A. (2013). On the cytokines produced by human neutrophils in tumors. *Semin. Cancer Biol.* 23: 159-170. doi: 10.1016/j.semcancer.2013.02.004.

Templeton, A. J., McNamara, M. G., Šeruga, B., Vera-Badillo, F. E., Aneja, P., Ocaña, A., et al., (2014). Prognostic role of neutrophil-to-lymphocyte ratio in solid tumors: A systematic review and meta-analysis. *J. Natl. Cancer Inst.* 106: dju124. doi: 10.1093/jnci/dju124.

Treffers, L. W., Ten Broeke, T., Rösner, T., Jansen, J. H. M., van Houdt, M., Kahle, S., et al., (2020). IgA-mediated killing of tumor cells by neutrophils is enhanced by CD47-SIRPα checkpoint inhibition. *Cancer Immunol. Res.* 8: 120-130. doi: 10.1158/2326-6066.CIR-19-0144.

Trellakis, S., Bruderek, K., Dumitru, C., Gholaman, H., Gu, X., Bankfalvi, A., et al., (2011). Polymorphonuclear granulocytes in human head and neck cancer: enhanced inflammatory activity, modulation by cancer cells and expansion in advanced disease. *Int. J. Cancer* 129: 2183-2193. doi: 10.1002/ijc.25892.

Uribe-Querol, E., and Rosales, C. (2015). Neutrophils in cancer: Two sides of the same coin. *J. Immunol. Res.* 2015: 983698. doi: 10.1155/2015/983698.

Uribe-Querol, E., and Rosales, C. (2021a). "Phagocytosis," in *Encyclopedia of Infection and Immunity*. (Amsterdam: Elsevier Inc.), 1-11. doi: 10.1016/B978-0-12-818731-9.00049-5.

Uribe-Querol, E., and Rosales, C. (2021b): The multiple roles of trogocytosis in immunity, the nervous system and development. *BioMed Res. Inter.* 2021: 1601565. doi: 10.1155/2021/1601565.

Ustyanovska Avtenyuk, N., Visser, N., Bremer, E., and Wiersma, V. R. (2020). The neutrophil: The underdog that packs a punch in the fight against cancer. *Int. J. Mol. Sci.* 21: 7820. doi: 10.3390/ijms21217820.

Valerius, T., Repp, R., de Wit, T. P., Berthold, S., Platzer, E., Kalden, J. R., et al., (1993). Involvement of the high-affinity receptor for IgG (Fc gamma RI; CD64) in enhanced tumor cell cytotoxicity of neutrophils during granulocyte colony-stimulating factor therapy. *Blood* 82: 931-939.

Valgardsdottir, R., Cattaneo, I., Klein, C., Introna, M., Figliuzzi, M., and Golay, J. (2017). Human neutrophils mediate trogocytosis rather than phagocytosis of CLL B cells opsonized with anti-CD20 antibodies. *Blood* 129: 2636-2644. doi: 10.1182/blood-2016-08-735605.

Van Coillie, E., Van Aelst, I., Wuyts, A., Vercauteren, R., Devos, R., De Wolf-Peeters, C., et al., (2001). Tumor angiogenesis induced by granulocyte chemotactic protein-2 as a countercurrent principle. *Am. J. Pathol.* 159: 1405-1414.

van Egmond, M., and Bakema, J. E. (2013). Neutrophils as effector cells for antibody-based immunotherapy of cancer. *Semin. Cancer Biol.* 23: 190-199. doi: 10.1016/j.semcancer.2012.12.002.

Vano, Y. A., Oudard, S., By, M. A., Têtu, P., Thibault, C., Aboudagga, H., et al., (2018). Optimal cut-off for neutrophil-to-lymphocyte ratio: Fact or fantasy? A prospective cohort study in metastatic cancer patients. *PLoS One* 13: e0195042. doi: 10.1371/journal.pone.0195042.

Veglia, F., Sanseviero, E., and Gabrilovich, D. I. (2021). Myeloid-derived suppressor cells in the era of increasing myeloid cell diversity. *Nat. Rev. Immunol.* 21: 485–498. doi: 10.1038/s41577-020-00490-y.

Vidarsson, G., Dekkers, G., and Rispens, T. (2014). IgG subclasses and allotypes: from structure to effector functions. *Front. Immunol.* 5: 520. doi: 10.3389/fimmu.2014.00520.

Voets, E., Paradé, M., Lutje Hulsik, D., Spijkers, S., Janssen, W., Rens, J., et al., (2019). Functional characterization of the selective pan-allele anti-SIRPα antibody ADU-1805 that blocks the SIRPα-CD47 innate immune checkpoint. *J. Immunother. Cancer* 7: 340. doi: 10.1186/s40425-019-0772-0.

Wada, Y., Yoshida, K., Hihara, J., Konishi, K., Tanabe, K., Ukon, K., et al., (2006). Sivelestat, a specific neutrophil elastase inhibitor, suppresses the growth of gastric carcinoma cells by preventing the release of transforming growth factor-alpha. *Cancer Sci.* 97: 1037-1043.

Wada, Y., Yoshida, K., Tsutani, Y., Shigematsu, H., Oeda, M., Sanada, Y., et al., (2007). Neutrophil elastase induces cell proliferation and migration by the release of TGF-alpha, PDGF and VEGF in esophageal cell lines. *Oncol. Rep.* 17: 161-167. doi: 10.3892/or.17.1.161.

Wagner, C., Iking-Konert, C., Denefleh, B., Stegmaier, S., Hug, F., and Hänsch, G. M. (2004). Granzyme B and perforin: constitutive expression in human polymorphonuclear neutrophils. *Blood* 103: 1099-1104. doi: 10.1182/blood-2003-04-1069.

Waight, J. D., Hu, Q., Miller, A., Liu, S., and Abrams, S. I. (2011). Tumor-derived G-CSF facilitates neoplastic growth through a granulocytic myeloid-derived suppressor cell-dependent mechanism. *PLoS One* 6: e27690. doi: 10.1371/journal.pone.0027690.

Walsh, S., Cook, E., Goulder, F., Justin, T., and Keeling, N. (2005). Neutrophil-lymphocyte ratio as a prognostic factor in colorectal cancer. *J. Surg. Oncol.* 91: 181-184. doi: 10.1002/jso.20329.

Walters, I., Austin, C., Austin, R., Bonnert, R., Cage, P., Christie, M., et al., (2008). Evaluation of a series of bicyclic CXCR2 antagonists. *Bioorg. Med. Chem. Lett.* 18: 798-803.

Wang, W., Erbe, A. K., Hank, J. A., Morris, Z. S., and Sondel, P. M. (2015). NK cell-mediated antibody-dependent cellular cytotoxicity in cancer immunotherapy. *Front. Immunol.* 6: 368. doi: 10.3389/fimmu.2015.00368.

Wilson, T. J., Nannuru, K. C., Futakuchi, M., Sadanandam, A., and Singh, R. K. (2008). Cathepsin G enhances mammary tumor-induced osteolysis by generating soluble receptor activator of nuclear factor-kappaB ligand. *Cancer Res.* 68: 5803-5811. doi: 10.1158/0008-5472.CAN-07-5889.

Wilson, T. J., Nannuru, K. C., Futakuchi, M., and Singh, R. K. (2010). Cathepsin G-mediated enhanced TGF-beta signaling promotes angiogenesis via upregulation of VEGF and MCP-1. *Cancer Lett.* 288: 162-169. doi: 10.1016/j.canlet.2009.06.035.

Wilson, T. J., Nannuru, K. C., and Singh, R. K. (2009). Cathepsin G-mediated activation of pro-matrix metalloproteinase 9 at the tumor-bone interface promotes transforming growth factor-beta signaling and bone destruction. *Mol. Cancer Res.* 7: 1224-1233. doi: 10.1158/1541-7786.MCR-09-0028.

Winterbourn, C. C., Kettle, A. J., and Hampton, M. B. (2016). Reactive oxygen species and neutrophil function. *Annu. Rev. Biochem.* 85: 765-792. doi: 10.1146/annurev-biochem-060815-014442.

Wirths, S., Bugl, S., and Kopp, H. G. (2014). Neutrophil homeostasis and its regulation by danger signaling. *Blood.* 123: 3563-3566. doi: 10.1182/blood-2013-11-516260.

Wislez, M., Rabbe, N., Marchal, J., Milleron, B., Crestani, B., Mayaud, C., et al., (2003). Hepatocyte growth factor production by neutrophils infiltrating bronchioloalveolar subtype pulmonary adenocarcinoma: role in tumor progression and death. *Cancer Res.* 63: 1405-1412.

Wu, P., Wu, D., Ni, C., Ye, J., Chen, W., Hu, G., et al., (2014). γδT17 cells promote the accumulation and expansion of myeloid-derived suppressor cells in human colorectal cancer. *Immunity* 40: 785-800. doi: 10.1016/j.immuni.2014.03.013.

Wu, Y., Zhao, Q., Peng, C., Sun, L., Li, X., and Kuang, D. (2011). Neutrophils promote motility of cancer cells via a hyaluronan-mediated TLR4/PI3K activation loop. *J. Pathol.* 225: 438-447. doi: 10.1002/path.2947.

Yan, J., Kloecker, G., Fleming, C., Bousamra, M., 2nd, Hansen, R., Hu, X., et al., (2014). Human polymorphonuclear neutrophils specifically recognize and kill cancerous cells. *Oncoimmunology* 3: e950163. doi: 10.4161/15384101.2014.950163.

Yang, F., Feng, C., Zhang, X., Lu, J., and Zhao, Y. (2017). The diverse biological functions of neutrophils, beyond the defense against Infections. *Inflammation* 40: 311-323. doi: 10.1007/s10753-016-0458-4.

Yazdani, H. O., Roy, E., Comerci, A. J., van der Windt, D. J., Zhang, H., Huang, H., et al., (2019). Neutrophil extracellular traps drive mitochondrial homeostasis in tumors to augment growth. *Cancer Res.* 79: 5626-5639. doi: 10.1158/0008-5472.CAN-19-0800.

Yui, S., Osawa, Y., Ichisugi, T., and Morimoto-Kamata, R. (2014). Neutrophil cathepsin G, but not elastase, induces aggregation of MCF-7 mammary carcinoma cells by a protease activity-dependent cell-oriented mechanism. *Mediators Inflamm.* 2014: 971409. doi: 10.1155/2014/971409.

Yvan-Charvet, L., and Ng, L. G. (2019). Granulopoiesis and neutrophil homeostasis: A metabolic, daily balancing act. *Trends Immunol.* 40: 598-612. doi: 10.1016/j.it.2019.05.004.

Zhang, W., Huang, Q., Xiao, W., Zhao, Y., Pi, J., Xu, H., et al., (2020). Advances in anti-tumor treatments targeting the CD47/SIRPα axis. *Front. Immunol.* 11: 18. doi: 10.3389/fimmu.2020.00018.

Zhao, X. W., van Beek, E. M., Schornagel, K., Van der Maaden, H., Van Houdt, M., Otten, M. A., et al., (2011). CD47-signal regulatory protein-α (SIRPα) interactions form a barrier for antibody-mediated tumor cell destruction. *Proc. Natl. Acad. Sci. USA*. 108: 18342-18347. doi: 10.1073/pnas.1106550108.

Zhou, G., Yu, L., Fang, L., Yang, W., Yu, T., Miao, Y., et al., (2018). $CD177^+$ neutrophils as functionally activated neutrophils negatively regulate IBD. *Gut* 67: 1052-1063. doi: 10.1136/gutjnl-2016-313535.

Zhu, Y. P., Padgett, L., Dinh, H. Q., Marcovecchio, P., Blatchley, A., Wu, R., et al., (2018). Identification of an early unipotent neutrophil progenitor with pro-tumoral activity in mouse and human bone marrow. *Cell Rep.* 24: 2329-2341. doi: 10.1016/j.celrep.2018.07.097.

Zindl, C. L., Lai, J. F., Lee, Y. K., Maynard, C. L., Harbour, S. N., Ouyang, W., et al., (2013). IL-22-producing neutrophils contribute to antimicrobial

defense and restitution of colonic epithelial integrity during colitis. *Proc. Natl. Acad. Sci. USA*. 110: 12768-12773. doi: 10.1073/pnas.1300318110.
Zivkovic, M., Poljak-Blazi, M., Egger, G., Sunjic, S., Schaur, R., and Zarkovic, N. (2005). Oxidative burst and anticancer activities of rat neutrophils. *Biofactors* 24: 305-312.
Zivkovic, M., Poljak-Blazi, M., Zarkovic, K., Mihaljevic, D., Schaur, R. J., and Zarkovic, N. (2007). Oxidative burst of neutrophils against melanoma B16-F10. *Cancer Lett.* 246: 100-108. doi: 10.1016/j.canlet.2006.02.002.

Section 3. Innate Immunity and Parasites

Chapter 7

Innate Immunity Against *Toxoplasma gondii* Infection

Carmen T. Gómez de León and Jorge Morales-Montor*

Departamento de Inmunología, Instituto de Investigaciones Biomédicas,
Universidad Nacional Autónoma de México,
Ciudad de México, México

Abstract

Toxoplasma gondii is a protozoan parasite that belongs to Apicomplexa phylum, it can infect almost all warm-blooded animals worldwide and one-third of human population is infected with this parasite. In immunocompromised individuals, such as pregnant women, AIDS patients, chemotherapy recipients and immunosuppressed tissue transplant patients, infection and reactivation can lead to miscarriage, malformations in the newborn, blindness and lethal encephalitis. Since 50's decade a mixture of sulfadiazine-pyrimethamine has been used as conventional therapy against toxoplasmosis. Although sulfadiazine – pyrimethamine are synergic it is known that present severe side effects and has a limited effectiveness, mainly on chronic stage disease, and there is not an available vaccine for human use. As an obligate intracelular opportunistic pathogen, *T. gondii* celular invassion is essential for its survival, which occurs almost immediately after the degradation of ingested cysts. For this reason the study of innate immune

* Corresponding Author's E-mail: jmontor66@biomedicas.unam.mx.

In: The Innate Immune System in Health and Disease
Editor: Jorge Morales-Montor
ISBN: 978-1-68507-507-1

response has been difficult, especially in humans. In this chapter we analyzed the innate immune response produced by *T. gondii*, focusing in the main pathway infection, ingestion of contaminated food with sporulated cysts.

Keywords: *Toxoplasma gondii*, toxoplasmosis, protozoan, apicomplexa, inflammation, tachizoyte

Introduction

Toxoplasma seroprevalence varys from 10 – 15% in USA to 60 – 80% in Central and South America, Europe, and some parts of the Mediterranian and Assia (Pappas Georgios 2009; Hernandez-Cortazar Ivonne B 2017). It is an obligate intracellular protozoan able to infect almost all warm-blooded animals. *T. gondii* infection is opportunistic in individuals with normal immunity. Conversely, in immunocompromised individuals, infection and reactivation can lead to chronic infection, potentially resulting in miscarriage, malformations in the newborn, blindness and lethal encephalitis. Moreover, toxoplasmosis it is between the first causes of death in AIDS patients (Howard Jonathan C 2011; Hernandez-Cortazar Ivonne B 2017). Specific treatmet or human use vaccines are not available and the conventional therapy consists of a mixture of sulfadiazine-pyrimethamine. This treatment present severe side effects and has a limited effectiveness (Wei Hai-Xia 2015). As an obligate intracelular pathogen, celular invassion is essential for *Toxoplasma* survival, which occurs almost immediately after the degradation of ingested cysts. For this reason the study of innate immune response has been difficult, especially in humans.

Life Cycle

Toxoplasma life cycle is complex, since different parasite stages are present between the different hosts and sexual or asexual reproduction also depends on them; this leads to several infective forms and, in consecuence, different transmission pathways. Terrestrial and aquatic mammals and birds are considered intermediate hosts because only asexual stages occur in them. The sexual stages are exclusively seen in the Felidae family members, which includes the domestic cat (Dubey 2009; Attias Márcia 2020). Therefore, they are considered to be the definitive hosts.

During the life-cycle of *Toxoplasma*, three developmental stages can infect cells: 1) tachyzoite, the rapid multiplication form, present in acute infections; 2) bradyzoite, the slow multiplication form present in chronic infection and that originates the tissue cysts; and 3) sporozoite, which is produced only in the definitive host during the sexual reproduction and released in the oocysts via felid feces (Attias Márcia 2020).

Intermediate hosts, including humans, can be infected via different pathways: 1) ingestion of water, vegetables and fruits contaminated with viable oocysts, sporulated after previous elimination in the feces of cats; 2) intake of uncooked or undercooked meat containing viable tissue cysts; 3) congenital transmission from the mother through the placenta; 4) blood transfusion; 5) organ transplantation, where the organs may contain cysts or tachyzoites (Attias Márcia 2020). Despite the different infection pathways, the ingestion of contaminated food is the most common form to acquire Toxoplasmosis, for this reason we are going to focus in the innate immune response that this pathway generates.

Immune Response at the Infection Site

When *Toxoplasma* cysts and oocytes are ingested, the intestinal mucosa, (where a group the innate lymphoid cells (ILCs) resides), is the first barrier that parasites cope with. ILCs have important functions in protecting epithelial barriers against infections; these cells has been clasified in three groups based on their functional characteristics (Elemam Noha Mousaad 2017). Group 1 (ILC1s) consist of ILC1 and conventional natural killer (NK) cells. Activation of ILC1s leads to the production of Th1 cytokines, such as IFN-γ and TNF-α. Group 2 (ILC2s) produce IL-4, IL-5, IL-9 and IL-13, which are Th2 cytokines. Group 3 (ILC3s) produce IL-17A and IL-22.

ILC1s are found in both, mouse and human intestine where they participate in type 1 inflammation (Klose Christoph S N 2014; K. E. Klose Christoph S N 2013) and has been reported that contribute with the immune responses against intracellular pathogens, such as *Toxoplasma gondii* and *Clostridium difficile* (Klose Christoph S N 2014; Abt Michael C 2015; Elemam Noha Mousaad 2017). A strong IL-12 induced type 1 immunity, IFN-γ dependent, is required for succesfully control of *Toxoplasma* replication; this response involved in part the NKp46+ cells (Denkers EY 1993; Gazzinelli R

T 1993; Suzuki Y 1988), which participate as well in the recruitment of inflammatory monocytes (Schulthess Julie 2012) that are also required for the infection control (Dunay Ildiko R 2008). A more exhaustive analysis of immune cell response in mice that were orally infected with *Toxoplasma*, showed that ILC1s were the main producers of IFN-γ and TNF in response to the infection, while cNK cells and NKp46+ NK1.1+ ILC3s only contributed to a smaller extent (Klose Christoph S N 2014). IFN-γ derived from ILC1 resulted to be essential for the mainteinance of inflammatory IRF8+ cDC1s, which leads in parasite clearance (López-Yglesias Américo H 2021). Nevrtheless, ILC1s failed to recruit inflammatory monocytes (Klose Christoph S N 2014) (Figure 1).

The classical members of group 3 ILCs are the lymphoid tissue inducer (LTi) cells, which are crucial for the formation of secondary lymphoid organs (Bar-Ephraïm Yotam E 2016). Several studies pointed out the presence of lymphoid cells present in the intestine that express NKp46 but do not resemble NK cells, the NKp46+ cells possess the transcription factor RORγt, lack cytotoxic effectors and produce the IL-22 but no IFN-γ or TNF (Sonnenberg Gregory F 2011; Buonocore Sofia 2010). They are distinct from LTi cells that secrete IL-22 and, during bacterial infection in the small intestine lamina propria, help to protect the mucosal surface by regulating inflammation (Malmberg Karl-Johan 2009) (Figure 1).

In humans, almost all ILC3s express CCR6 and CD117, and could be distinguished on the basis of NKp44 expression. Adult human ILC3s can be subdivided into NKp44+ and NKp44− subsets. NKp44+ ILC3s are the most prevalent cells in the human gut under homeostatic conditions. These cells produce IL-22 and consequently, maintain gut barrier function (Satoh-Takayama Naoko 2008). The number of ILC3s in the lamina propria of aryl hydrocarbon receptor (Ahr)-deficient mice is decreased during *Toxoplasma* infection, resulting in overactivation of CD4 T cells. Thus, ILC3s and ILC1s are indirectly involved in the host defense against *Toxoplasma*. Notably, the mechanism by which *Toxoplasma* infection activates ILCs remains unknown (Sasai Miwa, 2019). Immune repsonse against *Toxoplasma* in humans also involves IL-12 production, which mainly depend of neutrophils and monocytes (Bliss SK 1999; Aldebert D 2007). Unlike mouse macrophages (Robben PM 2004), phagocytosis of the parasite by human monocytes triggers the IL-12 response (Tosh KW 2016) leading in the production of IFN-γ that initiates protective type 1 immunity (Gazzinelli RT 1994; Däubener W 1995) (Figure 1).

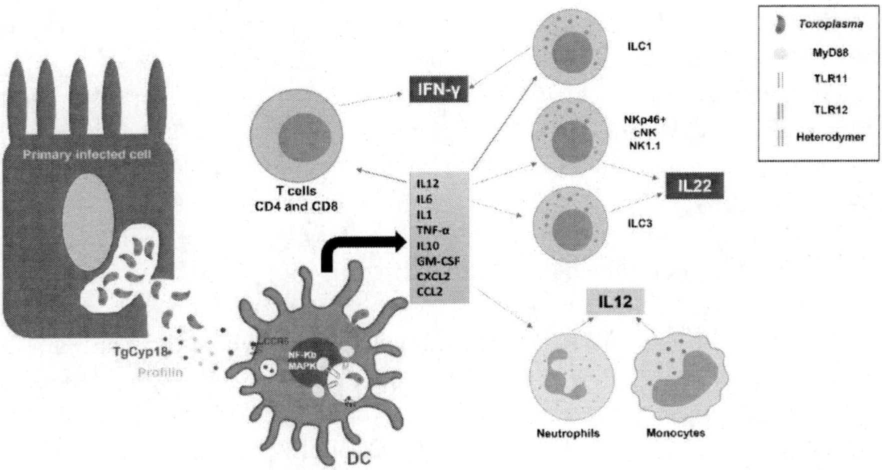

Figure 1. Primary response against *Toxoplasma*. When *Toxoplasma* reaches the gut, DCs can recognize it by CCR5 and TLR11/12, triggering a strong Th1 inflammatory response dependent of IL12 and IFN-γ.

For parasite survival, *Toxoplasma* actively infects host cells residing in a special membranous structure called parasitophorous vacuole (PVs), where *Toxoplasma* is able to proliferate. Host immune cells not only recognize *Toxoplasma*-derived ligands to activate acquired immunity but also destroy the PVs to remove *Toxoplasma* proliferation and block parasite expansion *in vivo*. However, *Toxoplasma* secretes various virulence factors into host cells to inhibit host acquired and innate immune responses (Sasai Miwa, Innate, adaptive, and cell-autonomous immunity against Toxoplasma gondii infection 2019).

Toxoplasma Recognition by Innate Sensors

Innate immunity is the first line of host defense that responds immediately and detects pathogen infection via pattern recognition receptors (PRRs), such as Toll-like receptors (TLRs), Nod-like receptors, and C-type lectins. Ligand recognition by PRRs induces proinflammatory cytokines, including TNF-α, interleukin-6 (IL-6) and IL-12, and plays a role in the subsequent activation of immune responses (Sasai Miwa 2013; Quinn Susan R 2011; Beutler Bruce 2006). TLRs are a group of 10 and 13 members in humans and mice, respectively, that differentially recognize pathogen-derived components and are involved in early pathogen detection.

In mice, macrophages are able to recognize *Toxoplasma* parasites by the interaction of TLR2 and TLR4 with the glycosylphosphatidylinositol (GPI) anchors parasite protein; TLR7 and TLR9 detect the parasite's nucleic acids (Debierre-Grockiego F 2007); whereas TLR11 and TLR12 from dendritic cells (DCs) has been shown to recognize *Toxoplasma* profilin, a typical pathogen associated molecular pattern (PAMP) that is essential for parasite invasion and that is common for other Apicomplexa pathogen members. TLR12 homodimer or TLR11-TLR12 heterodimer complex result in the induction of IL-12, IFN-γ, and CCL2, among others, via MyD88-dependent signaling pathway (NF-κB, and MAPK) (Koblansky A Alicia 2013) (Figure 1). Evenmore, TLR11$^{-/-}$ and TLR12$^{-/-}$ mice are highly sensitive to *Toxoplasma* infection and exhibit a defitient production of IL-12 and IFN-γ. Recent studies showed that when infected mice were treated with Imiquimod (an immune response modulated drug) TLR7, TLR11, and TLR12, were upregulated following by the interconversion from bradyzoites to tachyzoites. Consequently, MyD88 pathway was activated (Hamie M 2021). Interestingly TLR2 and TLR4 N-glycans from both, DCs and macrophages interact with *Toxopasma* surface adhesins microneme protein (MIC) 1 and 4 triggering IL-12 secretion (Mendonça-Natividade FC 2019; Sardinha-Silva A 2019).

In addition, *Toxoplasma*-derived cyclophilin-18 (TgCyp18) is detected by CCR5 (a chemokine receptor mainly expressed on DCs, macrophages, T cells and microglia) on DCs (Aliberti J 2011), which induces IL-12 production through activation of the G protein α I family and MAPK (Sher Alan 2003; Denkers Eric Y 2004). CCR5-dependent inflammation may be involved in fetal lethality and miscarriage when pregnant women undergo primary infection with *Toxoplasma*, since CCR5-deficient mice showed reduced embryonic lethality during pregnancy in comparison with that of wild-type mice (Nishimura Maki 2017). Although human CCR5 associates with TgCyp18, it is unclear whether CCR5 is critical for the production of IL-12 (Golding Hana 2003) (Figure 1).

Human monocytes produce IL-12 following *Toxoplasma* infection, suggesting that TLRs other than TLR11 and TLR12 may play a role in the detection of *Toxoplasma*-derived ligands. The ligand recognition mechanism of TLR11 and TLR12 in endosomes is similar to that of human TLR3, TLR7 and TLR9, which recognize nucleic acids, these TLRs, as well as TLR8, might be stimulated by *Toxoplasma*-derived nucleic acids in the endosome to induce IL-12 expression, which induces the production of IFN-γ that initiates protective type 1 immunity (Andrade Warrison A 2013). IFN-γ increases tryptophan degradation in human fibroblasts, inhibiting parasite replication (Pfefferkorn 1984). Evenmore, IFN-γ inducible proteins were found on the parasitophorous vacuole membrane (PVM). Interestingly, these immunity-related GTPases (IRGs) play an important role in cell-intrinsic antimicrobial defense in the mouse (Zhao Yang O 2009).

Mechanisms of Intracellular *Toxoplasma* Tachyzoites Clearance

IRGs are a group of 47–48 kDa proteins with a GTP binding domain; this group of proteins is classified into two subgroups: GKS-IRG and GMS-IRG (Boehm U 1998; Bekpen C 2005). In mice, GMS-IRG consists of Irgm1, Irgm2 and Irgm3 and they are mainly located in the endoplasmic reticulum and Golgi (Taylor GA 1997; Zerrahn J 2002; Zhao YO 2010). Except for GMS-IRG, all IRGs belong to GKS-IRGs, among which Irga6 and Irgb6 play major roles in the disruption of the parasitophorous vacuole membrane (PVM) (Bekpen C 2005; Hunn JP 2008; JD 2004). The effectiveness of Irga6 and Irgb6 depends of Irgm1 and Irgm3, since Irgm1 and Irgm3 deficient mice exhibited a reduced expression and misslocation of Irga6 and Irgb6, leading in highly susceptible to *Toxoplasma* infection mice (Henry SC 2009; Haldar AK 2013) (Figure 2).

A parallel IFN-γ-dependent mechanism of resistance in humans and mice consists of the guanylate binding proteins (GBPs), which are also recruited to the PVM and cause vacuolar membrane disruption and parasite clearance (Yamamoto Masahiro 2012; Degrandi Daniel 2013; Selleck Elizabeth M 2013). GBPs consist of 7 and 11 family members in humans and mice, respectively. In mice, GBP genes are aligned in two clusters located on chromosome 3 and 5 (Kim BH 2011; Selleck Elizabeth M 2013). GBPs in chromosome 3, such as GBP1, GBP2, GBP3, GBP5 and GBP7 are essential in *Toxoplamsa* host defense since mice that lacks of those GBPs exhibited a high susceptibility to the parasite with a defective disruption of the PVM (Yamamoto Masahiro 2012). In addition, macrophages lacking GBPchr3 showed defective Irgb6 loading on *Toxoplasma* PVM (Yamamoto Masahiro 2012). Although, GBP2 and GBP5 chimeric mutants showed that C-terminal isoprenylation of GBP2 regulated the recruitment of GBP2 to the PVM, the molecular mechanisms of how GTPases localize and recognize the PVM remain unknown. Besides isoprenylation alone cannot distinguish membranes of host organelles from PVs.

GBPs can participate in host defense without causing classical vacuolar membrane disruption since human GBP1 restricts replication of *Toxoplasma* in epithelial cells without targeting the parasitophorous vacuole (Johnston Ashleigh C 2016). Besides, ubiquitination on the PVM has also emerged as a key mechanism of parasite control that discriminates between host membrane and PVM. Ubiquitin recognition on intracellular bacteria triggers critical host effector mechanisms, for example in HeLa cells, ubiquitination of the PVM leads to non-canonical autophagy and parasite growth stunting (O. R. Selleck Elizabeth M 2015), and in umbilical vein endothelial cells cause endolysosomal fusion and parasite clearance (Clough Barbara 2016).

The types of ubiquitination on the *Toxoplasma* PVM have been identified as M1 linear, K48 and K63-linked polyubiquitin chains (Lee Y 2015; Haldar AK 2015; Clough B 2016). Participation of TRAF6 and TRIM21 in the regulation of PVM ubiquitination has been controversial, since *in vitro* assays showed that after IFN-γ treatment ubiquitination of PVM was regulated by TRAF6 and TRIM21, however TRIM21-deficient cells didn't failed in PVM ubiquitination nor disruption, which suggest that other E3 ligases may compensate TRAF6 (Foltz C 2017). The role of ubiquitin in the recruitment of GBPs and IRGs to the PVM has been also controversial and the biological significance of the recruitment of ubiquitin-binding effector p62/ Sqstm1 remains unknown since one study showed that p62 plays a role in IFN-γ-mediated *Toxoplasma* clearance (Haldar AK 2015), while another study demonstrated that p62 was not important in parasite clearance but was involved in presenting vacuolar antigens to CD8 T cells (Lee Y 2015).

Autophagy is another process that, as ubiquitination, can discriminate between host membranes and *Toxoplasma* PVM, is tightly controlled by autophagy-regulated (ATG) genes (Mizushima N 2011). In mammals Atg8 consists of the LC3 subfamily such as LC3A and LC3B; aditionally in humans LC3C and the GABARAP subfamily such as Gabarap, Gabarapl1 and Gate-16/Gabarapl2 are part of Atg8 (Slobodkin MR 2013). LC3B is a homologue of yeast Atg8, an ubiquitin-like protein required for the formation of autophagosomal membrane (Y 2001), that accumulates on autophagosomes to remove intracellular pathogens from cells. LC3-coated autophagosomes are able to fuse with lysosomes (Sumpter Jr R 2010) leading in xenophagy (Kudchodkar SB 2009). LC3 also accumulates on the *Toxoplasma* PVM when cells are activated via CD40, which induces the activation of phosphoinositide-3-class 3 (PIK3C3), Rab7 and vacuolar ATPase leading in the fussion of PVM with late endosomes or lysosomes (Andrade RM 2006). LC3 recruitment to the *Toxoplasma* PV in response to IFN-γ treatment is controversial (Sasai M 2017; Martens S 2005; Choi J 2014; Park S 2016) and even when autophagy is probably not functionally involved in anti-parasite immune responses mediated by IFN-γ, Atg proteins are able to regulate it. Cells lacking Atg5, an essential autophagy-related molecule, showed a defect in IFN-γ-mediated PV disruption (Choi J 2014). Evenmore, *Toxoplasma* PV and degradated parasites were not enveloped by the typically autophagy double membranes (Zhao Z 2008); besides, another Atg family members that belong to the Atg12 conjugation system, such as Atg3, Atg7 and Atg16L1, also participate in IFN-γ-mediated anti-*Toxoplasma* responses, but other Atg proteins that regulate the pathway do not. Suggesting that IFN-γ-mediated anti- *Toxoplasma* responses can be regulated by some Atg proteins in an autophagy-independent manner (O. R. Selleck Elizabeth M 2015; Haldar AK 2014; Ohshima J 2014) (Figure 2).

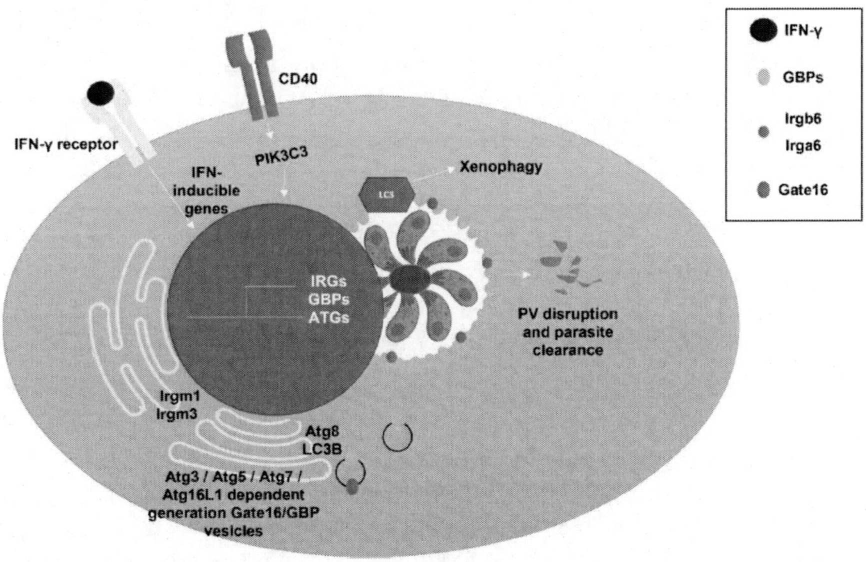

Figure 2. Mechanisms of intracellular *Toxoplasma* tachyzoites clearance. Disruption of PVM and parasite clearance may occur by IFN-γ regulated or independent manner.

Finally, a GABARAP subfamily member in the Atg8 family critically regulated the uniform cytosolic localization of GBPs, facilitating the efficient clearance of *Toxoplasma*. GBPs are located in cytoplasmic vesicular-like structures, this distribution doesn't seems to be affected when cells were stimulated with IFN-γ (F. S. Selleck Elizabeth M 2013). Among GABARAP subfamily members, Gate-16 demonstrated to be the most important for the distribution of GBPs, and its expression in LysM+ myeloid cells, mainly macrophages, neutrophils and monocytes, resulted essential for the survival of mice with acute toxoplasmosis (Sasai M 2017).

Toxoplasma Manipulation of Host Signaling Pathways

Parasites are able to modulate some host signaling pathways in order to survive and to avoid immune response. Although *Toxoplasma* resides within a vacuole in infected cells, it is capable to release effector proteins through its specialized secretory organelles, the rhoptries (ROP proteins) and the dense granules (GRA proteins), which manipulate host cell signaling and transcriptional responses.

Toxoplasma has three dominant clonal lineages that notably differ in their effects on host cells, mainly because they vary in the so called virulence factors, which includes several members of the ROP and GRA proteins. Type I and III, but not type II strains activate the signal transducer and activator of transcription 3 and 6 (STAT3 and STAT6) in human and mouse cells, thereby down-regulating IL-12 (Saeij J P J 2006). This effect has been related to the secretion of the rhoptry kinase ROP16 that phosphorylates and activates STAT3 and STAT6 (Saeij J P J 2007; Yamamoto M 2009; Ong Yi-Ching 2010). While the three types inhibit STAT1 transcriptional activity through a dense granule protein named *T. gondii* STAT1-dependent transcription (TgIST); which binds to activated STAT1 dimers and also to the chromatin-modifying Mi2/NuRD complex, preventing the dissociation of STAT1 from DNA, resulting in altered chromatin and blocking of IFN-γ-dependent transcription (Rosowski EE 2014; Gay G 2016; Olias P 2016).

Toxoplasma type I strain inhibites LPS-induced IL-1β production in primary human neutrophils, an effect that has been related to the inhibition of NF-κB signaling. In *Toxoplasma*-infected neutrophils, IκBα degradation and p65/RelA phosphorylation are reduced, as are transcripts for IL-1β and the inflammasome sensor NLRP3 (Lima TS 2018; Gov L 2013; Gov L 2017).

Recently has been reported that *Toxoplasma* is able to reprogram inflammatory responses interfereing with the β-catenin destruction complex. β-catenin is the main effector of the Wnt pathway, functioning as a coactivator of the T-cell factor/lymphoid enhancer factor (TCF/LEF) transcription factors

(Cadigan KM 2012). *Toxoplasma* GRA18 protein form complexes with host β-catenin, GSK3α/β, and the PP2A-B56 holoenzyme, promoting stabilization and nuclear translocation of β-catenin, which induces β-catenin-dependent gene expression associated with anti-inflammatory responses, such as CCL17 and CCL22 (He H 2018).

In phagocytes, such as neutrophils and macrophages, ROS production is an important antimicrobial response for the elimination of pathogens. However, ROS is not induced in *Toxoplasma* infected human macrophages (Wilson CB 1980) and non-phagocytic cells generate low levels of ROS (Bedard K 2007). In ARPE-19 cells, *Toxoplasma* targeted the main NADPH oxidase by reducing expression of Nox4 protein, which resulted in decreased intracellular ROS. The reduced Nox4 expression was associated with activation of PI3K/AKT signaling in infected cells (Zhou W 2013). Proliferation of type I *Toxoplasma* in murine inflammatory macrophages was also associated with decreased ROS production (Shrestha SP 2006). Mouse macrophages showed that clearance of type III, but not type I, *Toxoplasma* relies on NADPH activity, increased ROS production, and induction of GBP5, therefore virulent strains might block ROS production in order to parasite survival (Matta SK 2018).

Conclusion

As *Toxoplasma* is an intracelular parasite, the innate immune response must be quick and effective in order to clear the majority of ingested parasites that succesfully crossed the gut barrier and control the infection. ILC1's are the first cells that establishes a Th1 inlfammatory response, then ILC3's initiate a Th2 inflammatory response, leading to parasite clearance by phagocytosis or ubiquitination, therefore apoptosis. Nevertheless, parasites will develop from tachyzoite to bradyzoite inside of a cyst in a few hours, avoiding immune response; if host get immunocompromised, parasites will change into tachyzoites initiating the lytic cycle and diseminating.

References

Abt Michael C, Lewis Brittany B, Caballero Silvia, Xiong Huizhong, Carter Rebecca A, Sušac Bože, Ling Lilan, Leiner Ingrid, Pamer Eric G. 2015. "Innate Immune Defenses Mediated by Two ILC Subsets Are Critical for

Protection against Acute *Clostridium difficile* Infection." *Cell host & microbe* 18(1):27-37.

Aldebert D, Durand F, Mercier C, Brenier-Pinchart MP, Cesbron-Delauw MF, Pelloux H. 2007. "*Toxoplasma gondii* triggers secretion of interleukin-12 but low level of interleukin-10 from the THP-1 human monocytic cell line." *Cytokine* 37(3):206-11.

Aliberti J, Reis e Sousa C, Schito M, Hieny S, Wells T, Huffnagle G B, Sher A. 2011. "CCR5 provides a signal for microbial induced production of IL-12 by CD8 alpha+ dendritic cells." *Nature immunology* 1(1):83-7.

Andrade RM, Wessendarp M, Gubbels MJ, Striepen B, Subauste CS. 2006. "CD40 induces macrophage anti-*Toxoplasma gondii* activity by triggering autophagy-dependent fusion of pathogen-containing vacuoles and lysosomes." *The journal of clinical investigation* 116(9):2366-77.

Andrade Warrison A, Souza Maria do Carmo, Ramos-Martinez Espiridion, Nagpal Kamalpreet, Dutra Miriam S, Melo Mariane B, Bartholomeu Daniella C, Ghosh Sankar, Golenbock Douglas T, Gazzinelli Ricardo T. 2013. "Combined action of nucleic acid-sensing Toll-like receptors and TLR11/TLR12 heterodimers imparts resistance to *Toxoplasma gondii* in mice." *Cell host & microbe* 13(1):42-53.

Attias Márcia, Teixeira Dirceu E, Benchimol Marlene, Vommaro Rossiane C, Crepaldi Paulo Henrique, De Souza Wanderley. 2020. "The life-cycle of *Toxoplasma gondii* reviewed using animations." *Parasites and vectors* 13(1):588.

Bar-Ephraïm Yotam E, Mebius Reina E. 2016. "Innate lymphoid cells in secondary lymphoid organs." *Immunological reviews* 271(1):185-99.

Bedard K, Krause KH. 2007. "The NOX family of ROS-generating NADPH oxidases: physiology and pathophysiology." *Physiological reviews* 87(1):245-313.

Bekpen C, Hunn JP, Rohde C, Parvanova I, Guethlein L, Dunn DM, Glowalla E, Leptin M, Howard JC. 2005. "The interferon-inducible p47 (IRG) GTPases in vertebrates: loss of the cell autonomous resistance mechanism in the human lineage." *Genome biology* 6(11):R92.

Beutler Bruce, Jiang Zhengfan, Georgel Philippe, Crozat Karine, Croker Ben, Rutschmann Sophie, Du Xin, Hoebe Kasper. 2006. "Genetic analysis of host resistance: Toll-like receptor signaling and immunity at large." *Annual review of immunology* 24:353-89.

Bliss SK, Marshall AJ, Zhang Y, Denkers EY. 1999. "Human polymorphonuclear leukocytes produce IL-12, TNF-alpha, and the chemokines macrophage-inflammatory protein-1 alpha and -1 beta in response to *Toxoplasma gondii* antigens." *Journal of Immunology* 162(12):7369-75.

Boehm U, Guethlein L, Klamp T, Ozbek K, Schaub A, Fütterer A, Pfeffer K, Howard JC. 1998. "Two families of GTPases dominate the complex cellular response to IFN-gamma." *Journal of Immunology* 161(12):6715-23.

Buonocore Sofia, Ahern Philip P, Uhlig Holm H, Ivanov Ivaylo I, Littman Dan R, Maloy Kevin J, Powrie Fiona. 2010. "Innate lymphoid cells drive interleukin-23-dependent innate intestinal pathology." *Nature* 464(7293):1371-5.

Cadigan KM, Waterman ML. 2012. "TCF/LEFs and Wnt signaling in the nucleus." *Cold Spring Harbor Perspectives in Biology* 4(11):a007906.

Choi J, Park S, Biering SB, Selleck E, Liu CY, Zhang X, Fujita N, Saitoh T, Akira S, Yoshimori T, Sibley LD, Hwang S, Virgin HW. 2014. "The parasitophorous vacuole membrane of *Toxoplasma gondii* is targeted for disruption by ubiquitin-like conjugation systems of autophagy." *Immunity* 40(6):924-35.

Clough Barbara, Wright Joseph D, Pereira Pedro M, Hirst Elizabeth M, Johnston Ashleigh C, Henriques Ricardo, Frickel Eva-Maria. 2016. "K63-Linked Ubiquitination Targets *Toxoplasma gondii* for Endo-lysosomal Destruction in IFNγ-Stimulated Human Cells." *PLoS Pathogens* 12(11):e1006027.

Däubener W, Mackenzie C, Hadding U. 1995. "Establishment of T-helper type 1- and T-helper type 2-like human *Toxoplasma* antigen-specific T-cell clones." *Immunology* 86(1):79-84.

Debierre-Grockiego F, Campos MA, Azzouz N, Schmidt J, Bieker U, Garcia Resende M, Santos Mansur D, Weingart R, Schmidt RR, Golenbock DT, Gazzinelli RT, Schwarz RT. 2007. "Activation of TLR2 and TLR4 by glycosylphosphatidylinositols derived from *Toxoplasma gondii*." *Journal of Immunology* 179(2):1129-37.

Degrandi Daniel, Kravets Elisabeth, Konermann Carolin, Beuter-Gunia Cornelia, Klümpers Verena, Lahme Sarah, Wischmann Eva, Mausberg Anne K, Beer-Hammer Sandra, Pfeffer Klaus. 2013. "Murine guanylate binding protein 2 (mGBP2) controls *Toxoplasma gondii* replication." *Proceedings of the National Academy of Sciences of the United States of America* 110(1):294-9.

Denkers Eric Y, Butcher Barbara A, Del Rio Laura, Kim Leesun. 2004. "Manipulation of mitogen-activated protein kinase/nuclear factor-kappaB-signaling cascades during intracellular *Toxoplasma gondii* infection." *Immunological reviews* 201:191-205.

Denkers EY, Gazzinelli RT, Marin D, Sher A. 1993. "Emergence of NK1.1+ cells as effectors of IFN-gamma dependent immunity to *Toxoplasma gondii* in MHC class I-deficient mice." *The journal of experimental medicine* 178(5):1465-72.

Dubey, J P. 2009. "History of the discovery of the life cycle of *Toxoplasma gondii*." *International Journal for Parasitology* 39(8):877-82.

Dunay Ildiko R, Damatta Renato A, Fux Blima, Presti Rachel, Greco Suellen, Colonna Marco, Sibley L David. 2008. "Gr1(+) inflammatory monocytes are required for mucosal resistance to the pathogen *Toxoplasma gondii*." *Immunity* 29(2):306-17.

Elemam Noha Mousaad, Hannawi Suad, Maghazachi Azzam A. 2017. "Innate Lymphoid Cells (ILCs) as Mediators of Inflammation, Release of Cytokines and Lytic Molecules." *Toxins* 9(12):398.

Foltz C, Napolitano A, Khan R, Clough B, Hirst EM, Frickel EM. 2017. "TRIM21 is critical for survival of *Toxoplasma gondii* infection and localises to GBP-positive parasite vacuoles." *Scientific reports* 7(1):5209.

Gay G, Braun L, Brenier-Pinchart MP, Vollaire J, Josserand V, Bertini RL, Varesano A, Touquet B, De Bock PJ, Coute Y, Tardieux I, Bougdour A, Hakimi MA. 2016. "*Toxoplasma gondii* TgIST co-opts host chromatin repressors dampening STAT1-dependent gene regulation and IFN-γ-mediated host defenses." *The Journal of Experimental Medicine* 213(9):1779-98.

Gazzinelli R T, Hieny S, Wynn T A, Wolf S, Sher A. 1993. "Interleukin 12 is required for the T-lymphocyte-independent induction of interferon gamma by an intracellular parasite and induces resistance in T-cell-deficient hosts." *Proceedings of the national academy of sciences of the United States of America* 90(13):6115-9.

Gazzinelli RT, Wysocka M, Hayashi S, Denkers EY, Hieny S, Caspar P, Trinchieri G, Sher A. 1994. "Parasite-induced IL-12 stimulates early IFN-gamma synthesis and resistance during acute infection with *Toxoplasma gondii*." *Journal of Immunology* 153(6):2533-43.

Golding Hana, Aliberti Julio, King Lisa R, Manischewitz Jody, AndersenJohn, Valenzuela Jesus, Landau Nathaniel R, Sher Alan. 2003. "Inhibition of HIV-1 infection by a CCR5-binding cyclophilin from *Toxoplasma gondii*." *Blood* 102(9):3280-6.

Gov L, Karimzadeh A, Ueno N, Lodoen MB. 2013. "Human innate immunity to *Toxoplasma gondii* is mediated by host caspase-1 and ASC and parasite GRA15." *mBio* 4(4):e00255-13.

Gov L, Schneider CA, Lima TS, Pandori W, Lodoen MB. 2017. "NLRP3 and Potassium Efflux Drive Rapid IL-1β Release from Primary Human Monocytes during *Toxoplasma gondii* Infection." *Journal of Immunology* 199(8):2855-2864.

Haldar AK, Foltz C, Finethy R, Piro AS, Feeley EM, Pilla-Moffett DM, Komatsu M, Frickel EM, Coers J. 2015. "Ubiquitin systems mark pathogen-containing vacuoles as targets for host defense by guanylate binding proteins." *Proceedings of the national academy of Sciences of the United States of America* 112(41):E5628-37.

Haldar AK, Piro AS, Pilla DM, Yamamoto M, Coers J. 2014. "The E2-like conjugation enzyme Atg3 promotes binding of IRG and Gbp proteins to Chlamydia- and Toxoplasma-containing vacuoles and host resistance." *PLoS one* 9(1):e86684.

Haldar AK, Saka HA, Piro AS, Dunn JD, Henry SC, Taylor GA, Frickel EM, Valdivia RH, Coers J. 2013. "IRG and GBP host resistance factors target aberrant, "non-self" vacuoles characterized by the missing of "self" IRGM proteins." *PLoS Pathogens* 9(6):e1003414.

Hamie M, Najm R, Deleuze-Masquefa C, Bonnet PA, Dubremetz JF, Sabban ME, Hajj HE. 2021. "Imiquimod Targets Toxoplasmosis Through Modulating Host Toll-Like Receptor-MyD88 Signaling." *Frontiers in Immunology* 12:629917.

He H, Brenier-Pinchart MP, Braun L, Kraut A, Touquet B, Couté Y, Tardieux I, Hakimi MA, Bougdour A. 2018. "Characterization of a Toxoplasma effector uncovers an alternative GSK3/β-catenin-regulatory pathway of inflammation." *eLife* 7:e39887.

Henry SC, Daniell XG, Burroughs AR, Indaram M, Howell DN, Coers J, Starnbach MN, Hunn JP, Howard JC, Feng CG, Sher A, Taylor GA. 2009. "Balance of Irgm protein activities determines IFN-gamma-induced host defense." *Journal of leukocyte biology* 85(5):877-85.

Hernandez-Cortazar Ivonne B, Acosta-Viana Karla Y, Guzman-Marin Eugenia, Ortega-Pacheco Antonio, Segura-Correa Jose C, Jimenez-Coello Matilde. 2017. "Presence of *Toxoplasma gondii* in Drinking Water from an Endemic Region in Southern Mexico." *Foodborne pathogens and disease* (5):288-292.

Howard Jonathan C, Hunn Julia P, Steinfeldt Tobias. 2011. "The IRG protein-based resistance mechanism in mice and its relation to virulence in *Toxoplasma gondii*." *Current opinion in microbiology* (4):414-21.

Hunn JP, Koenen-Waisman S, Papic N, Schroeder N, Pawlowski N, Lange R, Kaiser F, Zerrahn J, Martens S, Howard JC. 2008. "Regulatory interactions between IRG resistance GTPases in the cellular response to *Toxoplasma gondii*." *The EMBO journal* 27(19):2495-509.

JD, MacMicking. 2004. "IFN-inducible GTPases and immunity to intracellular pathogens." *Trends in immunology* 25(11):601-9.

Johnston Ashleigh C, Piro Anthony, Clough Barbara, Siew Malvin, Virreira Winter Sebastian, Coers Jörn, Frickel Eva-Maria. 2016. "Human GBP1 does not localize to pathogen vacuoles but restricts *Toxoplasma gondii*." *Cellular microbiology* 18(8):1056-64.

Kim BH, Shenoy AR, Kumar P, Das R, Tiwari S, MacMicking JD. 2011. "A family of IFN-γ-inducible 65-kD GTPases protects against bacterial infection." *Science* 332(6030):717-21.

Klose Christoph S N, Flach Melanie, Möhle Luisa, Rogell Leif, Hoyler Thomas, Ebert Karolina, Fabiunke Carola, Pfeifer Dietmar, Sexl Veronika, Fonseca-Pereira Diogo, Domingues Rita G, Veiga-Fernandes Henrique, Arnold Sebastian J, Busslinger Meinrad. 2014. "Differentiation of type 1 ILCs from a common progenitor to all helper-like innate lymphoid cell lineages." *Cell* 157(2):340-356.

Klose Christoph S N, Kiss Elina A, Schwierzeck Vera, Ebert Karolina, Hoyler Thomas, d'Hargues Yannick, Göppert Nathalie, Croxford Andrew L, Waisman Ari, Tanriver Yakup, Diefenbach Andreas. 2013. "A T-bet gradient controls the fate and function of CCR6-RORγt+ innate lymphoid cells." *Nature* 494(7436):261-5.

Koblansky A Alicia, Jankovic Dragana, Oh Hyunju, Hieny Sara, Sungnak Waradon, Mathur Ramkumar, Hayden Matthew S, Akira Shizuo, Sher Alan, Ghosh Sankar. 2013. "Recognition of profilin by Toll-like receptor 12 is critical for host resistance to *Toxoplasma gondii*." *Immunity* 38(1):119-30.

Kudchodkar SB, Levine B. 2009. "Viruses and autophagy." *Reviews in medical virology* 19(6):359-78.

Lee Y, Sasai M, Ma JS, Sakaguchi N, Ohshima J, Bando H, Saitoh T, Akira S, Yamamoto M. 2015. "p62 Plays a Specific Role in Interferon-γ-Induced Presentation of a Toxoplasma Vacuolar Antigen." *Cell reports* 13(2):223-33.

Lima TS, Gov L, Lodoen MB. 2018. "Evasion of Human Neutrophil-Mediated Host Defense during *Toxoplasma gondii* Infection." *mBio* 9(1):e02027-17.

López-Yglesias Américo H, Burger Elise, Camanzo Ellie, Martin Andrew T, Araujo Alessandra M, Kwok Samantha F, Yarovinsky Felix. 2021. "T-bet-dependent ILC1- and NK cell-derived IFN-γ mediates cDC1-dependent host resistance against *Toxoplasma gondii*." *PLoS Pathogens* 17(1):e1008299.

Malmberg Karl-Johan, Ljunggren Hans-Gustaf. 2009. "Spotlight on IL-22-producing NK cell receptor-expressing mucosal lymphocytes." *Nature immunology* 10(1):11-2.

Martens S, Parvanova I, Zerrahn J, Griffiths G, Schell G, Reichmann G, Howard JC. 2005. "Disruption of *Toxoplasma gondii* parasitophorous vacuoles by the mouse p47-resistance GTPases." *PLoS pathogens* 1(3):e24.

Matta SK, Patten K, Wang Q, Kim BH, MacMicking JD, Sibley LD. 2018. "NADPH Oxidase and Guanylate Binding Protein 5 Restrict Survival of Avirulent Type III Strains of *Toxoplasma gondii* in Naive Macrophages." *mBio* 9(4):e01393-18.

Mendonça-Natividade FC, Duque Lopes C, Ricci-Azevedo R, Sardinha-Silva A, Figueiredo Pinzan C, Paiva Alegre-Maller AC, Nohara LL, Carneiro AB, Panunto-Castelo A, Almeida IC, Roque-Barreira MC. 2019. "Receptor Heterodimerization and Co-Receptor Engagement in TLR2 Activation

Induced by MIC1 and MIC4 from *Toxoplasma gondii*." *International Journal of Molecular Science* 20(20):5001.

Mizushima N, Yoshimori T, Ohsumi Y. 2011. "The role of Atg proteins in autophagosome formation." *Annual review of cell and develop-ment biology* 27:107-32.

Nishimura Maki, Umeda Kousuke, Suwa Masayuki, Furuoka Hidefumi, Nishikawa Yoshifumi. 2017. "CCR5 Is Involved in Interruption of Pregnancy in Mice Infected with *Toxoplasma gondii* during Early Pregnancy." *Infection and immunology* 85(9):e00257-17.

Ohshima J, Lee Y, Sasai M, Saitoh T, Ma JS, Kamiyama N, Matsuura Y, Pann-Ghill S, Hayashi M, Ebisu S, Takeda K, Akira S, Yamamoto M. 2014. "Role of mouse and human autophagy proteins in IFN-γ-induced cell-autonomous responses against *Toxoplasma gondii*." *Journal of immunology* 192(7):3328-35.

Olias P, Etheridge RD, Zhang Y, Holtzman MJ, Sibley LD. 2016. "Toxoplasma Effector Recruits the Mi-2/NuRD Complex to Repress STAT1 Transcription and Block IFN-γ-Dependent Gene Expression." *Cell Host & Microbe* 20(1):72-82.

Ong Yi-Ching, Reese Michael L, Boothroyd John C. 2010. "Toxoplasma rhoptry protein 16 (ROP16) subverts host function by direct tyrosine phosphorylation of STAT6." *The Journal of Biological Chemistry* 285(37):28731-40.

Pappas Georgios, Roussos Nikos, Falagas Matthew E. 2009. "Toxoplasmosis snapshots: global status of *Toxoplasma gondii* seroprevalence and implications for pregnancy and congenital toxoplasmosis." *International Journal for Parasitology* 39(12):1385-94.

Park S, Choi J, Biering SB, Dominici E, Williams LE, Hwang S. 2016. "Targeting by AutophaGy proteins (TAG): Targeting of IFNG-inducible GTPases to membranes by the LC3 conjugation system of autophagy." *Autophagy* 12(7):1153-67.

Pfefferkorn, ER. 1984. "Interferon gamma blocks the growth of *Toxoplasma gondii* in human fibroblasts by inducing the host cells to degrade tryptophan." *Proceedings of the National Academy of Sciences of the United Stas of America* 81(3):908-12.

Quinn Susan R, O'Neill Luke A. 2011. "A trio of microRNAs that control Toll-like receptor signalling." *International Immunology* 23(7):421-5.

Robben PM, Mordue DG, Truscott SM, Takeda K, Akira S, Sibley LD. 2004. "Production of IL-12 by macrophages infected with *Toxoplasma gondii* depends on the parasite genotype." *Journal of Immunology* 172(6):3686-94.

Rosowski EE, Nguyen QP, Camejo A, Spooner E, Saeij JP. 2014. "*Toxoplasma gondii* inhibits gamma interferon (IFN-γ)- and IFN-β induced host cell

STAT1 transcriptional activity by increasing the association of STAT1 with DNA." *Infection and immunity* 82(2):706-19.

Saeij J P J, Boyle J P, Coller S, Taylor S, Sibley L D, Brooke-Powell E T, Ajioka J W, Boothroyd J C. 2006. "Polymorphic secreted kinases are key virulence factors in toxoplasmosis." *Science* 314(5806):1780-3.

Saeij JPJ, Coller S, Boyle JP, Jerome ME, White MW, Boothroyd JC. 2007. "Toxoplasma co-opts host gene expression by injection of a polymorphic kinase homologue." *Nature* 445(7125):324-7.

Sardinha-Silva A, Mendonça-Natividade FC, Pinzan CF, Lopes CD, Costa DL, Jacot D, Fernandes FF, Zorzetto-Fernandes ALV, Gay NJ, Sher A, Jankovic D, Soldati-Fav D, Grigg ME, Roque-Barreira MC. 2019. "The lectin-specific activity of *Toxoplasma gondii* microneme proteins 1 and 4 binds Toll-like receptor 2 and 4 N-glycans to regulate innate immune priming." *PLoS Pathogens* 15(6):e1007871.

Sasai M, Sakaguchi N, Ma JS, Nakamura S, Kawabata T, Bando H, Lee Y, Saitoh T, Akira S, Iwasaki A, Standley DM, Yoshimori T, Yamamoto M. 2017. "Essential role for GABARAP autophagy proteins in interferon-inducible GTPase-mediated host defense." *Nature immunology* 18(8):899-910.

Sasai Miwa, Yamamoto Masahiro. 2019. "Innate, adaptive, and cell-autonomous immunity against *Toxoplasma gondii* infection." *Experimental & molecular medicine* 51(12):1-10.

Sasai Miwa, Yamamoto Masahiro. 2013. "Pathogen recognition receptors: ligands and signaling pathways by Toll-like receptors." *International reviews of immunology* 32(2):116-33.

Satoh-Takayama Naoko, Vosshenrich Christian A J, Lesjean-Pottier Sarah, Sawa Shinichiro, Lochner Matthias, Rattis Frederique, Mention Jean-Jacques, Thiam Kader, Cerf-Bensussan Nadine, Mandelboim Ofer, Eberl Gerard, Di Santo James P. 2008. "Microbial flora drives interleukin 22 production in intestinal NKp46+ cells that provide innate mucosal immune defense." *Immunity* 29(6):958-70.

Schulthess Julie, Meresse Bertrand, Ramiro-Puig Emma, Montcuquet Nicolas, Darche Sylvie, Bègue Bernadette, Ruemmele Frank, Combadière Christophe, Di Santo James P, Buzoni-Gatel Dominique, Cerf-Bensussan Nadine. 2012. "Interleukin-15-dependent NKp46+ innate lymphoid cells control intestinal inflammation by recruiting inflammatory monocytes." *Immunity* 37(1):108-21.

Selleck Elizabeth M, Fentress Sarah J, Beatty Wandy L, Degrandi Daniel, Pfeffer Klaus, Virgin 4th Herbert W, Macmicking John D, Sibley L David. 2013. "Guanylate-binding protein 1 (Gbp1) contributes to cell-autonomous immunity against *Toxoplasma gondii.*" *PLoS Pathogens* 9(4):e1003320.

Selleck Elizabeth M, Orchard Robert C, Lassen Kara G, Beatty Wandy L, Xavier Ramnik J, Levine Beth, Virgin Herbert W, Sibley L David. 2015. "A Noncanonical Autophagy Pathway Restricts *Toxoplasma gondii* Growth in a Strain-Specific Manner in IFN-γ-Activated Human Cells." *mBio* 6(5):e01157-15.

Sher Alan, Collazzo Carmen, Scanga Charles, Jankovic Dragana, Yap George, Aliberti Julio. 2003. "Induction and regulation of IL-12-dependent host resistance to *Toxoplasma gondii*." *Immunologic research* 27(2-3):521-8.

Shrestha SP, Tomita T, Weiss LM, Orlofsky A. 2006. "Proliferation of *Toxoplasma gondii* in inflammatory macrophages in vivo is associated with diminished oxygen radical production in the host cell." *International journal of parasitology* 36(4):433-41.

Slobodkin MR, Elazar Z. 2013. "The Atg8 family: multifunctional ubiquitin-like key regulators of autophagy." *Essays in biochemistry* 55:51-64.

Sonnenberg Gregory F, Monticelli Laurel A, Elloso M Merle, Fouser Lynette A, Artis David. 2011. "CD4(+) lymphoid tissue-inducer cells promote innate immunity in the gut." *Immunity* 34(1):122-34.

Sumpter Jr R, Levine B. 2010. "Autophagy and innate immunity: triggering, targeting and tuning." *Seminars in cell and developmental biology* 21(7):699-711.

Suzuki Y, Orellana M A, Schreiber R D, Remington J S. 1988. "Interferon-gamma: the major mediator of resistance against *Toxoplasma gondii*." *Science* 240(4851):516-8.

Taylor GA, Stauber R, Rulong S, Hudson E, Pei V, Pavlakis GN, Resau JH, Vande Woude GF. 1997. "The inducibly expressed GTPase localizes to the endoplasmic reticulum, independently of GTP binding." *The Journal of Biological Chemistry* 272(16):10639-45.

Tosh KW, Mittereder L, Bonne-Annee S, Hieny S, Nutman TB, Singer SM, Sher A, Jankovic D. 2016. "The IL-12 Response of Primary Human Dendritic Cells and Monocytes to *Toxoplasma gondii* Is Stimulated by Phagocytosis of Live Parasites Rather Than Host Cell Invasion." *Journal of Immunology* 196(1):345-56.

Wei Hai-Xia, Wei Shan-Shan, Lindsay David S, Peng Hong-Juan. 2015. "A Systematic Review and Meta-Analysis of the Efficacy of Anti-*Toxoplasma gondii* Medicines in Humans." *PloS One* 10(9): e0138204.

Wilson CB, Tsai V, Remington JS. 1980. "Failure to trigger the oxidative metabolic burst by normal macrophages: possible mechanism for survival of intracellular pathogens." *The journal of experimental medicine* 151(2):328-46.

Y, Ohsumi. 2001. "Molecular dissection of autophagy: two ubiquitin-like systems." *Nature reviews. Molecular cell biology* 2(3):211-6.

Yamamoto M, Standley DM, Takashima S, Saiga H, Okuyama M, Kayama H, Kubo E, Ito H, Takaura M, Matsuda T, Soldati-Favre D, Takeda K. 2009. "A single polymorphic amino acid on *Toxoplasma gondii* kinase ROP16 determines the direct and strain-specific activation of Stat3." *The Journal of Experimental Medicine* 206(12):2747-60.

Yamamoto Masahiro, Okuyama Megumi, Ma Ji Su, Kimura Taishi, Kamiyama Naganori, Saiga Hiroyuki, Ohshima Jun, Sasai Miwa, Kayama Hisako, Okamoto Toru, Huang David C S, Soldati-Favre Dominique, Horie Kyoji, Takeda Junji, Takeda Kiyoshi. 2012. "A cluster of interferon-γ-inducible p65 GTPases plays a critical role in host defense against *Toxoplasma gondii*." *Immunity* 37(2):302-13.

Zerrahn J, Schaible UE, Brinkmann V, Guhlich U, Kaufmann SHE. 2002. "The IFN-inducible Golgi- and endoplasmic reticulum- associated 47-kDa GTPase IIGP is transiently expressed during listeriosis." *Journal of Immunology* 168(7):3428-36.

Zhao Yang O, Khaminets Aliaksandr, Hunn Julia P, Howard Jonathan C. 2009. "Disruption of the *Toxoplasma gondii* parasitophorous vacuole by IFNgamma-inducible immunity-related GTPases (IRG proteins) triggers necrotic cell death." *PLoS Pathogens* 5(2):e1000288.

Zhao YO, Könen-Waisman S, Taylor GA, Martens S, Howard JC. 2010. "Localisation and mislocalisation of the interferon-inducible immunity-related GTPase, Irgm1 (LRG-47) in mouse cells." *PLoS one* 5(1):e8648.

Zhao Z, Fux B, Goodwin M, Dunay IR, Strong D, Miller BC, Cadwell K, Delgado MA, Ponpuak M, Green KG, Schmidt RE, Mizushima N, Deretic V, Sibley LD, Virgin HW. 2008. "Autophagosome-independent essential function for the autophagy protein Atg5 in cellular immunity to intracellular pathogens." *Cell host & microbe* 4(5):458-69.

Zhou W, Quan JH, Lee YH, Shin DW, Cha GH. 2013. "*Toxoplasma gondii* Proliferation Require Down-Regulation of Host Nox4 Expression via Activation of PI3 Kinase/Akt Signaling Pathway." *PLoS one* 8(6):e66306.

Chapter 8

The Role of the Innate Immune Response during Helminth Infections

Víctor Hugo Del Río-Araiza[1,*], Karen Elizabeth Nava-Castro[2] and Jorge Morales-Montor[3,†]

[1]Departamento de Parasitología, Facultad de Medicina Veterinaria y Zootecnia, Universidad Nacional Autónoma de México, Ciudad de México, México
[2]Departamento de Ciencias Ambientales. Centro de Ciencias de la Atmósfera, Universidad Nacional Autónoma de México, Ciudad de México, México
[3]Departamento de Inmunología, Instituto de Investigaciones Biomédicas, Universidad Nacional Autónoma de México, Ciudad de México, México

Abstract

Helminthosis not only affect a large number of people in the world (more than 1.5 billion people are estimated to be infected by these parasites), they also affect a large number of animal species, and many of them are zoonotic. This implies great economic losses, not only in the cost of the different treatments, but also due to the decrease in the different productive and reproductive parameters in other animal species. Although many of them are asymptomatic infections, in some cases

[*] Corresponding Author's E-mail: victordelrioa@fmvz.unam.mx.
[†] Corresponding Author's E-mail: jmontor66@iibiomedicas.unam.mx.

In: The Innate Immune System in Health and Disease
Editor: Jorge Morales-Montor
ISBN: 978-1-68507-507-1

serious health problems can occur, as well as a predisposition to different comorbidities or co-infections. In controlling a helminth infection, the immune response plays an important role. In particular, the innate response is one of the first and main defense mechanisms of the organism that are activated when coming into contact with these parasites. It has long been known that different cells of the innate immune response participate through different mechanisms, in the control and elimination of helminths. However, both new cells and new strategies implemented by the innate immune system to combat these infections have recently been described. In this chapter, we will discuss the participation of cells of the innate immune response, as well as the different effector mechanisms that are activated during helminth infections, and how they can act, either directly on parasites or by modulating the future adaptive immune response.

Keywords: estradiol, immune response, parasite, parasite infection

Introduction

Helminths are metazoan organisms that can be divided into different phylums. The two of greatest medical and veterinary importance are Platyhelminthes (trematodes and cestodes), also known as flatworms, and Nematoda (nematodes), known as cylindrical worms. Many of these organisms cause diseases known as helminthosis. Such diseases can affect both man and many animal species. In the case of animal health, diseases caused by helminths produce economic losses equivalent to several billion dollars worldwide, whether due to the death of animals, decrease in production parameters, seizure of organs, treatments, etc. To cite a few examples, in Mexico it has been estimated that the economic losses caused by swine cysticercosis can reach US$19.5 million annually (Bhattarai et al. 2019), while the cost of treating bovine fasciolosis is around an estimated US$26.02 million annually in Nigeria alone (Odeniran, Omolabi, and Ademola 2020). Both, very important diseases because they are considered zoonotic in nature. In terms of public health, it is estimated that more than 2 billion people in the world are infected with this type of parasite, which cause malnutrition, stunting, cognitive deficiencies and chronic immunopathologies (James et al. 2018; Jourdan et al. 2018). Soil-transmitted helminthosis (STH) are infections considered neglected tropical diseases (NTDs), which occur most frequently in developing countries (it is estimated that there are at least 120 endemic countries) and have been associated with conditions of marginalization and poverty. Among the most important parasites that cause disease we can mention some such as *Ascaris lumbricoides, Necantor americanus, Ancylostoma duodenale, Trichuris* spp., *Toxocara canis*, etc (Ásbjörnsdóttir et al. 2017). Although these diseases are rarely fatal, they have a high morbidity. It is estimated that they represent more than 5 million disability-adjusted life years (DALY) and decreased production in people infected with this type of parasite (de Vlas et al. 2016). For the control of helminthosis, the immune response plays a very important role. Particularly, innate immunity participates in the initiation of the response against this type of pathogens, as well as in the subsequent assembly of the adaptive response. During this chapter we will address the role of innate immunity cells, as well as the different mechanisms that occur during infections caused by helminths.

Innate Immune Response to Helminths

The cells of the innate response and their effector mechanisms are indispensable for the initiation of an immune response and subsequently, for the assembly of a protective adaptive response, carried out by type 2 (Th2) cooperative T cells (Anthony et al. 2007). This communication that takes place between the innate and adaptive response is carried out through the production of cytokines that participate in the activation and proliferation of different cell types. Although the immune response is compartmentalized, many of these cells that are present in the different organs and tissues will participate in a similar and coordinated way during helminthosis in order to control and eliminate these parasites. During these infections, the damage caused by parasites, or the products secreted by them, have an important role and participate in the induction of innate and adaptive immune responses (Motran et al. 2018). These secretions are known as excretion/secretion antigens (ESAg) and refer to molecules that the parasite produces, releases, and are recognized by the immune system (Hewitson, Grainger, and Maizels 2009; White and Artavanis-Tsakonas 2012; Harnett 2014).

Within the first line of defense during infections caused by helminths, we have epithelial cells and innate immunity cells (dendritic cells, alternatively activated macrophages (AAMs or M2), neutrophils, eosinophils, basophils, mast cells and type 2 innate lymphoid cells (ILCs2)) (Allen and Maizels 2011; Babu and Nutman 2019).

Epithelial Cells

Epithelial cells are one of the body's first defense barriers that most helminths face. In response to the damage generated by helminths, these cells will produce chemokines and cytokines to attract and activate other cells of the immune response (Hammad and Lambrecht 2015). At the beginning of the infection in the mucous membranes, there will be an increase in the contraction of smooth muscle, hypersecretion of mucus by goblet cells, hyperproliferation of epithelial cells, and the production of cytokines such as IL-25, IL-33 and thymic stromal lymphoprotein (TSLP) (Allen and Maizels 2011; Babu and Nutman 2019). In addition to these cytokines, epithelial cells will produce chemokines such as CCL11, CCL17, CCL22, CCL24, CCL26 and prostaglandin D2, which will act by recruiting ILCs2, dendritic cells, basophils, eosinophils and barley cells among others (Xue et al. 2014; Hammad and Lambrecht 2015). Also, the cytokines mentioned above will favor the activation of these cells. The cytokines IL25, IL-33, TSLP will activate cells of the innate response such as ILCs2, eosinophils, basophils, mast cells and dendritic cells. The latter will also promote the future adaptive response Th2 (Allen and Maizels 2011; Halim et al. 2012; Hammad and Lambrecht 2015) (Figure 1).

Recently, a type of epithelial cells, called "Tuft cells" have been described. These cells possess chemosensory-secretory capabilities. It has been described that they can secrete cytokines such as IL-25, TSLP, eicosanoids (Cysteinyl leukotrienes and prostaglandin D2) and the neurotransmitter acetylcholine. Although all their functions have not been described, these characteristics position them as important cells in the development of anthelmintic immunity (Schneider, O'Leary, and Locksley 2019) (Figure 1).

Another mechanism by which epithelial cells can regulate type 2 immune responses is by producing damage-associated molecular patterns (DAMPs) or endogenous alarmins such as uric acid, IL-1α, IL-33, High Mobility Group Box 1 (HMGB1) and S100 family proteins (Hammad and Lambrecht 2015).

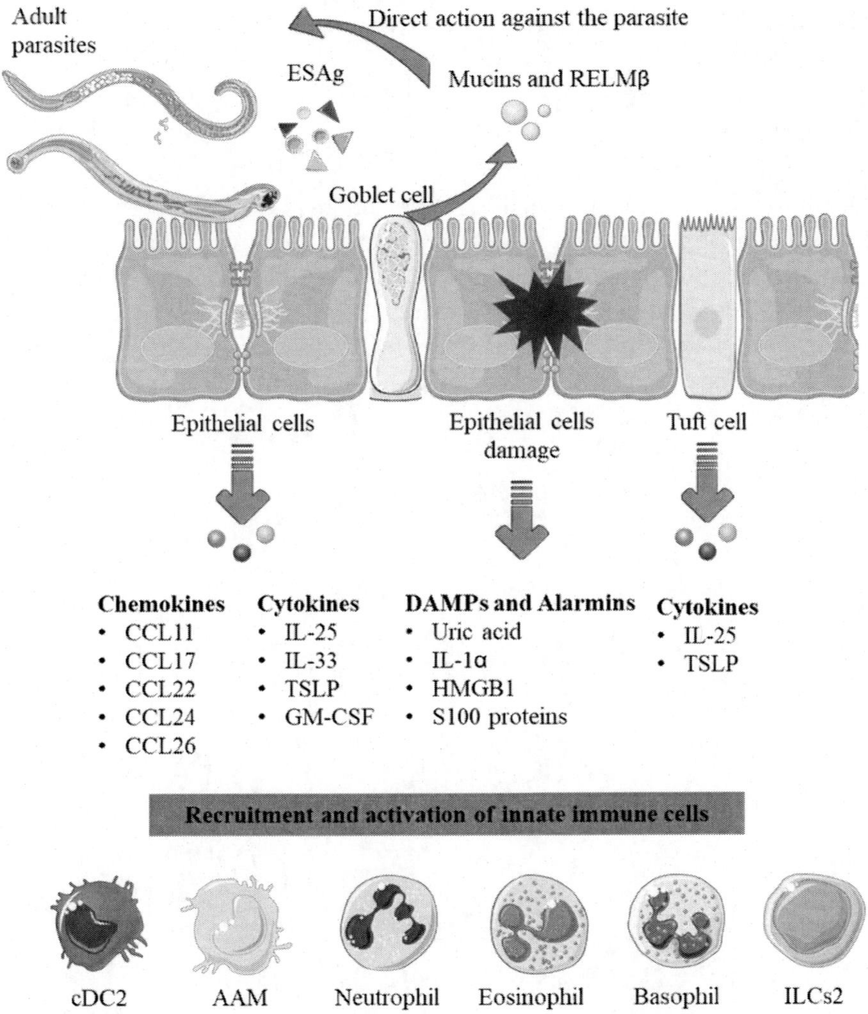

Figure 1. Epithelial cells participation during helmints infection.

Finally, cytokines such as IL-13 will favor hyperplasia of goblet cells, which will produce mucins and Resistin-like molecule-β (RELMβ). These molecules have different functions depending on the site where they are (pulmonary or intestinal epithelium) and regulate the production of mucus, favor the recruitment of leukocytes and the deposition of collagen, as well as the remodeling of tissues. They can also act directly on helminths by inhibiting their feeding and facilitating their expulsion (Allen and Maizels 2011; Hammad and Lambrecht 2015; Babu and Nutman 2019).

Dendritic Cells

Dendritic cells (DCs) are cells specialized in capturing, processing, and presenting antigens to T lymphocytes during the immunological synapse. Overall, three populations of DCs have been described: conventional dendritic cells (cDCs), which in turn divide into 2 types (cDC1 and cDC2); and plasmocytoid dendritic cells (pDC), which are characterized by secreting interferon type I (J. Y. Shin et al. 2020).

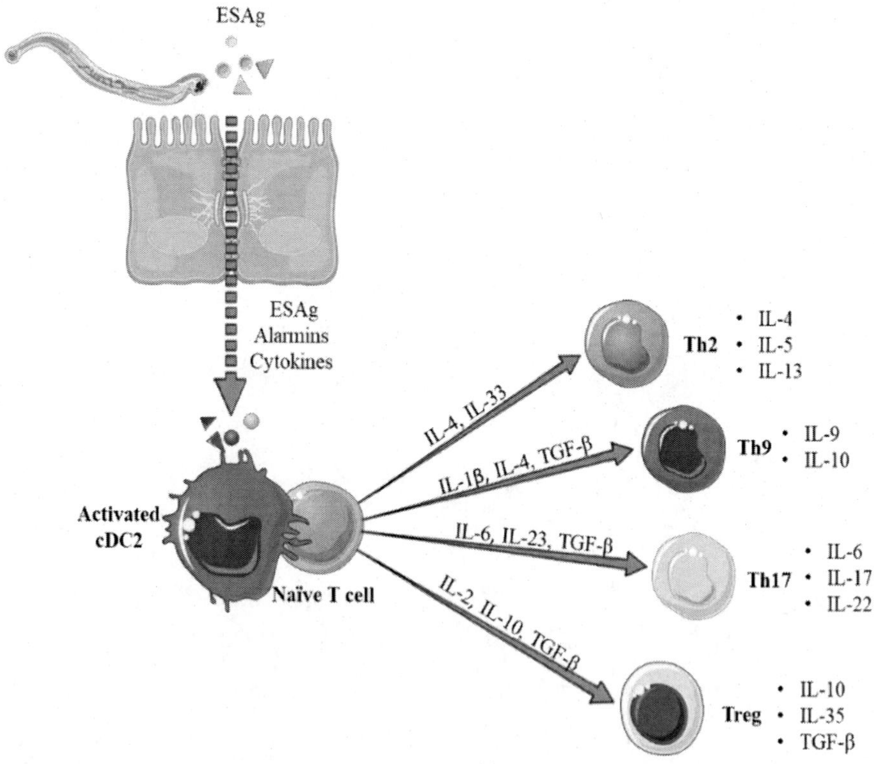

Figure 2. cDC2 activation during helminth infection and its role in the initiation of the adaptive immune response.

During infections caused by helminths, cDC2 are responsible for orchestrating and mediating a subsequent adaptive response of type Th2. They recognize esAg and surface antigens of helminths through pattern recognition receptors (PRRs) such as: Tool-type receptors (TLRs), c-type lectin receptors, and NOD-type receptors (White and Artavanis-Tsakonas 2012; Motran et al. 2018). The characteristic markers of cDC2 are: CD301b, PDL2 and CD11b, CD172, in addition to receptors for cytokines such as IL-4 (IL-4R), IL-13 (IL-13R), IL-25 (IL-25R), IL-33 (IL-33R) and TSLP (TSLP-R) (Babu and Nutman 2019). These cells are so important for the polarization of a Th2-type response that it has been described that a depletion of them suppresses these types of responses. In addition, DCs stimulated with helminth antigens rapidly induce responses of this type (MacDonald et al. 2001; Balic et al. 2004; Leech and Grencis 2006; Carvalho et al. 2009). Some of the cytokines produced by epithelial cells such as IL-1α, IL-33, TSLP and GM-CSF are responsible for programming DCs to perform Th2 (cDC2) responses. They suppress the secretion of IL-12 (cytokine involved in the polarization of Th1-like responses), induce the expression of OX40L and Notch receptor ligands (Schuijs, Hammad, and Lambrecht 2019). Once the cDC2 are activated, they will begin with the antigenic presentation and the triggering of the adaptive immune response, favoring the differentiation towards Th2, Th9, Th17 and Treg profiles depending on the cytokine micro-environment that is present (Kumar et al. 2019) (Figure 2).

Macrophages

Macrophages are another kind of professional antigen-presenting cells that play an important role in the development of the immune response. Two main types of these cells have been described: classically activated macrophages (CAM), which are present in the Th1-type response; and alternately activated macrophages (AAMs), important during a Th2-type response (immunity to helminths) (Kreider et al. 2007). Additionally, it has been reported that AAMs have anti-inflammatory and pro-angiogenic functions, during wound repair processes (since they have improved phagocytic capacities), through the production of immunomodulatory cytokines such as IL-10, extracellular matrix components, as well as different chemotactic and angiogenic factors (Gordon and Martinez 2010). AAMs, unlike CAM, are not activated by the IFN-γ. In addition, they have high levels of expression of Arginase-1, an enzyme that is responsible for the metabolism of arginine, Ym1 and Ym2 and the resistin-like molecule (RELM-α) (Loke et al. 2002).

Polarization of AAMs is mediated by the presence of cytokines such as IL-4 and IL-13 through the IL-4Rα signaling and the Signal transducer and activator of transcription-6 (STAT-6) (Van Dyken and Locksley 2013; Gundra et al. 2014). These cytokines required for AAMs polarization may be secreted from other cells of innate immunity such as ILCs2 (Tait Wojno and Artis 2016; Bouchery, Le Gros, and Harris 2019), basophils (van Panhuys et al. 2011; Inclan-Rico et al. 2020) and eosinophils (Weller and Spencer 2017; Coakley and Harris 2020). Helminths can also induce polarization and activation of AAMs by producing ESAg (Maizels, Smits, and McSorley 2018), either being directly recognized by PRRs present in macrophages (mannose receptor (MR)) (van Die and Cummings 2017) or in conjunction with IL-4R.

Another cytokine that has gained relevant importance in the polarization of AAMs is the macrophage migration inhibitor factor (MIF). Although it is a pleiotropic cytokine that has been associated with Th1-type responses; during helminth infections it has been observed that this molecule can have a dichotomous effect associated with the antigenic challenge used. Damle et. al., (2017), evaluated the role of MIF using *Nippostrongylus brasiliensis* as an antigenic challenge. They reported that at 7 days post-infection, MIF-deficient mice showed a reduction in parasite loads due to a more intense Th2-like response, where they observed increased expression of the transcription factor GATA3, as well as increased IL-13 production in mesenteric lymph nodes (Damle et al. 2017). On the other hand, other researchers highlight the importance of MIF in the polarization of the Th2 response. First, Stravitsky et. al., (2003), mention that the blockade of the MIF during infection by *Schistosoma japonicum* (4.5 to 6.5 weeks post-infection) increase parasite load and reduce parasite fertility (Stavitsky et al. 2003). Filbey et. al., (2019), they also demonstrated that MIF plays a critical role in the polarization of AAMs, as well as in the protective immune response at the intestinal level against the parasite *Heligmosomoides polygirus* (Filbey et al. 2019).

Finally, AAMs may activate T lymphocytes and induce their differentiation to a Th2 profile. These, in turn, will allow to maintain the phenotype of AAMs through the production of cytokines such as IL-4 and IL-13, which allow feedback to AAMs. All these effector mechanisms allow AAMs to repair damaged tissues through the secretion of different molecules such as RELM-α, RELM-β, Arginase 1, Ym1, Ym2, among others. This act directly on the parasite and participate in the healing of wounds caused by helminths, chymotaxes, activation of cells of the immune system and regulation of the immune response (Allen and Maizels 2011; Coakley and Harris 2020) (Figure 3).

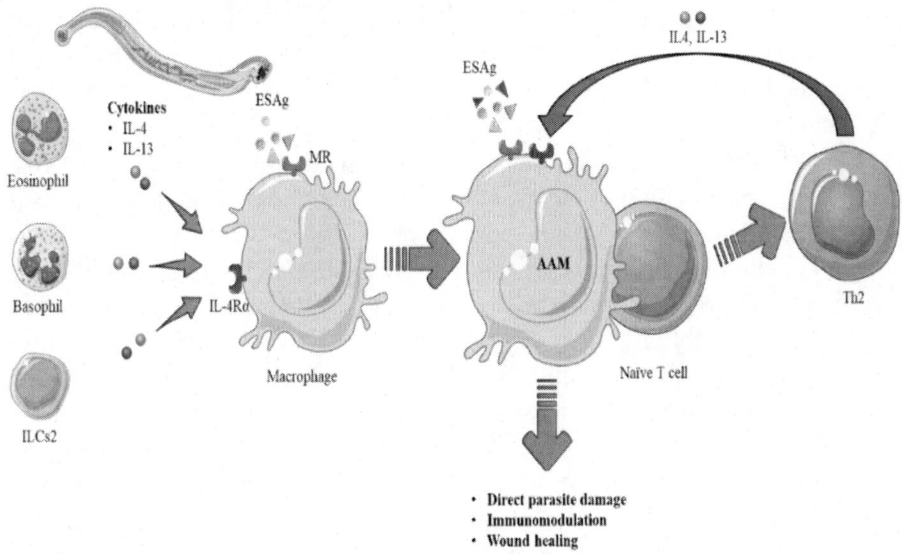

Figure 3. Differentication, activation and maintaining of AAMs during helmith infections.

Granulocytes

Granulocytes, or also known as polymorphonuclear leukocytes, are cells of the innate immune response that have as a characteristic to present granules in their cytoplasm. These cells can divide into four subpopulations: eosinophils, basophils, neutrophils and mast cells. All of them have been shown to play a fundamental role in the control and elimination of helminth parasites (Figure 4). Neutrophils are the most abundant granulocytes and the first immune response cells to respond to and reach sites of infection, damage or inflammation. Recently it has been shown that there are different subpopulations of neutrophils, which have different functions either if they are in homeostasis or pathological conditions (Silvestre-Roig et al. 2019). However, their role during helminth infections has been poorly studied since they have typically been considered important for infections caused by bacteria and fungi. Among the main effector mechanisms that neutrophils use to fight pathogens are phagocytosis, reactive oxygen species production and the release of soluble antimicrobials (granular proteins) contained in their granules. A few years ago, it was described that activated neutrophils can generate the release of nucleic acids with histones and granular proteins, which have been called extracellular neutrophil traps (NETs), and the process by which they perform it as NETosis (Brinkmann and Zychlinsky 2012; Mutua and Gershwin 2020). These traps work by capturing and destroying various pathogens such as viruses, fungi, bacteria and protozoan parasites. Regarding the role of neutrophils during nematode infections, different groups have reported the formation of NETs, although the functional role of NETs has not been fully demonstrated (Díaz-Godínez and Carrero 2019).

Neutrophils participate in conjunction with AAMs to immobilize and eliminate parasites. Bonne-Année et al., (2013, 2014), demonstrated that there is a collaboration between neutrophils and AAMs that is dependent on complement proteins, in addition to the NETs, to kill the larvae of the parasite *Strongyloides stercolaris* (Bonne-Année et al. 2013; 2014). Chen et al., (2014), reported that during a primary infection by *Nippostrongylus brasiliensis*, lung resident neutrophils show a "N2 phenotype" which shows a positive regulation of IL-13, IL-33, IgF1, RELM-α and Ym1. In addition, these "N2s" can instruct macrophages to develop a phenotype of AAMs during a secondary infection to mediate damage and elimination of the parasite. It is worth mentioning that this cooperation is dependent on the expression of IL-4Rα in macrophages (Chen et al. 2014). Finally, another mechanism by which neutrophils can kill helminths is by producing and releasing the enzyme myeloperoxidase (O'Connell et al. 2011).

It has been recently described that some helminths could evade and degrade NETs by using different mechanisms. Bouchery et al., (2020), showed that during the infection with *Nippostrongylus brasiliensis,* neutrophils were rapidly recruited and deployed NETs around the larvae located in the skin of infected mice. However, larvae released a deoxyribonuclease (Nb-DNase II), as part of their ESAg, to degrade the DNA structure of NETs. In fact, they also described that neutrophils can kill larvae in *in vitro* studies, and that this effect was enhanced by Nb-DNase II neutralization (Bouchery et al. 2020). Chauhan et al., (2021), reported that ESAg of *Mesocestoides corti*, may modulate the response of neutrophils by inhibiting the formation of ROS-induced NETs by blocking the activation of the Transient Receptor Potential Melastatin 2 (TRPM2) (Chauhan et al. 2021).

Eosinophils account for about 5% of leukocytes in peripheral blood in healthy people and animals. However, during infections caused by helminths, a very marked eosinophilia occurs in both blood and tissues. Cytokines such as IL-5, GM-CSF, and IL-3 accelerate the growth, maturation, survival, and activation of eosinophils (Obata-Ninomiya, Domeier, and Ziegler 2020). IL-5 is a cytokine that is critical for inducing the recruitment and activation of eosinophils to sites of infection (L. Huang and Appleton 2016). Eosinophils are also capable of producing cytokines and chemokines such as IL-4, IL-5, IL-8, IL-10, IL-12, IL-13, IL-16, IL-18, TGF-β, TNF-α, CCL5 and CCL11. These are stored in their granules and released at the time of degranulation (Stone, Prussin, and Metcalfe 2010). In addition, these granules contain proteins such as eosinophil peroxididase (EPO), eosinophil-derived neurotoxin (EDN), major basic protein (MBP), and eosinophilic cationic protein (ECP), all with cytotoxic qualities for parasites (Acharya and Ackerman 2014). The protective role of these proteins in infections caused by helminths has been observed in diverse studies. Spetch et al., (2006) determined that the lack of MBP or EPO during *Litomosoides sigmodontis* infection has a significant increase in parasite loads with respect to control mice. In addition, they observed that these proteins participate in the regulation of the immune response since the absence of MBP had an increase in the production of IL-10. Moreover, EPO-deficient mice showed an increase in IL-5 levels, as well as increased eosinophilia (Specht et al. 2006). O'Connell et al., (2011), evaluated the mechanisms used by eosinophils to control the *Strongyloides stercolaris* infection. They determine that eosinophils kill parasite larvae through an MBP-dependent mechanism produced during the primary immune response, when other effector cells are absent (O'Connell et al. 2011). Yang et al., (2008), propose that the EDN acts as an alarmin that interacts with the TLR2-MYD88 of the DCs, promoting an increase in the production of cytokines Th2 (IL-5, IL-6, IL13), Treg (IL-10) and antibodies (IgG1) all these very important cytokines during helminth infections (Yang et al. 2008).

While eosinophils can kill helminths directly, it is not yet clear whether they can carry out this action *in vivo*. In different studies where eosinophil-deficient mice have been used, little to no differences are shown during helminthosis, suggesting that the response of these cells to parasites may be situational and redundant (LeMessurier and Samarasinghe 2019). In another study, Fabre et al., (2009), demonstrated an immunomodulatory role of eosinophils during chronic infection with *Trichinella spiralis*, since their presence was fundamental for the survival of the parasite in the muscular phase. They observed that when there was an absence of eosinophils, muscle larvae died, and this correlated with an increase in IFN-γ and a decrease in IL-4 production. In addition, larval survival improved when mice were treated with an inducible nitric oxide synthase (iNOS) enzyme inhibitor, which is indicative that the nitric oxide (NO) pathway plays an important role in eliminating the parasite (Fabre et al. 2009). In another report, Gebreselassie et al., (2012), show that the death of *Trichinella spiralis* larvae in mice lacking eosinophils correlates with a decrease in Th2 lymphocyte recruitment (IL-4+), and an increase in the recruitment of iNOS-producing neutrophils to infected muscles and iNOS in tissue-resident macrophages (Gebreselassie et al. 2012). Finally, Strandmark et al., (2017), using eosinophil-deficient mice that were infected with *Heligmosomoides polygyrus*, reported a decrease in parasite fertility. Also, this lack of eosinophils generated an increase in the expression of the transcription factor GATA-3 and IL-4 producing cells during acute infection. This finding was strictly limited to Peyer's plaques. These results suggest a new role of eosinophils in regulating the immune response during helminth intestinal infections, during which they restrict Th cell responses and isotype change to IgG in Peyer's plaques, in order to maintain local IgA production (Strandmark et al. 2017).

Basophils represent a proportion of less than 0.5% of leukocytes in the blood and during Th2-type responses a high number of them are observed in the periphery. IL-3 has been shown to be necessary for the survival and proliferation of basophils during nematode infections (Shen et al. 2008). These basophils at the time of being activated secrete large amounts of IL-4, which promotes a Th2 response. This response has been described in other infections caused by helminths such as *Trichinella spiralis* (Giacomin et al. 2012), *Heligmosomoides polygyrus* (Schwartz et al. 2014) and *Litomosoides sigmodontis* (Torrero et al. 2010). They can also produce cytokines such as IL-6, TNF-α and IL-13, inflammatory mediators such as histamine and platelet activating factor, bioactive lipids such as prostaglandins and leukotrienes, and growth fractors such as amphiergulin and macrophage colony-stimulating factor (M-CSF) (Voehringer 2013). These cells have been shown to play an important role in the immune response against helminths (Karasuyama et al. 2011), although apparently, this response appears to be parasite-specific.

Syracuse et al., (2011) used a known model of helminth infection, in which protective immunity is dependent on TSLP-TSLPR signaling to determine the role of this signaling pathway in basophil response. They showed that *Trichuris muris* infection in wild-type mice results in an increase in basophils in the periphery, but the Tslpr-/- mice did not show this induced basophilia. In addition, TSLP-induced basophils isolated from Tslpr+/+ mice and transferred to Tslpr-/- mice susceptible to *Trichuris muris* infection were sufficient to restore Th2 response, IgE production, goblet cell hyperplasia,

RELM-β secretion by goblet cells, and significantly reduced parasite loads (Siracusa et al. 2011). Webb et al., (2019) demonstrated that during infection by *Trichuris muris,* intestinal basophils positively regulate the expression of the Notch receptor. *In vitro*, Notch inhibition decreased the production of cytokines from basophils. In addition, Notch signaling is necessary to produce basophilic cytokines induced by IL-33. They also observed that Notch regulates gene expression in basophils when faced with helminth infection, promoting optimal elimination of the parasite triggered by a Th2 response and basophil positioning in the cecum during *Trichuris muris* infection (Webb et al. 2019). In another study, Inclan-Rico et al., (2020) reported that infection with *Nippostrongylus brasiliensis* induced accumulation of basophils in the lungs after the exit of the larva from the tissue, suggesting that its role is to contribute to the protection of the host by promoting the healing of tissues affected by larvae during their migration rather than acting directly on parasites. They also showed that the absence of basophils during infection causes an exacerbated ILCs2 response, resulting in increased inflammation and reduced lung function. In addition, hyperreactive ILCs from mice with basophil depletion showed reduced expression of neuromedin B receptor (NMBR). Treatment with exogenous NMB reduced the ILC2 response, eosinophilia and mucus production after infection with *Nippostrongylus brasiliensis*. Finally, the ILCs2 of mice with depleted basophils did not respond to the stimulus with NMB, suggesting that basophils may inhibit type 2 responses through increased NMBR expression in ILCs2, concluding that basophils are not always inducers or potentiators of type-2 immune responses (Inclan-Rico et al. 2020).

Mast cells are inflammatory cells that respond to different stimuli of innate and adaptive immunity through the immediate and delayed release of inflammatory mediators. These cells express high-affinity Fcε receptors that are sensitized to parasitic antigen-specific IgE and can be activated by ESAg. Like basophils, they contribute to the production of IL-4 and IL-13 during the early stages of helminth infections, either in response to ESAg, cytokines or alarmins (IL-3, TSLP, IL-33, IL-18) or anaphyllotoxins. In secondary infections, these cells can be additionally activated by binding specific antibodies against the parasite (Voehringer 2013). Ohnmacht and Voehringer (2010), observed that mice deficient in mast cells, had lower levels of IgG compared to wid-type. In addition, after infection with *Nippostrongylus brasiliensis* the elimination kinetics of the parasite were delayed only during primary infection (Ohnmacht and Voehringer 2010). Hashimoto et al., (2010), make an inverse correlation between the fequanity of *Heligmosomoides polygyrus* and the number of mast cells (Hashimoto et al. 2010). In the case of *Trichinella spiralis*, Knight et al., (2000), reported that the murine mast cell protease-1 (mMCP-1) was required to limit the deposition of *Trichinella spiralis* larvae in muscle tissue and to expel adult parasites at the intestinal level. They described that mMCP-1 gene deletion caused a delay in the expulsion of adult parasites of *Trichinella spiralis*, as well as an increased deposition of muscle larvae, despite a normal and sometimes greater number of mast cells (Knight et al. 2000). Shin et al., (2008), and Gurish et al., (2004), during chronic trichiellosis, report that the secretion of mMCP-6 by mast cells contributes to the recruitment of eosinophils to the site where the larva was located. This defect in eosinophil recruitment was demonstrated in IgE-deficient mice (K. Shin et al. 2008; Gurish et al. 2004). Linnemann et al., (2020), using mice deficient in mast cells ($Cpa3^{cre}$ mice) during *Litomosoides sigmodontis* infection, demonstrate that mast cells are not mandatory for the initiation of an immune response, nor for the defense of the host during infection by *Litomosoides sigmodontis*.

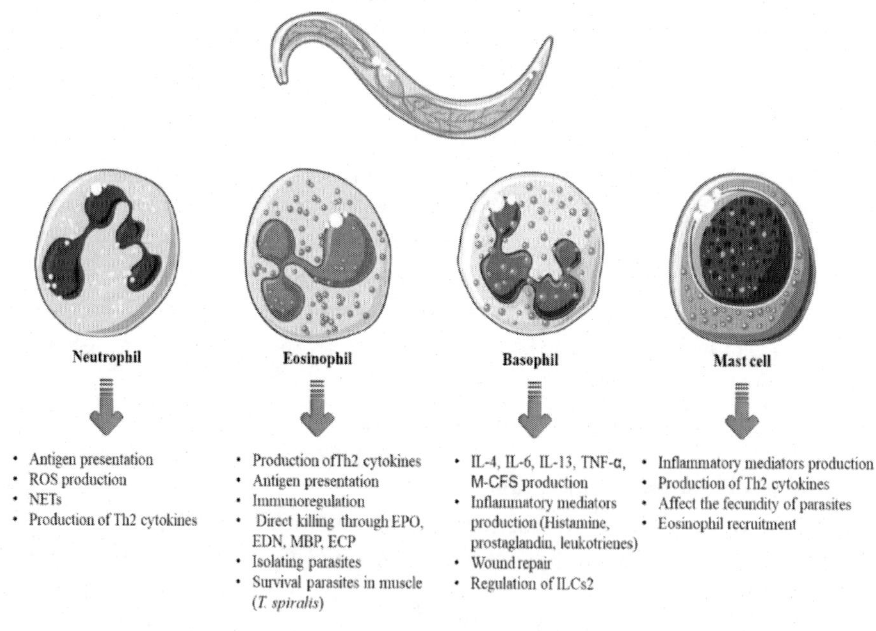

Figure 4. Granulocye functions during helmith infections.

This was concluded because they did not find differences in the elimination kinetics of microfilariae nor differences in the response of cytokines and antibodies against the parasite between wt mice and mast cells-deficient mice (Linnemann et al. 2020).

Innate Lymphoid Cells Type 2

ILCs2 are tissue-resident cells that are abundant in the mucous barriers and participate as immunological and physiological mediators. They promote homeostasis, integrity and repair of these tissues. These cells were discovered and described in studies using helminths as an antigenic challenge and found that at the intestinal level, they produced Th2 cytokines such as IL-4, IL-5 and IL-13 (Neill et al. 2010; Moro et al. 2010). Their presence has been reported in several tissues and they are the first line of defense during helminth infections, before the development of the adaptive response (Inclan-Rico and Siracusa 2018). ILCs2 are sensitive to the effect of alarmins and cytokines such as IL-25, IL-33 and TSLP, since they express receptors for these molecules. These cytokines, as we mentioned earlier, are released as a response to damage to the epithelial caused by helminth infections. When interacting with ILCs2, they produce cytokines such as IL-4, IL-5 and IL-13, which will promote type 2 immune responses (Vivier et al. 2018) (Figure 5). Fallon et al., (2006), first reported the role of ILCs2 (before they were identified as innate lymphocytes) in the elimination of *Nippostrongylus brasiliensis*. They observed that IL-25 deficient mice eliminated parasites in a deficient way, in addition, there was a decreased production of Th2 cytokines compared to wild-type mice. With the above, they established that IL-25 participates by regulating a new subpopulation of cells that were neither B nor T cells (NBNT cells), responsible for producing Th2 cytokines (IL-4, IL-5, IL-13) during infections caused by helminths. In *in vivo* experiments, when recombinant IL-25 was administrated, they observed these NBNT cells were increased, which lead to a rapid exposure of the parasite. This effect was independent of the presence of T or B cells, but dependent on Th2 cytokines (Fallon et al. 2006). After this experiment, Neill et al., (2010), demonstrated that another important cytokine to induce rapid elimination of *Nippostrongylus brasiliensis* was IL-33. This cytokine in addition to IL-25 are early activators of ILCs2 (Neill et al. 2010).

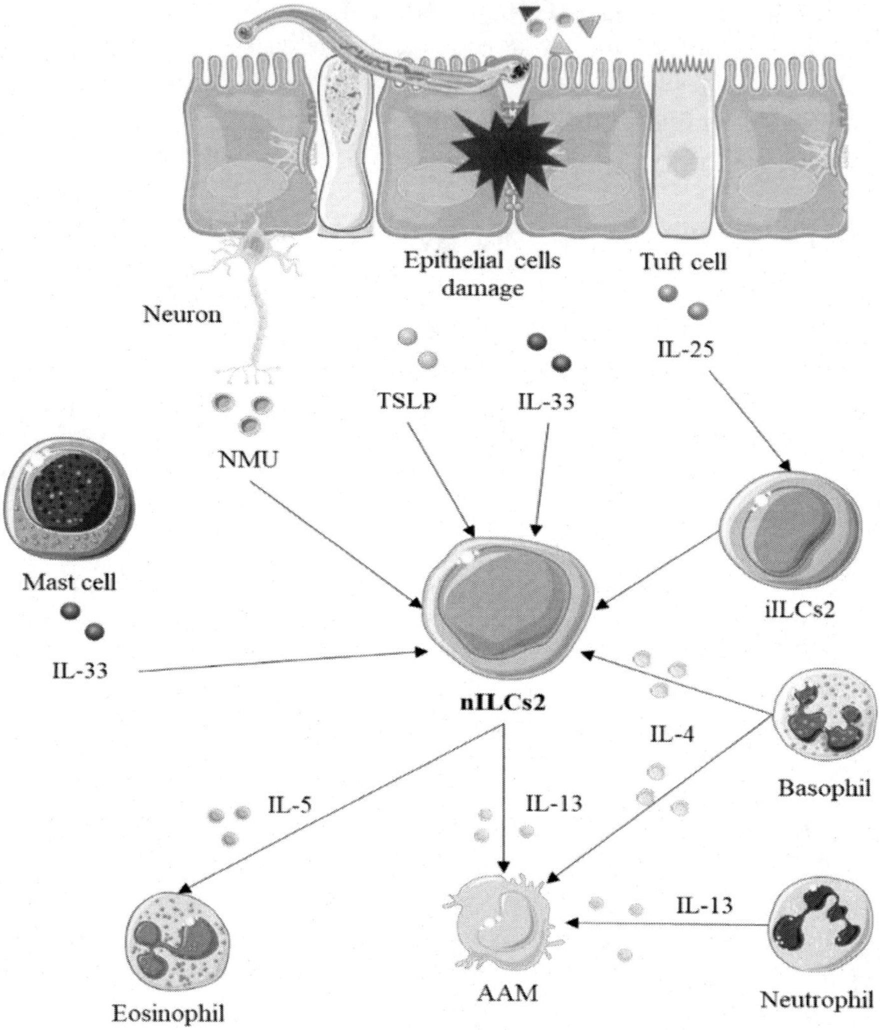

Figure 5. Interaction between ILCS2 and other innate immunity cells during helminth infections.

Although these studies define ILCs2 as a single subpopulation, Huang et al., (2015), described different subsets of ILCs2: natural ILCs (tissue residents) (nILC2), which respond to IL-33; and inflammatory ILCs (iILC2), which respond to IL-25. In addition, they propose that iILC2 are transient progenitors of ILCs, which are mobilized by inflammatory or infectious processes and that develop in nILC2, which will contribute to the development of immunity against helminths (Y. Huang et al. 2015) (Figure 5). In addition to immunological mediators, it has been shown that ILCs2 can be positively or negatively regulated by the effect of neurotransmitters such as neuromedin U (NMU) (Cardoso et al. 2017; Wallrapp et al. 2017; Klose et al. 2017) and noradrenaline (Moriyama et al. 2018), both molecules can influence the functioning of ILCs2 and the elimination of helminths (Figure 5). Finally, Roberts et al., (2021), demonstrated that pulmonary ILCs2 can synthesize and release acetylcholine during *Nippostrongylus brasiliensis* infection and that this feature is induced by IL-25 and IL-33. In addition, they demonstrated that transgenic mice that have ILCs2 that do not synthesize acetylcholine have an impaired immune response against *Nippostrongylus brasiliensis*, as well as reduced expression of IL-5 and IL-13, mucins Muc5b and Muc5ac in the lung, and an altered gut barrier response, demonstrating that acetylcholine-producing ILCs2 are an important factor in promoting type 2 immunity (Roberts et al. 2021).

Conclusion

In the last decade, there have been important advances in the knowledge about how helminth parasites are recognized by innate immune cells. Also, how other innate cells are necessary to generate a specific immune response to eliminate them. These network of molecules, include cytokines, soluble factors, and danger signals arising from the damage produced by parasite migration, which promotes a Th2-type response. This response in turn drives the alternative activation of Mo, and the activation of Eos, basophils, and mast cells, which contribute to the expulsion of helminths. Among the innate immune cells, macrophages, that are differenciated to a phenotype M2, play a key role in providing factors for rapid and efficient tissue repair and to complete the process of wound healing. There is also involvement of neutrophils, eosinophils, basophils and mast cells in death mechanisms against helminths. However, there is growing evidence showing the importance of ILCs in the mechanisms of protection against parasitic helminths. Some interactions between phagocytes with other cell types (e.g., ILC2) during helminth infection have been demonstrated. However, further studies are still required to elucidate the factors that modulate the maturation of DCs to promote the Th2 response in T cells and possible interaction between other cells such as Mo, neutrophils, basophils, and mast cells during infection by helminth parasites.

References

Acharya, K. Ravi, and Steven J. Ackerman. 2014. "Eosinophil Granule Proteins: Form and Function." *Journal of Biological Chemistry*. J Biol Chem. https://doi.org/10.1074/jbc.R113.546218.

Allen, Judith E, and Rick M Maizels. 2011. "Diversity and Dialogue in Immunity to Helminths." *Nature Reviews Immunology* 11 (6): 375–88. https://doi.org/10.1038/nri2992.

Anthony, Robert M., Laura I. Rutitzky, Joseph F. Urban, Miguel J. Stadecker, and William C. Gause. 2007. "Protective Immune Mechanisms in Helminth Infection." *Nature Reviews Immunology*. Nature Publishing Group. https://doi.org/10.1038/nri2199.

Ásbjörnsdóttir, Kristjana H., Arianna R. Means, Marleen Werkman, and Judd L. Walson. 2017. "Prospects for Elimination of Soil-Transmitted Helminths." *Current Opinion in Infectious Diseases* 30 (5): 482–88. https://doi.org/10.1097/QCO.0000000000000395.

Babu, Subash, and Thomas B. Nutman. 2019. "Immune Responses to Helminth Infection." In *Clinical Immunology*, 437-447.e1. Elsevier. https://doi.org 10.1016/b978-0-7020-6896-6.00031-4.

Balic, Adam, Yvonne Harcus, Martin J. Holland, and Rick M. Maizels. 2004. "Selective Maturation of Dendritic Cells by Nippostrongylus Brasiliensis-Secreted Proteins Drives Th2 Immune Responses." *European Journal of Immunology* 34 (11): 3047–59. https://doi.org/10.1002/eji.200425167.

Bhattarai, Rachana, Hélène Carabin, Jefferson V. Proaño, Jose Flores-Rivera, Teresa Corona, Ana Flisser, Leith León-Maldonado, and Christine M. Budke. 2019. "The Monetary Burden of Cysticercosis in Mexico." *PLOS Neglected Tropical Diseases* 13 (7): e0007501. https://doi.org/10.1371/JOURNAL.PNTD.0007501.

Bonne-Année, Sandra, Laura A. Kerepesi, Jessica A. Hess, Amy E. O'Connell, James B. Lok, Thomas J. Nolan, and David Abraham. 2013. "Human and Mouse Macrophages Collaborate with Neutrophils to Kill Larval Strongyloides Stercoralis." *Infection and Immunity* 81 (9): 3346–55. https://doi.org/10.1128/IAI.00625-13.

Bonne-Année, Sandra, Laura A. Kerepesi, Jessica A. Hess, Jordan Wesolowski, Fabienne Paumet, James B. Lok, Thomas J. Nolan, and David Abraham. 2014. "Extracellular Traps Are Associated with Human and Mouse Neutrophil and Macrophage Mediated Killing of Larval Strongyloides Stercoralis." *Microbes and Infection* 16 (6): 502–11. https://doi.org/10.1016/J.MICINF.2014.02.012.

Bouchery, Tiffany, Graham Le Gros, and Nicola Harris. 2019. "ILC2S—Trailblazers in the Host Response against Intestinal Helminths." *Frontiers in Immunology*. Frontiers Media S.A. https://doi.org/10.3389/fimmu.2019.00623.

Bouchery, Tiffany, Mati Moyat, Javier Sotillo, Solomon Silverstein, Beatrice Volpe, Gillian Coakley, Theodora Dorita Tsourouktsoglou, et al. 2020. "Hookworms Evade Host Immunity by Secreting a Deoxyribonuclease to Degrade Neutrophil Extracellular Traps." *Cell Host and Microbe* 27 (2): 277-289.e6. https://doi.org/10.1016/j.chom.2020.01.011.

Brinkmann, Volker, and Arturo Zychlinsky. 2012. "Neutrophil Extracellular Traps: Is Immunity the Second Function of Chromatin?" *Journal of Cell Biology*. J Cell Biol. https://doi.org/10.1083/jcb.201203170.

Cardoso, Vania, Julie Chesné, Hélder Ribeiro, Bethania Garcia-Cassani, Tania Carvalho, Tiffany Bouchery, Kathleen Shah, Nuno L. Barbosa-Morais, Nicola Harris, and Henrique Veiga-Fernandes. 2017. "Neuronal Regulation of Type 2 Innate Lymphoid Cells via Neuromedin U." *Nature* 549 (7671): 277–81. https://doi.org/10.1038/nature23469.

Carvalho, Lucas, Jie Sun, Colleen Kane, Fraser Marshall, Connie Krawczyk, and Edward J. Pearce. 2009. "Review Series on Helminths, Immune Modulation and the Hygiene Hypothesis: Mechanisms Underlying Helminth Modulation of Dendritic Cell Function." *Immunology* 126 (1): 28–34. https://doi.org/10.1111/j.1365-2567.2008.03008.x.

Chauhan, Arun, Atul Sharma, Jitendra K. Tripathi, Yuyang Sun, Pramod Sukumran, Brij B. Singh, Bibhuti B. Mishra, and Jyotika Sharma. 2021. "Helminth Derived Factors Inhibit Neutrophil Extracellular Trap Formation and Inflammation in Bacterial Peritonitis." *Scientific Reports* 11 (1): 12718. https://doi.org/10.1038/s41598-021-92001-9.

Chen, Fei, Wenhui Wu, Ariel Millman, Joshua F. Craft, Eunice Chen, Nirav Patel, Jean L. Boucher, Joseph F. Urban, Charles C. Kim, and William C. Gause. 2014. "Neutrophils Prime a Long-Lived Effector Macrophage Phenotype That Mediates Accelerated Helminth Expulsion." *Nature Immunology* 15 (10): 938–46. https://doi.org/10.1038/ni.2984.

Coakley, Gillian, and Nicola Laraine Harris. 2020. "Interactions between Macrophages and Helminths." *Parasite Immunology*. Blackwell Publishing Ltd. https://doi.org/10.1111/pim.12717.

Damle, S. R., R. K. Martin, J. V. Cross, and D. H. Conrad. 2017. "Macrophage Migration Inhibitory Factor Deficiency Enhances Immune Response to Nippostrongylus Brasiliensis." *Mucosal Immunology* 10 (1): 205–14. https://doi.org/10.1038/mi.2016.29.

Díaz-Godínez, César, and Julio C. Carrero. 2019. "The State of Art of Neutrophil Extracellular Traps in Protozoan and Helminthic Infections." *Bioscience Reports* 39 (1): 1–19. https://doi.org/10.1042/BSR20180916.

Die, Irma van, and Richard D. Cummings. 2017. "The Mannose Receptor in Regulation of Helminth-Mediated Host Immunity." *Frontiers in Immunology*. Front Immunol. https://doi.org/10.3389/fimmu.2017.01677.

Dyken, Steven J. Van, and Richard M. Locksley. 2013. "Interleukin-4-and Interleukin-13-Mediated Alternatively Activated Macrophages: Roles in Homeostasis and Disease." *Annual Review of Immunology*. NIH Public Access. https://doi.org/10.1146/annurev-immunol-032712-095906.

Fabre, Valeria, Daniel P. Beiting, Susan K. Bliss, Nebiat G. Gebreselassie, Lucille F. Gagliardo, Nancy A. Lee, James J. Lee, and Judith A. Appleton. 2009. "Eosinophil Deficiency Compromises Parasite Survival in Chronic Nematode Infection." *The Journal of Immunology* 182 (3): 1577–83. https://doi.org/10.4049/jimmunol.182.3.1577.

Fallon, Padraic G., Sarah J. Ballantyne, Niamh E. Mangan, Jillian L. Barlow, Ayan Dasvarma, Duncan R. Hewett, Ann McIlgorm, Helen E. Jolin, and Andrew N.J. McKenzie. 2006. "Identification of an Interleukin (IL)-25-Dependent Cell Population That Provides IL-4, IL-5, and IL-13 at the Onset of Helminth Expulsion." *Journal of Experimental Medicine* 203 (4): 1105–16. https://doi.org/10.1084/jem.20051615.

Filbey, Kara J., Fumi Varyani, Yvonne Harcus, James P. Hewitson, Danielle J. Smyth, Henry J. McSorley, Alasdair Ivens, et al. 2019. "Macrophage Migration Inhibitory Factor (MIF) Is Essential for Type 2 Effector Cell Immunity to an Intestinal Helminth Parasite." *Frontiers in Immunology* 10 (OCT). https://doi.org/10.3389/fimmu.2019.02375.

Gebreselassie, Nebiat G., Andrew R. Moorhead, Valeria Fabre, Lucille F. Gagliardo, Nancy A. Lee, James J. Lee, and Judith A. Appleton. 2012. "Eosinophils Preserve Parasitic Nematode Larvae by Regulating Local Immunity." *The Journal of Immunology* 188 (1): 417–25. https://doi.org/10.4049/jimmunol.1101980.

Giacomin, Paul R., Mark C. Siracusa, Kevin P. Walsh, Richard K. Grencis, Masato Kubo, Michael R. Comeau, and David Artis. 2012. "Thymic Stromal Lymphopoietin-Dependent Basophils Promote Th2 Cytokine Responses Following Intestinal Helminth Infection." *The Journal of Immunology* 189 (9): 4371–78. https://doi.org/10.4049/jimmunol.1200691.

Gordon, Siamon, and Fernando O. Martinez. 2010. "Alternative Activation of Macrophages: Mechanism and Functions." *Immunity* 32 (5): 593–604. https://doi.org/10.1016/j.immuni.2010.05.007.

Gundra, Uma Mahesh, Natasha M. Girgis, Dominik Ruckerl, Stephen Jenkins, Lauren N. Ward, Zachary D. Kurtz, Kirsten E. Wiens, et al. 2014.

"Alternatively Activated Macrophages Derived from Monocytes and Tissue Macrophages Are Phenotypically and Functionally Distinct." *Blood* 123 (20): e110. https://doi.org/10.1182/BLOOD-2013-08-520619.

Gurish, Michael F., Paul J. Bryce, Hong Tao, Alison B. Kisselgof, Elizabeth M. Thornton, Hugh R. Miller, Daniel S. Friend, and Hans C. Oettgen. 2004. "IgE Enhances Parasite Clearance and Regulates Mast Cell Responses in Mice Infected with Trichinella Spiralis." *The Journal of Immunology* 172 (2): 1139–45. https://doi.org/10.4049/jimmunol.172.2.1139.

Halim, Timotheus Y.F., Ramona H. Krauß, Ann C. Sun, and Fumio Takei. 2012. "Lung Natural Helper Cells Are a Critical Source of Th2 Cell-Type Cytokines in Protease Allergen-Induced Airway Inflammation." *Immunity* 36 (3): 451–63. https://doi.org/10.1016/J.IMMUNI.2011.12.020.

Hammad, Hamida, and Bart N. Lambrecht. 2015. "Barrier Epithelial Cells and the Control of Type 2 Immunity." *Immunity* 43 (1): 29–40. https://doi.org/10.1016/j.immuni.2015.07.007.

Harnett, William. 2014. "Secretory Products of Helminth Parasites as Immunomodulators." *Molecular and Biochemical Parasitology* 195 (2): 130–36. https://doi.org/10.1016/J.MOLBIOPARA.2014.03.007.

Hashimoto, K., R. Uchikawa, T. Tegoshi, K. Takeda, M. Yamada, and N. Arizono. 2010. "Immunity-Mediated Regulation of Fecundity in the Nematode Heligmosomoides Polygyrus - The Potential Role of Mast Cells." *Parasitology* 137 (5): 881–87. https://doi.org/10.1017/S0031182009991673.

Hewitson, James P., John R. Grainger, and Rick M. Maizels. 2009. "Helminth Immunoregulation: The Role of Parasite Secreted Proteins in Modulating Host Immunity." *Molecular and Biochemical Parasitology*. Mol Biochem Parasitol. https://doi.org/10.1016/j.molbiopara.2009.04.008.

Huang, Lu, and Judith A. Appleton. 2016. "Eosinophils in Helminth Infection: Defenders and Dupes." *Trends in Parasitology* 32 (10): 798–807. https://doi.org/10.1016/j.pt.2016.05.004.

Huang, Yuefeng, Liying Guo, Jin Qiu, Xi Chen, Jane Hu-Li, Ulrich Siebenlist, Peter R. Williamson, Joseph F. Urban, and William E. Paul. 2015. "IL-25-Responsive, Lineage-Negative KLRG1hi Cells Are Multipotential 'Inflammatory' Type 2 Innate Lymphoid Cells." *Nature Immunology* 16 (2): 161–69. https://doi.org/10.1038/ni.3078.

Inclan-Rico, Juan M., John J. Ponessa, Nuriban Valero-Pacheco, Christina M. Hernandez, Chandler B. Sy, Alexander D. Lemenze, Aimee M. Beaulieu, and Mark C. Siracusa. 2020. "Basophils Prime Group 2 Innate Lymphoid Cells for Neuropeptide-Mediated Inhibition." *Nature Immunology* 21 (10): 1181–93. https://doi.org/10.1038/s41590-020-0753-y.

Inclan-Rico, Juan M., and Mark C. Siracusa. 2018. "First Responders: Innate Immunity to Helminths." *Trends in Parasitology* 34 (10): 861–80. https://doi.org/10.1016/j.pt.2018.08.007.

James, Spencer L, Degu Abate, Kalkidan Hassen Abate, Solomon M Abay, Cristiana Abbafati, Nooshin Abbasi, Hedayat Abbastabar, et al. 2018. "Global, Regional, and National Incidence, Prevalence, and Years Lived with Disability for 354 Diseases and Injuries for 195 Countries and Territories, 1990–2017: A Systematic Analysis for the Global Burden of Disease Study 2017." *The Lancet* 392 (10159): 1789–1858. https://doi.org/10.1016/S0140-6736(18)32279-7.

Jourdan, Peter Mark, Poppy H.L. Lamberton, Alan Fenwick, and David G. Addiss. 2018. "Soil-Transmitted Helminth Infections." *The Lancet* 391 (10117): 252–65. https://doi.org/10.1016/S0140-6736(17)31930-X.

Karasuyama, Hajime, Kaori Mukai, Kazushige Obata, Yusuke Tsujimura, and Takeshi Wada. 2011. "Nonredundant Roles of Basophils in Immunity." *Annual Review of Immunology* 29: 45–69. https://doi.org/10.1146/annurev-immunol-031210-101257.

Klose, Christoph S.N., Tanel Mahlakõiv, Jesper B. Moeller, Lucille C. Rankin, Anne Laure Flamar, Hiroki Kabata, Laurel A. Monticelli, et al. 2017. "The Neuropeptide Neuromedin U Stimulates Innate Lymphoid Cells and Type 2 Inflammation." *Nature* 549 (7671): 282–86. https://doi.org/10.1038/nature23676.

Knight, Pamela A., Steven H. Wright, Catherine E. Lawrence, Yvonne Y.W. Paterson, and Hugh R.P. Miller. 2000. "Delayed Expulsion of the Nematode Trichinella Spiralis in Mice Lacking the Mucosal Mast Cell-Specific Granule Chymase, Mouse Mast Cell Protease-1." *Journal of Experimental Medicine* 192 (12): 1849–56. https://doi.org/10.1084/jem.192.12.1849.

Kreider, Timothy, Robert M. Anthony, Joseph F. Urban, and William C. Gause. 2007. "Alternatively Activated Macrophages in Helminth Infections." *Current Opinion in Immunology* 19 (4): 448–53. https://doi.org/10.1016/J.COI.2007.07.002.

Kumar, Sunil, Yideul Jeong, Muhammad Umer Ashraf, and Yong Soo Bae. 2019. "Dendritic Cell-Mediated Th2 Immunity and Immune Disorders." *International Journal of Molecular Sciences* 20 (9). https://doi.org/10.3390/ijms20092159.

Leech, Melanie D., and Richard K. Grencis. 2006. "Induction of Enhanced Immunity to Intestinal Nematodes Using IL-9-Producing Dendritic Cells." *The Journal of Immunology* 176 (4): 2505–11. https://doi.org/10.4049/jimmunol.176.4.2505.

LeMessurier, Kim S., and Amali E. Samarasinghe. 2019. "Eosinophils: Nemeses of Pulmonary Pathogens?" *Current Allergy and Asthma Reports*. Current Medicine Group LLC 1. https://doi.org/10.1007/s11882-019-0867-1.

Linnemann, Lara Christine, Martina Reitz, Thorsten B. Feyerabend, Minka Breloer, and Wiebke Hartmann. 2020. "Limited Role of Mast Cells during Infection with the Parasitic Nematode Litomosoides Sigmodontis." *PLoS Neglected Tropical Diseases* 14 (7): 1–14. https://doi.org/10.1371/journal.pntd.0008534.

Loke, P'ng, Meera G. Nair, John Parkinson, David Guiliano, Mark Blaxter, and Judith E. Allen. 2002. "IL-4 Dependent Alternatively-Activated Macrophages Have a Distinctive in Vivo Gene Expression Phenotype." *BMC Immunology* 3 (July): 1–11. https://doi.org/10.1186/1471-2172-3-7.

MacDonald, Andrew S., Amy D. Straw, Beverley Bauman, and Edward J. Pearce. 2001. "CD8 − Dendritic Cell Activation Status Plays an Integral Role in Influencing Th2 Response Development." *The Journal of Immunology* 167 (4): 1982–88. https://doi.org/10.4049/jimmunol.167.4.1982.

Maizels, Rick M, Hermelijn H Smits, and Henry J. McSorley. 2018. "Modulation of Host Immunity by Helminths: The Expanding Repertoire of Parasite Effector Molecules." *Immunity* 49 (5): 801–18. https://doi.org/10.1016/j.immuni.2018.10.016.

Moriyama, Saya, Jonathan R. Brestoff, Anne Laure Flamar, Jesper B. Moeller, Christoph S.N. Klose, Lucille C. Rankin, Naomi A. Yudanin, et al. 2018. "B2-Adrenergic Receptor-Mediated Negative Regulation of Group 2 Innate Lymphoid Cell Responses." *Science* 359 (6379): 1056–61. https://doi.org/10.1126/science.aan4829.

Moro, Kazuyo, Taketo Yamada, Masanobu Tanabe, Tsutomu Takeuchi, Tomokatsu Ikawa, Hiroshi Kawamoto, Jun Ichi Furusawa, Masashi Ohtani, Hideki Fujii, and Shigeo Koyasu. 2010. "Innate Production of TH 2 Cytokines by Adipose Tissue-Associated c-Kit+ Sca-1+ Lymphoid Cells." *Nature* 463 (7280): 540–44. https://doi.org/10.1038/nature08636.

Motran, Claudia Cristina, Leonardo Silvane, Laura Silvina Chiapello, Martin Gustavo Theumer, Laura Fernanda Ambrosio, Ximena Volpini, Daiana Pamela Celias, and Laura Cervi. 2018. "Helminth Infections: Recognition and Modulation of the Immune Response by Innate Immune Cells." *Frontiers in Immunology*. Frontiers Media S.A. https://doi.org/10.3389/fimmu.2018.00664.

Mutua, Victoria, and Laurel J. Gershwin. 2020. "A Review of Neutrophil Extracellular Traps (NETs) in Disease: Potential Anti-NETs Therapeutics." *Clinical Reviews in Allergy & Immunology 2020*, August, 1–18. https://doi.org/10.1007/S12016-020-08804-7.

Neill, Daniel R., See Heng Wong, Agustin Bellosi, Robin J. Flynn, Maria Daly, Theresa K.A. Langford, Christine Bucks, et al. 2010. "Nuocytes Represent a New Innate Effector Leukocyte That Mediates Type-2 Immunity." *Nature* 464 (7293): 1367–70. https://doi.org/10.1038/nature08900.

O'Connell, Amy E., Jessica A. Hess, Gilberto A. Santiago, Thomas J. Nolan, James B. Lok, James J. Lee, and David Abraham. 2011. "Major Basic Protein from Eosinophils and Myeloperoxidase from Neutrophils Are Required for Protective Immunity to Strongyloides Stercoralis in Mice." *Infection and Immunity* 79 (7): 2770–78. https://doi.org/10.1128/IAI.00931-10.

Obata-Ninomiya, Kazushige, Phillip P. Domeier, and Steven F. Ziegler. 2020. "Basophils and Eosinophils in Nematode Infections." *Frontiers in Immunology*. Frontiers Media S.A. https://doi.org/10.3389/fimmu.2020.583824.

Odeniran, Paul Olalekan, Kehinde Foluke Omolabi, and Isaiah Oluwafemi Ademola. 2020. "Economic Model of Bovine Fasciolosis in Nigeria: An Update." *Tropical Animal Health and Production* 52 (6): 3359–63. https://doi.org/10.1007/s11250-020-02367-7.

Ohnmacht, Caspar, and David Voehringer. 2010. "Basophils Protect against Reinfection with Hookworms Independently of Mast Cells and Memory Th2 Cells." *The Journal of Immunology* 184 (1): 344–50. https://doi.org/10.4049/jimmunol.0901841.

Panhuys, Nicholas van, Melanie Prout, Elizabeth Forbes, Booki Min, William E. Paul, and Graham Le Gros. 2011. "Basophils Are the Major Producers of IL-4 during Primary Helminth Infection." *The Journal of Immunology* 186 (5): 2719–28. https://doi.org/10.4049/jimmunol.1000940.

Roberts, Luke B, Corinna Schnoeller, Rita Berkachy, Matthew Darby, Jamie Pillaye, Menno J Oudhoff, Naveen Parmar, et al. 2021. "Acetylcholine Production by Group 2 Innate Lymphoid Cells Promotes Mucosal Immunity to Helminths." *Science Immunology* 6 (57): 359. https://doi.org/10.1126/sciimmunol.abd0359.

Schneider, Christoph, Claire E. O'Leary, and Richard M. Locksley. 2019. "Regulation of Immune Responses by Tuft Cells." *Nature Reviews Immunology* 19 (9): 584–93. https://doi.org/10.1038/s41577-019-0176-x.

Schuijs, Martijn J., Hamida Hammad, and Bart N. Lambrecht. 2019. "Professional and 'Amateur' Antigen Presenting Cells In Type 2 Immunity." *Trends in Immunology* 40 (1): 22. https://doi.org/10.1016/J.IT.2018.11.001.

Schwartz, Christian, Adriana Turqueti-Neves, Susanne Hartmann, Philipp Yu, Falk Nimmerjahn, and David Voehringer. 2014. "Basophil-Mediated Protection against Gastrointestinal Helminths Requires IgE-Induced Cytokine Secretion." *Proceedings of the National Academy of Sciences of the*

United States of America 111 (48): E5169–77. https://doi.org/10.1073/pnas.1412663111.

Shen, Tao, Sohee Kim, Jeong Su Do, Lu Wang, Chris Lantz, Joseph F. Urban, Graham Le Gros, and Booki Min. 2008. "T Cell-Derived IL-3 Plays Key Role in Parasite Infection-Induced Basophil Production but Is Dispensable for in Vivo Basophil Survival." *International Immunology* 20 (9): 1201–9. https://doi.org/10.1093/intimm/dxn077.

Shin, Jenn Yeu, Chen Yu Wang, Chi Chien Lin, and Ching Liang Chu. 2020. "A Recently Described Type 2 Conventional Dendritic Cell (CDC2) Subset Mediates Inflammation." *Cellular and Molecular Immunology* 17 (12): 1215–17. https://doi.org/10.1038/s41423-020-0511-y.

Shin, Kichul, Gerald F. M. Watts, Hans C. Oettgen, Daniel S. Friend, Alan D. Pemberton, Michael F. Gurish, and David M. Lee. 2008. "Mouse Mast Cell Tryptase MMCP-6 Is a Critical Link between Adaptive and Innate Immunity in the Chronic Phase of Trichinella Spiralis Infection." *The Journal of Immunology* 180 (7): 4885–91. https://doi.org/10.4049/jimmunol.180.7.4885.

Silvestre-Roig, Carlos, Zvi G. Fridlender, Michael Glogauer, and Patrizia Scapini. 2019. "Neutrophil Diversity in Health and Disease." *Trends in Immunology* 40 (7): 565–83. https://doi.org/10.1016/j.it.2019.04.012.

Siracusa, Mark C., Steven A. Saenz, David A. Hill, Brian S. Kim, Mark B. Headley, Travis A. Doering, E. John Wherry, et al. 2011. "TSLP Promotes Interleukin-3-Independent Basophil Haematopoiesis and Type 2 Inflammation." *Nature* 477 (7363): 229–33. https://doi.org/10.1038/nature10329.

Specht, Sabine, Michael Saeftel, Manuela Arndt, Elmar Endl, Bettina Dubben, Nancy A. Lee, James J. Lee, and Achim Hoerauf. 2006. "Lack of Eosinophil Peroxidase or Major Basic Protein Impairs Defense against Murine Filarial Infection." *Infection and Immunity* 74 (9): 5236–43. https://doi.org/10.1128/IAI.00329-06.

Stavitsky, Abram B., Christine Metz, Shifan Liu, Jia Xianli, and Richard Bucala. 2003. "Blockade of Macrophage Migration Inhibitory Factor (MIF) in Schistosoma Japonicum-Infected Mice Results in an Increased Adult Worm Burden and Reduced Fecundity." *Parasite Immunology* 25 (7): 369–74. https://doi.org/10.1046/j.1365-3024.2003.00641.x.

Stone, Kelly D., Calman Prussin, and Dean D. Metcalfe. 2010. "IgE, Mast Cells, Basophils, and Eosinophils." *Journal of Allergy and Clinical Immunology* 125 (2): S73–80. https://doi.org/10.1016/J.JACI.2009.11.017.

Strandmark, J., S. Steinfelder, C. Berek, A. A. Kühl, S. Rausch, and S. Hartmann. 2017. "Eosinophils Are Required to Suppress Th2 Responses in Peyer's

Patches during Intestinal Infection by Nematodes." *Mucosal Immunology* 10 (3): 661–72. https://doi.org/10.1038/mi.2016.93.

Tait Wojno, Elia D., and David Artis. 2016. "Emerging Concepts and Future Challenges in Innate Lymphoid Cell Biology." *Journal of Experimental Medicine*. The Rockefeller University Press. https://doi.org/10.1084/jem.20160525.

Torrero, Marina N., Marc P. Hübner, David Larson, Hajime Karasuyama, and Edward Mitre. 2010. "Basophils Amplify Type 2 Immune Responses, but Do Not Serve a Protective Role, during Chronic Infection of Mice with the Filarial Nematode Litomosoides Sigmodontis." *The Journal of Immunology* 185 (12): 7426–34. https://doi.org/10.4049/jimmunol.0903864.

Vivier, Eric, David Artis, Marco Colonna, Andreas Diefenbach, James P. Di Santo, Gérard Eberl, Shigeo Koyasu, et al. 2018. "Innate Lymphoid Cells: 10 Years On." *Cell*. Cell. https://doi.org/10.1016/j.cell.2018.07.017.

Vlas, Sake J. de, Wilma A. Stolk, Epke A. le Rutte, Jan A.C. Hontelez, Roel Bakker, David J. Blok, Rui Cai, et al. 2016. "Concerted Efforts to Control or Eliminate Neglected Tropical Diseases: How Much Health Will Be Gained?" *PLoS Neglected Tropical Diseases* 10 (2). https://doi.org/10.1371/journal.pntd.0004386.

Voehringer, David. 2013. "Protective and Pathological Roles of Mast Cells and Basophils." *Nature Reviews Immunology* 13 (5): 362–75. https://doi.org/10.1038/nri3427.

Wallrapp, Antonia, Samantha J. Riesenfeld, Patrick R. Burkett, Raja Elie E. Abdulnour, Jackson Nyman, Danielle Dionne, Matan Hofree, et al. 2017. "The Neuropeptide NMU Amplifies ILC2-Driven Allergic Lung Inflammation." *Nature* 549 (7672): 351–56. https://doi.org/10.1038/nature24029.

Webb, Lauren M., Oyebola O. Oyesola, Simon P. Früh, Elena Kamynina, Katherine M. Still, Ravi K. Patel, Seth A. Peng, et al. 2019. "The Notch Signaling Pathway Promotes Basophil Responses during Helminth-Induced Type 2 Inflammation." *Journal of Experimental Medicine* 216 (6): 1268–79. https://doi.org/10.1084/jem.20180131.

Weller, Peter F., and Lisa A. Spencer. 2017. "Functions of Tissue-Resident Eosinophils." *Nature Reviews Immunology*. Nature Publishing Group. https://doi.org/10.1038/nri.2017.95.

White, Rhiannon R., and Katerina Artavanis-Tsakonas. 2012. "How Helminths Use Excretory Secretory Fractions to Modulate Dendritic Cells." *Virulence*. Virulence. https://doi.org/10.4161/viru.22832.

Xue, Luzheng, Maryam Salimi, Isabel Panse, Jenny M. Mjösberg, Andrew N.J. McKenzie, Hergen Spits, Paul Klenerman, and Graham Ogg. 2014. "Prostaglandin D2 Activates Group 2 Innate Lymphoid Cells through Chemoattractant Receptor-Homologous Molecule Expressed on TH2 Cells." *Journal of Allergy and Clinical Immunology* 133 (4): 1184–94. https://doi.org/10.1016/j.jaci.2013.10.056.

Yang, De, Qian Chen, Bo Su Shao, Ping Zhang, Kahori Kurosaka, Rachel R. Caspi, Suzanne M. Michalek, Helene F. Rosenberg, Ning Zhang, and Joost J. Oppenheim. 2008. "Eosinophil-Derived Neurotoxin Acts as an Alarmin to Activate the TLR2-MyD88 Signal Pathway in Dendritic Cells and Enhances Th2 Immune Responses." *Journal of Experimental Medicine* 205 (1): 79–90. https://doi.org/10.1084/jem.20062027.

About the Editor

Jorge Morales-Montor. His Doctoral Thesis was recognized with the Lola and Igo Flisser PUIS Award as the best graduate thesis at the national level in the field of parasitology. He received a fellowship from the Fogarty Foundation to perform a Postdoctoral Research stay at the University of Georgia. His scientific production already has 153 articles in international indexed journals, with more than 3,100 citations, and an H Index of 35. He has also edited several books and published more than 55 chapters. He is a member of the Mexican Academy of Sciences, Academy New York Sciences, American Society of Immunologists, Latin American Academy of Sciences and The National Academy of Medicine, among others. He has more than 35 awards and 30 distinctions, has graduated more than 35 Bachelor's, 8 Master's and 12 Doctorate students, and has directed more than 7 Post-Doctoral Researchers. He was President of the Mexican Society of Parasitology and Mexican Society of Neuroimmunoendocrinology, and he has obtained more than 20 research grants in his career.

Index

D

K

Q

R

S

T

U

V

W

β